MICROBIOLOGICAL
RESEARCH AND DEVELOPMENT
FOR THE FOOD INDUSTRY

MICROBIOLOGICAL
RESEARCH AND DEVELOPMENT
FOR THE FOOD INDUSTRY

Edited by **Peter J. Taormina**

CRC Press
Taylor & Francis Group
Boca Raton   London   New York

CRC Press is an imprint of the
Taylor & Francis Group, an **informa** business

CRC Press
Taylor & Francis Group
6000 Broken Sound Parkway NW, Suite 300
Boca Raton, FL 33487-2742

First issued in paperback 2016

Version Date: 20120418

ISBN 13: 978-1-138-19920-0 (pbk)
ISBN 13: 978-1-4398-3483-1 (hbk)

Library of Congress Cataloging-in-Publication Data

Microbiological research and development for the food industry / edited by Peter J. Taormina.
    p. cm.
    Includes bibliographical references and index.
    ISBN 978-1-4398-3483-1 (hardback)
    1. Food--Microbiology. 2. Food industry and trade. I. Taormina, Peter J., 1973-

QR115.M465 2012
579'.16--dc23   2012009375

Visit the Taylor & Francis Web site at
http://www.taylorandfrancis.com

and the CRC Press Web site at
http://www.crcpress.com

Contents

Foreword

The intent of this book is to describe the purposes and processes for conducting microbiological research and development for companies involved in food, beverage, and ingredient production and distribution, as well as for the many food-associated industries, including processing plant sanitation and food testing. The book covers a broad range of topics of importance to practicing microbiologists in the food industry. Included are the basics of setting up a food microbiology laboratory; procedures for validating the efficacy of process and product food safety controls; practices and protocols for developing effective food preservatives, sanitizers, and biocides; approaches to respond to food safety emergencies such as food recalls or in-plant pathogen contamination; predicting survival and growth of microbes in foods through modeling, identifying, and applying appropriate assays for bacterial pathogen detection in foods and identification; and approaches to meaningful communication of food microbiology research outcomes. Examples of successful research projects from industrial food microbiology laboratories are included throughout the chapters. The authors of each chapter are experts on their respective topics and are an excellent mix of industrial and academic scientists.

This book is a terrific primer and subsequent reference for industrial food microbiologists, who typically have to garner this information by on-the-job experience or through a consultant as many of these topics are not sufficiently addressed in university courses. To my knowledge, this book is the first of its kind. I know of none other that addresses food microbiological research and development from an industry perspective. I applaud Peter Taormina and his colleagues for undertaking this initiative as the book provides useful information that might not otherwise be available.

Michael P. Doyle
Center for Food Safety
University of Georgia
Griffin, Georgia

Contributors

Bledar Bisha
Center for Meat Safety and
Quality, Food Safety Cluster
Colorado State University
Fort Collins, Colorado

Scott L. Burnett
Malt-O-Meal Company
Lakeville, Minnesota

Mark Carter
QC Laboratories
Southampton, Pennsylvania

Hari P. Dwivedi
bioMérieux, Inc.
Hazelwood, Missouri

Ruth Eden
BioLumix, Inc.
Ann Arbor, Michigan

Paul A. Gibbs
Microbiology Department
Leatherhead Food Research
Leatherhead, Surrey
United Kingdom

Lawrence Goodridge
Center for Meat Safety and
Quality, Food Safety Cluster
Colorado State University
Fort Collins, Colorado

Margaret D. Hardin
IEH Laboratories and Consulting
Group
Lake Forest Park, Washington

Evangelia Komitopoulou
Food Safety Research Department
Leatherhead Food Research
Leatherhead, Surrey
United Kingdom

Jeffrey L. Kornacki
Kornacki Microbiology Solutions,
Inc.
Madison, Wisconsin

Junzhong Li
Ecolab Research Center
Eagan, Minnesota

John C. Mills
bioMérieux, Inc.
Hazelwood, Missouri

Keila L. Perez
Department of Animal Science
Texas A&M University
College Station, Texas

Patricia Rule
bioMérieux, Inc.
Hazelwood, Missouri

Gerard Ruth
Charm Sciences, Inc.
Lawrence, Massachusetts

Peter J. Taormina
John Morrell Food Group
Cincinnati, Ohio

T. Matthew Taylor
Department of Animal Science
Texas A&M University
College Station, Texas

Peter Wareing
Food Safety Research Department
Leatherhead Food Research
Leatherhead, Surrey
United Kingdom

chapter one

The case for microbiological research and development

Paul A. Gibbs, Peter J. Taormina, and
Evangelia Komitopoulou

Contents

1.1   What is food microbiological research and development?

Food microbiology is a relatively new applied science that continues to grow in size and scope. Although the study of microorganisms in foods and beverages can be traced back many years, contemporary food microbiology as an applied discipline is more recent and has roots in dairy, food, and environmental sanitation (Jay et al. 2005). Food microbiology encompasses safety and quality aspects of foods and food processing, but it can also involve antimicrobials and biocides, detection and enumeration methodology, fermentation optimization, and many other topics. Food microbiology also influences and informs public health activities like outbreak prevention and detection and trace-back investigation, principally through the use of laboratory data and results of on-site investigations (Guzewich et al. 1997). Food microbiology continues to evolve rapidly in consort with new food product and ingredient development as well

1

as with advancement of clinical diagnosis of disease, epidemiology, and pathogen detection technology.

All along the way, food microbiology as an applied scientific discipline has always fundamentally involved research and development. This seems to be the case principally because it is the study of dynamic organisms in complex systems at a very small scale and at a very large scale simultaneously. By way of example, consider a bacterium on a piece of lettuce. The survival, death, or multiplication of this single bacterial cell on a lettuce leaf can be influenced by the intrinsic microenvironment of the lettuce tissue, by prior conditions to which the cell was exposed, and by extrinsic factors to which it becomes exposed (e.g., atmosphere, temperature, native microflora, and presence of antimicrobials or sanitizers). At the same time, these microscopic interacting factors are occurring many times over at a large scale—thousands, if not millions, of times on other pieces of lettuce in the same container or perhaps (to expand the scale and scope of the example) within multiple lots of production. It is this form of complexity that requires a research and development focus to food microbiology. Taking the bacterium on lettuce example further, first one must research and perhaps even develop the best visualization, detection, or enumeration methods for studying this biological system. Then, an impetus may exist to research the best antimicrobial system or sanitizing chemical to reduce or eliminate this particular type of bacterium from this particular food system. Then, one might find it necessary to utilize biotechnology to develop a safe surrogate bacterium so that this system can be studied and validated within processing environments. Perhaps this work would lead to the ability to develop statistically based sampling schemes and to perform risk assessment. This illustrates how dual-scale complexity colors the study of microorganisms in foods and underscores the need for the scientific method for understanding these systems better.

Much of the increased interest in food microbiology has been driven by the increasing concern about food safety as well as the defense of food and agricultural commodities from intentional contamination. This concern has opened up new opportunities for food safety technologies—from pathogen, toxin, and allergen detection assays to novel processing technologies to destroy pathogens in new products and in new ways. From a quality standpoint, new food product and packaging developments and new food distribution channels continually drive new microbiological research and development of technologies to process and preserve food in such a way that it limits spoilage.

Microbiological dogma and basic research findings must always be adapted and validated when applied to the behavior of microorganisms in complex food systems and dynamic agricultural and food-processing environments. Microorganisms are exposed to myriad selective pressures in these settings that warrant detailed thoughtful study. As such, food microbiology research and development is fundamental to food safety and quality systems. To undertake such research and development, one must possess the knowledge about both the microorganism and the food system or environment itself to generate meaningful, relevant information.

In many instances, there is a lack of understanding of the microbial to food relationship that warrants specific targeted research and experimentation to validate hypotheses. Sometimes, outcomes of basic microbial research conducted *in vitro* may not hold true *in situ*, and assumptions must be checked by introducing bacterial, fungal, or viral cultures to food systems and environments. Of course, the very methodologies for recreating these introductions of microorganisms to food-related conditions or environments can themselves be a whole realm of study and research. The proper tracking of microorganisms through the use of cultural, polymerase chain reaction (PCR), or immunological methods often requires validation since food constituents can interfere with enumeration and detection (Swaminathan and Feng 1994). Sometimes, there is a lack of fundamental knowledge in the food science arena regarding microorganisms that had already been researched within basic or clinical microbiology. In other cases, newly discovered or otherwise lesser-known microorganisms are brought to the forefront of microbial sciences due to their selective advantage, and consequently their relevance, in food systems or in the human gastrointestinal tract. A microorganism brought to the forefront of interest in this manner may be the subject of further basic research to better understand its fundamental nature. It is the job of food microbiologists to bridge these gaps between the basic fundamental sciences and the food world. This can be achieved by thoughtful and insightful research and development.

1.2   Microbial research: Foundation of food safety and quality systems

Food safety microbiology is often considered anything pertaining to the detection, control, and elimination of bacterial, fungal, parasitic, and viral pathogens in foods and food-processing environments. Microbial food quality could similarly be defined except that the work would focus on nonpathogenic microorganisms. In some cases, there is no real practical distinction between interventions against foodborne pathogens and those targeting spoilage microorganisms. Often, strategies to detect, control, or eliminate pathogens will have an impact on nonpathogens and vice versa.

Perhaps now more than ever food industry executives are willing to consider significant financial commitment to food safety and consumer protection efforts. With high-profile outbreaks and recalls occurring almost monthly, and with increasing accountability being placed on food company executives by governments, food safety is at the forefront of the minds of most people who spend their days producing, regulating, researching, testing, or writing about foods. A 2009 "top-of-mind" survey of 596 chief executive officers (CEOs) and senior executives who replied anonymously between December 2008 and January 2009 revealed that "food safety" was the second most prominent concern for CEOs, behind corporate social responsibility (CIES–The Food Business Forum 2009). Food safety was still a top five concern in 2010 according to a follow-up survey of a similar group (Consumer Goods Forum 2010). A survey of U.S. consumers revealed that food safety is a concern, and that nearly half of consumers are confident in the safety of the food supply (International Food Information Council 2010). This implies roughly half of U.S. consumers lack confidence in the safety of the food supply. Considering these findings, it would seem that making the case for investment in food microbiology research and development would be easy.

However, all too commonly in industry, investment in food safety means little more than passing third-party audits, acquiring certificates of analysis (COAs) from suppliers, or obtaining negative test results from routine environmental and finished product pathogen and allergen testing. It is not surprising because, after all, the finished product testing is the most easily understood aspect of all that comprises a food safety system. In short, test results are easily communicated to and understood by laypersons. Entities that buy food commodities often require finished product testing results in the form of a COA, which helps their belief, justified or not, that a product was produced safely. Even some regulators prefer reviewing testing data rather than process control documentation as the primary means to make food safety assessments. However, the limitations of finished product testing in terms of the statistical probability of actually detecting low levels of contamination with a high degree of confidence have been thoroughly described (Dahms 2004; van Schothorst et al. 2009), as have the limitations and inherent uncertainty of the microbial assays themselves (Corry et al. 2007) in terms of sensitivity and specificity. Food microbiological research and development is a slow and complex endeavor compared to product testing and auditing. It requires experts to design studies that are meaningful and to interpret and apply the findings. While research projects can take months or even years to complete, an audit or pathogen testing procedure can take days or weeks, and business decisions are commonly made based on these results rather than on sound scientific research. Lack of rapidity and complexity can be obstacles to making a strong case for such research.

Food scientists and food microbiologists in particular understand these limitations and tend to be advocates for the hazard analysis and critical control point (HACCP) system as an overarching approach toward controlling food safety hazards. The HACCP approach relies more on

controlling the food production process as a means to reducing the biological, chemical, and physical hazards identified (Joint FAO/WHO Food
Standards Programme Codex Alimentarius Commission 2001; Pierson
and Corlett 1992) and incorporates finished product testing merely as a
tool to verify that the HACCP system is working not as a means to truly
assess and control risk. A HACCP system is only as good as the supporting science used to guide the many decisions made during its development and ongoing reassessments and will only succeed in truly reducing
risk if the supporting science is relevant and applied correctly to the specific food production process. This is where robust food safety and quality research and development on actual products and processes become
fundamental to effective risk assessment and risk reduction. Microbial
research that is designed to validate prerequisite programs, critical control points (CCPs), and control points (CPs) is essential to the proper
implementation of HACCP systems. Audits, COAs, and finished product testing, while necessary, unfortunately can produce a false sense of
security if they are the only such measures employed in a food safety
and quality system. Companies that produce agricultural, ingredient,
food, and beverage products must incorporate sound science into food
safety and quality systems and must understand the performance of
food-processing systems and interventions to truly determine and manage risk of foodborne illness. Wherever scientific information is lacking,
research and development must ensue to fill those gaps. While audits,
COAs, and finished product testing can be understood by laypersons, a
truly effective system must reach beyond perception and make research
and development investments in food safety and quality systems, even if
competitors are not. This does not ensure absence of critical failures (i.e.,
outbreaks, recalls, and spoilage), but it does reduce risk of adverse events
occurring, sometimes even with a quantifiable degree of confidence.

Thoughtful and insightful collection, analysis, and interpretation of
data can provide much better detail and confidence about the quality
and safety of manufactured food products. The application and effective
implementation of good manufacturing practices (GMPs) and HACCP
systems represent efficacious means of control of food quality and safety.
This leads to better understanding of risk and better decisions for consumer and brand protection, especially when coupled with a strong
microbial research and development program.

1.3   Understanding products and processes

Food microbiologists work at the interface between the microbial sciences and food science, so it is critical that they understand how microbes
behave when interacting with food products and processes. For example,
the quantitative efficacy of lethal processes, such as cooking, may vary

depending on the strain of bacterium and by components of the food substrates. Attributes like water activity (a_w) and lipid content have been shown to have considerable effects on the heat resistance of microorganisms. When a food process does not go the way it was expected and spoilage or contamination ensues, food microbiology researchers should be there ready to provide answers and to develop and implement solutions.

Also, since there is increasing commercial activity in the development of "new" food and beverage products in response to or anticipation of consumer demands, there is a corresponding need for ongoing research and validation of the microbial safety and quality of new food products. New food product developments can range from considerable cutting-edge developments that greatly change the microenvironment, to simple line or flavor extensions that are, in reality, only slight modifications of previous products having inconsequential effects on microorganisms. Sometimes, new developments in ingredient technologies or food processes lead to new food and beverage product developments that can have drastic or minimal effects. It is the role of food microbiology researchers to understand if and how these new developments will change the behavior of foodborne microorganisms. Indeed, without proper research and validation of new ingredients or new food products, these amendments to formulas and processes can have unfortunate consequences for producers and consumers.

There are some rather stinging examples of problems occurring when product developments went into commerce without ample supporting scientific research. Substitution of simple ingredients, such as sucrose by glucose/fructose syrups, has given rise to spoilage problems with respect to fermentation by wild-type, sucrose-negative, strains of *Zygosaccharomyces bailii*, with unfortunate explosive results. Many strains of this yeast are now exhibiting preference for fructose as well as resistance to preservatives (Stratford et al. 2000). Similarly, utilization of a lower DE (dextrose equivalent) value starch hydrolysate, providing a slight rise in water activity (a_w), led to fermentation and adaptation of a yeast for growth and gas production at the lower, original a_w.

In heat processing of foods with low a_w, several research reports have identified that microorganisms in the vegetative phase become much more heat resistant, culminating in extreme thermotolerance at a_w values near or below about 0.20; examples include *Salmonella* in chocolate (Barrile and Cone 1970; Barrile et al. 1970) or on dry nuts (Doyle and Mazzotta 2000) or in peanut butter (Ma et al. 2009; Shachar and Yaron 2006). However, the effect is not always directly correlated with a_w *per se* as the solutes involved (e.g., sugars, polyols, salt) also have marked and different effects on heat resistance (Corry 1974; Mattick et al. 2001). Thermal processing of such foods should therefore be based on D-value of microorganisms under relevant intrinsic conditions, such as low a_w or high fat.

Fundamental lack of understanding of behavior of microorganisms in food products has led to foodborne disease outbreaks, such as transmission of *Salmonella* through consumption of peanut butter or peanut paste as ingredients in numerous snack foods (Anonymous 2007, 2009). However, in this instance microbiological research had already demonstrated the ability of the pathogen to survive in various peanut products under common conditions of storage (Burnett et al. 2000). So it was perhaps an apparent lack of risk communication or failure of processors and auditors to ascribe risk of this biological hazard properly during the HACCP plan development and subsequent reassessments. It remains to be seen whether the follow-up research on survival of the pathogen (Park et al. 2008) will be considered in future risk analysis for peanut butter pastes and spreads. Nonetheless, literature reviews, an early and critical facet of research projects, would have uncovered *Salmonella* as a biological hazard in this scenario.

Many years ago, raw milk on the farm was cooled in aluminum cans with running cold water (to about 10–12°C) before transport to the dairy at ambient temperatures. The characteristic spoilage of such milk was caused by lactic acid bacteria. In the 1960s and 1970s, bulk tank chilling of raw milk on the farm was installed, which led to a change in the developing microflora to a pseudomonad type, with very different physiological and metabolic properties than lactic acid bacteria. This pseudomonad-type predominance of the microflora gave rise to problems in the old "methylene blue" or "rezasurin" test for the hygienic quality of raw milk. It was also found that on ultrahigh temperature (UHT—141°C < 3 seconds) treatment of milk contaminated with large numbers of pseudomonads, the lipase and protease enzymes produced by these microorganisms before heat treatment were not totally destroyed. Therefore, the sterile UHT milk gradually developed a soft, gelled structure and "soapy" taints.

A potential food safety issue was avoided in the reformulation of cured meats. On cooking meats containing nitrite, nitrosamines can be formed. As there was evidence that nitrosamines can elicit cancerous cell development and proliferation, there was a call for the reduction on the levels of nitrite used. However, it was realized that the levels of salt, nitrite, and phosphates were extremely important in preventing growth and toxin production by *Clostridium botulinum* (Perigo and Roberts 1968; Perigo et al. 1967), and a large research program on the formulation of cured meats effectively obviated the development of low-salt and nitrite-cured meats, almost certainly avoiding cases of botulism. This was an example of when a consortium of industry, academic, and government and nongovernment organizations collaborated on food microbiological research to solve a real issue. Even today, there are occasional cases of botulism caused by poorly cured hams, usually by home curing; the salt and nitrite does not penetrate rapidly or sufficiently, and the temperatures are too high to inhibit

C. *botulinum* spores from germinating, growing, and forming toxin. It was only later that the specific and synergistic physiological actions of salt, nitrite, and phosphates on clostridial metabolism, growth, and toxin production were elucidated (Woods et al. 1981). Today, a similar debate rages about another important food ingredient, salt. As with the nitrite debate, the debate over safe levels of salt consumption creates conflicting views between those concerned with human nutrition and physiology and human cardiovascular disease against those in charge of prevention of foodborne microbial disease and spoilage. It has been suggested that a reduction in salt content of foods without proper research and validation could lead to an increase in human foodborne illness (Taormina 2010).

It is quite clear from these few examples that product development teams must be made aware of the possible consequences of formulation and process changes in relation to potential microbiological safety and shelf life problems before progressing too far with such product modifications. It is too late if the microbiologists are asked to evaluate the safety and shelf life of a "new" product just 1 month before launch. There are few guarantees in biology, and food microbiology is no different. Applying proper research and validation will not guarantee that a product will never cause a single illness or an isolated spoilage event, but neither can sampling 99% of a lot as there would still be that 1% of the lot in question. However, proper research, development, and validation can create a strong level of confidence in the safety and wholesomeness of food. A rush to market with a product without proper research can lead to market withdrawals, recalls, and worse—human illness and even death.

1.4　Finding and mitigating risks

Foodborne disease outbreaks cause substantial economic impact through medical costs and costs related to loss of productivity and quality of life (Shin et al. 2010). There are also many important social costs that are typically underestimated, such as the value of pain, suffering, and functional disability. Governments weigh the cost of food safety prevention and control regulations against the estimated benefits to the population of reducing foodborne disease to determine net benefits so that governments have information to allocate funds among competing programs efficiently (Buzby and Roberts 2009). In like manner, the food industry must run the cost-to-benefit models to see if investment in food safety systems above and beyond regulatory compliance is justified for the benefit of further risk reduction.

Outbreaks of food poisoning or foodborne infections, especially of the more serious diseases such as Shiga toxin-producing *Escherichia coli* (STEC), listeriosis, or botulism that are traceable to a food source can cause irreparable harm to people. Outbreaks can also rapidly destroy the reputation and financial viability of a food business, which also affects people, albeit in a different but real way. Exposure of a food business to these types of risk is an extreme hazard to the viability of the company in itself. In most cases of foodborne illness, it is now possible to predict the hazards associated with a particular food product, both from historical records and from knowledge of the ecological niches occupied by the hazardous organisms and their survival and growth characteristics. Thus, a food manufacturer should be able to predict the risks associated with producing particular foods and take all reasonable steps to control the hazards in formulations and processes.

However, this is not always the case since there have been some notable developments in the hazards and risks associated with some common foods. Such is the case with the STEC organisms that are a relatively recent arrival among the ranks of hazardous bacteria. In what appears to be its normal host animal, cattle, STEC causes little if any problem, and thus is not recognized as a zoonotic organism, as is the case with *Salmonella*. Similarly, *Listeria monocytogenes* seems to have taken advantage of the ecological niche of chilled foods with little other means of preservation (e.g., soft cheeses, cold-smoked fish, patés, coleslaw, etc.), and listeriosis was not recognized as a significant human foodborne infection until there was a large outbreak of listeriosis (from coleslaw in Canada in 1981; Schlech et al., 1983). Since then, there have been several outbreaks traced to various food sources and sporadic cases not traceable to a food source. Exacerbating factors are the increase in the elderly population and immunocompromised persons, increased demand for convenience foods ready to eat (RTE) foods, decreased use of preservatives, and increasing consumption of fresh produce.

Another such case is that of botulism from garlic in oil as a condiment (in 1985 in Vancouver, Canada, and in 1988 in St. Louis and other places). This problem arose from not recognizing the hazards of *C. botulinum* in soil-grown vegetables or recognizing that garlic is not an item with a low a_w and is contained in an anaerobic environment. However, there were no previous records of botulism from garlic in oil that could be used to indicate the risks.

Risk analysis is not yet a precise science owing to the variability regarding the occurrence of most of the pathogens in foods, infective/toxic dose in specific food types, numbers of cases caused by each specific hazard, exposure of susceptible populations, and so on. However, from the point of view of the food business, it is essential to reduce risks to the lowest possible level by taking precautions in the formulations of foods and in processes and storage conditions applied to foods. This requires setting limits with validated parameters (i.e., CCPs) of formulations and processes and monitoring those parameters (e.g., pH, a_w, times, and temperatures). These parameters work best when designed to achieve a specific food safety objective (FSO).

1.5   Competitive aspects: Intellectual property and return on investment

Some of the major areas of microbiological research and development that have an impact on the food and beverage industry include the following: fermentations that improve food and food ingredients; process technologies, biocides, or antimicrobial agents that control, reduce, or eliminate microorganisms from food systems or processing environments; and methodologies to detect, track, and study microorganisms in food and beverage products and related environments. Development of each of these technical advancements requires different approaches and types of experimentation for proof of concept, optimization, and validation phases. The end result for the inventor or developer is a novel technology that can be monetized as a return on the investment in the research and development efforts.

Microbiological methods for enumeration and detection of microorganisms continue to be in high demand, given all the reasons described. Many new developments have been reported over the past several years, and with public and private research funding continuing to fuel research, new developments will surely ensue. A private market research company issued a report summarizing the growing market for food safety testing in the United States; the report valued the market size at $3.4 billion in 2010, with a projected climb to $4.7 billion by 2015 (Gainer 2010). It was revealed that bacterial pathogen testing represented the majority of that market, dwarfing the markets for both toxin testing and genetically modified organism (GMO) testing (Figure 1.1). A survey from the early 2000s

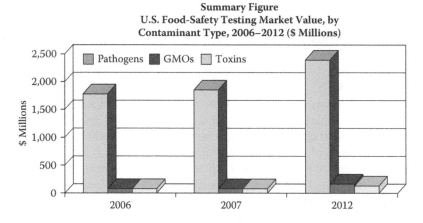

Figure 1.1 U.S. food safety testing market value estimation by contamination type, 2006–2012. (Reprinted from Gainer, K. 2010. *Food safety testing: technologies and markets.* Wellesley, MA: BCC Research. With permission from BCC Research.)

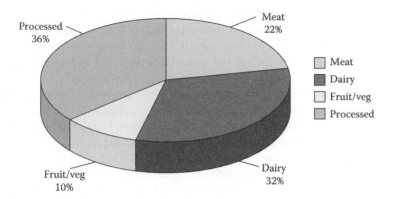

Figure 1.2 Total microbial tests used by the food industry by sector. (Reprinted from Alocilja, E. C., and S. M. Radke. 2003. Market analysis of biosensors for food safety. *Biosensors and Bioelectronics* 18 (5–6):841–846. With permission from Elsevier.)

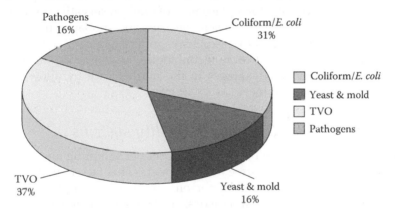

Figure 1.3 Total microbial tests used by the food industry by assay. TVO = total viable organisms. (Reprinted from Alocilja, E. C., and S. M. Radke. 2003. Market analysis of biosensors for food safety. *Biosensors and Bioelectronics* 18 (5–6):841–846. With permission from Elsevier.)

(Figure 1.2) indicated a somewhat even distribution of testing in this area among food commodities as well as among target microorganism groups (Figure 1.3) (Alocilja and Radke 2003).

Areas for new development in microbiological detection from foods will most likely center around sensitivity, specificity, time to completion of assay, and cost per assay. Also, new tests will be needed for emerging pathogens and spoilage microorganisms.

Research toward new methods of preservation and processing of foods is expensive, and there must be a financial return for the inventor, by means of protected intellectual property (i.e., by patents, licensing royalties, or specific competitive advantages in the marketplace that are created by technology). Several new processing technologies have been researched in depth, but to date not all of them have proven totally satisfactory when applied on a large scale or acceptable to the wider public—the ultimate customer. As a case in point, irradiation is an effective, and only recently permitted and acceptable, treatment for the elimination of specific organisms (i.e., verocytotoxic *Escherichia coli* [VTEC] in ground beef) when all other control measures seem to have failed in the control of this very serious pathogen. However, irradiation generally has not been accepted by the consumer, even though extensive research has shown unequivocally the safety, benefits, and theoretical consumer acceptance (Bruhn 1995) of this relatively inexpensive process. High hydrostatic pressure (HHP) processing has also been demonstrated to be very effective in elimination of *Listeria monocytogenes* from packaged RTE meats and is now accepted and used routinely by some manufacturers. HHP provides a trendy solution to the consumer demand for "clean label" food products since it offers the potential for replacement of food preservatives otherwise used for controlling growth of *L. monocytogenes* and spoilage microflora in RTE meats. However, it is rather expensive in terms of capital equipment and running costs. Both of these technologies, as examples, have given their users competitive advantages in the marketplace by allowing the sale of pathogen-free products.

1.6   Proactive food safety and quality systems

It has been well established that finished product sampling and analyses cannot guarantee pathogen-free food products. This fact, however, is not recognized by most media and laypersons, and it is not simple to communicate to those not educated in statistics, microbiology, or general sciences. Typically, the very first bit of information mentioned in media reports about outbreaks of foodborne illness is the summary of product testing results—whether tests were done or what percentage of results was positive. The distribution of organisms in foods is not homogeneous, and random sampling and analysis, even of quite large numbers of samples, is subject to large statistical variation and low statistical confidence of detection of a low level of contamination in a production lot.* Thus, for the U.S. space program, the Pillsbury Corporation developed the approach to safe food production referred to as HACCP, which is essentially an extension and formalization of GMPs. When implemented effectively, HACCP

---

* For more details on the statistical aspects of food testing for microorganisms, including distribution of microorganisms in food and statistical sampling, see the work of Jarvis (1989).

identifies the biological hazards (i.e., microbial pathogens or their toxins) associated with production of a given food product, and the steps (CCPs) in the processing that eliminate or control that hazard. The CCPs must be validated with respect to efficacy in controlling the hazards and must be monitored (e.g., measurements of times, temperatures, pH, and a_w values) on a regular basis (e.g., for each batch of food produced) and the results recorded. It is essential that any changes, however apparently minor, result in a reevaluation of the hazard analysis and the efficacy of the process steps, the CCPs. Examples may include changes in sugars used (discussed previously); exchanging citric acid for acetic acid (e.g., in pickles, sauces, and mayonnaises); levels of salt and nitrite in cured meat products; omission of propionate from bread to provide a clean label that will result in germination and growth of surviving *Bacillus* spp. spores and production of "ropy" bread or food poisoning; and so on. Process and product validation in support of HACCP decisions are discussed in detail in Chapters 3 and 4, respectively.

Just as important is the auditing of suppliers of raw materials since these may be sourced from different parts of the world with quite different types and levels of pathogens. Examples of the latter include the introduction of "new" serotypes of *Salmonella* into a country via raw materials (e.g., dried egg from China in the 1950s) or from finished products (e.g., *Salmonella* Napoli into the United Kingdom in chocolate snack bars from Italy). It is also essential not to forget those hazardous organisms that occur and cause infections or intoxications only rarely (e.g., botulism) and the control measures necessary to control the access of the organism to foods and processing plants and to apply correct processes and formulations to minimize contamination and growth.

One effective starting point for analyzing hazards and setting of CCPs is to use the various microbiological predictive (or computational) modeling programs, such as Growth Predictor, freely available from the Institute of Food Research (Norwich, UK). This is far more effective than many challenge experiments of new products or processes (although not all growth or death conditions may be available for particular organisms). Another example is the Pathogen Modeling Program (PMP 7.0). Predictive microbiological modeling is discussed in Chapter 8.

Food manufacturers and the microbiologists in charge of food safety and quality systems must keep abreast of the current literature with regard to the appearance of new hazards in their raw materials or environment and evaluate their current CCPs for efficacy in eliminating or controlling the growth of these emergent or reemergent pathogens. The value of Internet warnings from government agencies or newsletters and e-mail listserves for rapid dissemination of data should not be underestimated as sources of such information. Alternatively, or in addition, it is helpful to keep in contact with one of the food research institutes, food research associations, or the like for up-to-date information.

1.7 Outsourcing microbial research and development to contract labs, universities, and consultants

Maintaining and staffing a dedicated microbiological laboratory "in-house" can be an expensive endeavor that may not be justifiable within certain organizations and companies. In some cases, there is a compelling financial argument for outsourcing analytical work on an "as-needed" basis. One meat company that was headquartered near the Chicago stockyards in the 1970s had extensive research laboratories complete with laboratory animal testing until a new president no longer saw the need and dissolved the entire group (W. L. Brown, personal communication, 2003). The head of that laboratory group, Dr. William Brown, later founded a successful research microbiological and analytical laboratory that provides services to the food industry to this day.* There are many laboratories capable of microbiological and analytical contractual work, charging competitive rates for routine analyses (e.g., for aerobic and anaerobic plate counts, yeast molds, coliforms and *E. coli, Salmonella,* and *Listeria,* to name a few). As mentioned, the statistics of distribution and sampling for microorganisms do not lead to a large degree of statistical confidence in the results obtained, particularly for the presence or absence of pathogens. Also, the business model of contract analytical laboratories is that of rapid turnaround and low prices, not necessarily in developing and applying the most appropriate methods or in problem solving. Such laboratories may or may not be able to offer any advice based on the results or suggest more relevant follow-up analyses. Further, the specialized research and development projects on food ingredients, food products, technologies, and so on may be beyond the scope of contract testing labs. The authors of this chapter have worked in roles on every side of the possible working relationships between contract lab, university lab, consultant, and food processor or allied food industry products and services. Each situation should be evaluated differently as there is a variety of different strengths and weaknesses as part of the character of each of the available outsourced lab options. Matching projects to strengths of these available resources is essential.

---

* In the United States, there are several examples of food microbiologists taking their industry knowledge and founding successful third-party research and testing laboratories: Dr. John H. Silliker founded Silliker Corporation in 1967; Dr. Robert H. Deibel founded Deibel Labs.

Sometimes, research can be guided from afar, thanks to modern communications and travel possibilities, and therefore an outsourced research partner may be an attractive alternative. In many cases, the methods applied are crucial in coming to conclusions about the shelf life or safety risks of a batch of food, so high-level expertise might be one critical factor in choosing outside services. Examples are the choice of incubation temperatures for spoilage organisms for fish and vacuum-packaged (VP) meats; *Photobacterium phosphoreum*, a potent spoilage organism for ice-stored cod especially, and other marine fish, is a strict psychrophile, not surviving above about 28°C and therefore not detected in pour plates. Similarly, the strictly psychrotrophic clostridia responsible for spoilage of VP meats, *C. laramie* and *estertheticum*, are also not detected in pour plates as they are killed by the temperature of molten cooled agar. The methods of detecting sublethally damaged cells of pathogens, for example, also require knowledge and application of modifications of the standard selective methods and media to obtain correct results. The fruit juice spoilage organism *Alicyclobacillus acidoterrestris* also requires an appropriate medium for detection as it does not tolerate high levels of amino nitrogen compounds. It should be noted that there are several commercial food microbiology laboratories throughout the world that staff well-educated and trained scientists who can devise thoughtful research to address real needs in a cost-effective way. It is not our place to point them out in this text, but we rather leave it to the reader to seek these laboratories with this book in hand as a reference.

Microbiological problem solving in the food and beverage industries does seem to require considerable experience and an ability to "think outside the box," as well as familiarity with formulations and processes. While some contract laboratories are known for their expertise in these areas, still others may not have the depth of expertise to investigate problems as they arise or may charge quite heavily for such work as they subcontract it to an expert not on their staff.

Another possible avenue for outsourcing research and solving problems is university faculty experts in food microbiology. Sometimes, dealing with an academic research group can take longer than commercial laboratories since projects can be delayed while a student is identified for the research and becomes familiar with the topic. Another obstacle to successful microbiological research and development projects with universities is the area of grants and contracts. Sometimes, universities will not be willing to enter into a research agreement unless the university retains at least some of the rights to new developments. Neither of these options seems acceptable to food businesses, which often need immediate advice and solutions to problems. However, universities offer research depth, detail, instrumentation, and expertise that few private laboratories can match. This makes universities excellent options for long-term, complex, and large studies. The food research associations or institutes would appear to be among the best options for the food and beverage industries as they have a wide variety of experienced scientists able to combine their expertise for resolving problems quickly, many of which they have probably seen previously. One aspect of the food and beverage industries that these laboratories can help companies with particularly is that of reformulation of products since they have in-house product innovation teams who can interact with the microbiologists with respect to processing for food safety and shelf life issues, long before a product is ready to launch.

Whether food microbiological research and development is outsourced to contract laboratories, consultants, universities, or food research institutes, the most important consideration is to approach the work in a partnering mentality with a thought given to the long-term benefits rather than a series of short, quick studies. Inevitably, if some entity recognizes the need to perform research externally, the need will continue to rise and grow. Many of the issues and questions that warrant research have a long-term effect. As such, establishing a working partnership with an outside entity on a contractual basis or even a less-formal arrangement is highly recommended.

1.8   Conclusion

The following chapters contain much of the information one would need to conduct food microbiological research and development for a variety of needs and in a variety of settings. Although there is a special focus on research and development that has an impact on the food and beverage industry, the content of this book as a whole is meant to be useful to anyone involved with food microbiological research and development. Government and academic researchers working in this area will see the benefit and usefulness of the information presented here, and so will those who do research directly for the food industry. Whether efforts are focused on methods development; troubleshooting contamination; developing new preservatives, sanitizers, or biocides; or validating the behavior of microorganisms in food systems, the information in subsequent chapters should have plenty of relevance and utility. For those not actually conducting the research and development but rather managing others who do so, this book will assist with understanding the entire process and ultimately aid in managing and guiding others to the expected completion of the work.

Food microbiology as an applied scientific field is growing in importance and scope. Research and development activities will continue to be fundamental to the advancement of the understanding of foodborne microorganisms. It is hoped this book will be an essential guide to those involved with any aspect of food microbiology research and development.

References

Alocilja, E. C., and S. M. Radke. 2003. Market analysis of biosensors for food safety. *Biosensors and Bioelectronics* 18 (5–6):841–846.

Anonymous. 2007. Multistate outbreak of *Salmonella* serotype Tennessee infections associated with peanut butter—United States, 2006–2007. *Morbidity and Mortality Weekly Report* 56 (21):521–524.

Anonymous. 2009. Multistate outbreak of *Salmonella* infections associated with peanut butter and peanut butter-containing products—United States, 2008–2009. *Morbidity and Mortality Weekly Report* 58 (4):85–90.

Barrile, J. C., and J. F. Cone. 1970. Effect of added moisture on the heat resistance of *Salmonella anatum* in milk chocolate. *Applied Microbiology* 19 (1):177–178.

Barrile, J. C., J. F. Cone, and P. G. Keeney. 1970. A study of salmonellae survival in milk chocolate. *Manufacturing Confectioner* 50 (9):34–39.

Bruhn, C. M. 1995. Consumer attitudes and market response to irradiated food. *Journal of Food Protection* 58:175–181.

Burnett, S. L., E. R. Gehm, W. R. Weissinger, and L. R. Beuchat. 2000. Survival of *Salmonella* in peanut butter and peanut butter spread. *Journal of Applied Microbiology* 89 (3):472–477.

Buzby, J. C., and T. Roberts. 2009. The economics of enteric infections: human foodborne disease costs. *Gastroenterology* 136 (6):1851–1862.

CIES–The Food Business Forum. 2009. Top of mind 2009. Available from http://www.ciesnet.com/pfiles/press_release/2009-02-02-PR-TOM.pdf, accessed November 12, 2010.

Consumer Goods Forum. 2010. Top of mind survey 2010. Available from http://www.theconsumergoodsforum.com; accessed November 12, 2010.

Corry, J. E. L. 1974. The effect of sugars and polyols on the heat resistance of salmonellae. *Journal of Applied Bacteriology* 37:31–43.

Corry, J. E. L., B. Jarvis, S. Passmore, and A. Hedges. 2007. A critical review of measurement uncertainty in the enumeration of food micro-organisms. *Food Microbiology* 24 (3):230–253.

Dahms, S. 2004. Microbiological sampling plans—statistical aspects. *Mitteilungen aus Lebensmitteluntersuchung und Hygiene* 95:32–44.

Doyle, M. E., and A. S. Mazzotta. 2000. Review of studies on the thermal resistance of salmonellae. *Journal of Food Protection* 63 (6):779–795.

Gainer, K. 2010. *Food safety testing: technologies and markets*. Wellesley, MA: BCC Research.

Guzewich, J. J., F. L. Bryan, and E. C. D. Todd. 1997. Surveillance of foodborne disease I. Purposes and types of surveillance systems and networks. *Journal of Food Protection* 60 (5):555–566.

International Food Information Council. 2010. 2010 food and health survey: consumer attitudes toward food safety, nutrition and health 2010. Available from http://www.foodinsight.org/Resources/Detail.aspx?topic=2010_Food_Health_Survey_Consumer_Attitudes_Toward_Food_Safety_Nutrition_Health; accessed November 12, 2010.

Jarvis, B. 1989. *Statistical aspects of the microbiological analysis of foods*. 25 vols. Vol. 21, *Progress in industrial microbiology*. Oxford, UK: Elsevier.

Jay, J. M., M. J. Loessner, and D. A. Golden. 2005. History of microorganisms in food. In *Modern food microbiology*. New York: Springer. pp. 1–10.

Joint FAO/WHO Food Standards Programme Codex Alimentarius Commission. 2001. Hazard analysis and critical control point (HACCP) system and guidelines for its application in the CODEX Alimentarius Commission and the FAO/WHO Food Standards Programme (ed. FAO/WHO). Rome: FAO/WHO.

Ma, L., G. Zhang, P. Gerner-Smidt, V. Mantripragada, I. Ezeoke, and M. P. Doyle. 2009. Thermal inactivation of *Salmonella* in peanut butter. *Journal of Food Protection* 72 (8):1596–1601.

Mattick, K. L., F. Jorgensen, P. Wang, J. Pound, M. H. Vandeven, L. R. Ward, J. D. Legan, H. M. Lappin-Scott, and T. J. Humphrey. 2001. Effect of challenge temperature and solute type on heat tolerance of *Salmonella* serovars at low water activity. *Applied and Environmental Microbiology* 67 (9):4128–4136.

Park, E. J., S. W. Oh, and D. H. Kang. 2008. Fate of *Salmonella* Tennessee in peanut butter at 4 and 22 degrees C. *Journal of Food Science* 73 (2):M82–M86.

Perigo, J. A., and T. A. Roberts. 1968. Inhibition of clostridia by nitrite. *Journal of Food Technology* 3:91–94.

Perigo, J. A., E. Whiting, and T. E. Bashford. 1967. Observations of the inhibition of vegetative cells of *Clostridium sporogenes* by nitrite which has been autoclaved in a laboratory medium, discussed in the context of sublethally processed cured meats. *Journal of Food Technology* 2:377–397.

Pierson, M. D., and D. A. Corlett. 1992. *HACCP: principles and applications* (ed. Institute of Food Technologists). New York: Van Nostrand Reinhold.

Schlech, W. F., P. M. Lavigne, R. A. Bortolussi, A. C. Allen, E. V. Haldane, A. J. Wort, A. W. Hightower, S. E. Johnson, S. H. King, E. S. Nicholls, and C. V. Broome. 1983. Epidemic listeriosis: evidence for transmission by food. *New England Journal of Medicine* 308:203–206.

Shachar, D., and S. Yaron. 2006. Heat tolerance of *Salmonella enterica* serovars Agona, Enteritidis, and Typhimurium in peanut butter. *Journal of Food Protection* 69 (11):2687–2691.

Shin, H., S. Lee, J. S. Kim, J. Kim, and K. H. Han. 2010. Socioeconomic costs of foodborne disease using the cost-of-illness model: applying the QALY method. *Journal of Preventative Medicine and Public Health* 43 (4):352–361.

Stratford, M., P. D. Hofman, and M. B. Cole. 2000. Fruit juices, fruit drinks, and soft drinks. In *The microbiological safety and quality of food* (ed. B. M. Lund, T. C. Baird-Parker, and G. W. Gould). Gaithersburg, MD: Aspen, pp. 836–869.

Swaminathan, B., and P. Feng. 1994. Rapid detection of food-borne pathogenic bacteria. *Annual Review of Microbiology* 48 (1):401–426.

Taormina, P. J. 2010. Implications of salt and sodium reduction on microbial food safety. *Critical Reviews in Food Science and Nutrition* 50 (3):209–227.

van Schothorst, M., M. H. Zwietering, T. Ross, R. L. Buchanan, and M. B. Cole. 2009. Relating microbiological criteria to food safety objectives and performance objectives. *Food Control* 20 (11):967–979.

Woods, L. F. J., J. M. Wood, and P. A. Gibbs. 1981. The involvement of nitric oxide in the inhibition of the phosphoroclastic system in *Clostridium sporogenes* by sodium nitrite. *Journal of General Microbiology* 125:399–406.

chapter two

Building research and development capabilities

Peter J. Taormina

Contents

2.1 Introduction

One cannot realistically consider the implementation of research ideas at the bench and performing of research projects until the mechanisms to plan, conduct, and report research are in place. This chapter provides the framework for the startup process for a food microbiology research laboratory. The intent of this chapter is to pose questions that should be considered when starting up or restarting microbial research activities related to food, beverage, ingredient, or microbial methods. As one undertakes this process, he or she will find that there are numerous possible choices at each step. The overriding goal of the research and development (R&D) will largely dictate how these decisions are made. Also, the work style of the lead researcher, support staff, and organizational hierarchy will most likely influence decisions, leading to an eventual workflow and data-reporting mechanism that matches the same. This chapter discusses considerations for selecting the team; securing the funding; setting up the laboratory; stocking the laboratory with equipment, materials, and supplies; developing external research partnerships; and enabling an effective reporting mechanism.

Once a decision is made to conduct microbiological R&D within an organization, the process of building infrastructure, staffing positions, and developing workflow systems must begin. Rather than a sequential step-by-step process, such an endeavor is more likely to be a long-term project moving forward in all aspects toward suitable and sustainable levels of productivity. The laboratory cannot function without a team, but the team cannot function without a laboratory. Similarly, the laboratory needs equipment, materials, supplies, and systems for workflow and reporting, but such cannot be made of use without a team of laboratorians or technicians to work and then produce results. In some organizations, it may be necessary to have a data-reporting mechanism in place to help justify expenditures on team, laboratory, and supplies. In other instances, seed money may be required to get the project up and running. The development of team, laboratory, equipment/supplies, contract partnerships, and reporting mechanisms, concurrently in many cases, may be required by the organization and can be the most difficult aspect of building microbial research capabilities. In short, getting started requires lots of planning, preparation, and hard work.

Early success is critical to surviving the startup phase and establishing a successful research microbiology laboratory. Fortunately for bacteriologists, obtaining initial data can be done quickly in many instances due to growth rates of most bacteria and speed of most molecular microbiology methods. Mycologists may find data generation takes a bit longer than for bacteriologists. Virologists and parasitologists may have more difficulty in obtaining results quickly due to relative difficulty in propagation of these microorganisms and obtaining sufficient material for study. As such, R&D on viruses and parasites might require a much more substantial commitment in terms of time to produce deliverables and up-front monetary investment or long-term funding. However, the reward for researchers in foodborne virus and parasitic disease who establish laboratories in these areas could be a much less-crowded field of competition for scientific discovery and competitive funding.

Regardless of specialization, the need for food microbiological R&D for all pathogen types and commensal microorganisms exists. As mentioned in Chapter 1, this need should continue to expand as the intricacy of the food system increases, human populations increase, and the science of epidemiology uncovers previously unrecognized routes of foodborne illness. From development of novel methods of detection to strategies to control and eliminate microorganisms from foods and beverages, technological advancements that anticipate or respond to these growing needs will be critical.

2.2   The team

Before the first pipette tip touches microbial culture, a team of people must exist. In so-called bootstrapping situations, this may be a team of one person for a time, but a plan to assemble a larger team must be quickly put into place. For early stages of operating a microbiological research lab, a team may consist of or include contract or part-time technical support. It is not advisable that the laboratory leader (whether manager, director, or other) attempt to perform daily laboratory activities as well as oversee the management of the lab. His or her time will be better spent managing operations, overseeing projects, and managing the team while ensuring compliance with all pertinent safety and accreditation criteria. The person overseeing laboratory operations (manager, director, principal investigator, etc.) will be ultimately responsible for biological and chemical safety and environmental compliance, but may also oversee work schedules and performance reviews, conflict resolutions, and client or customer relations. Perhaps most importantly, the lead researcher must finalize the reports and explain and promote the findings of the research. This is the aspect of the process that largely determines future funding opportunity. Examples of typical organizational structures for industry, academic, and government research groups are shown in Figure 2.1.

Every team needs a leader. Obviously, the principal investigator or research leader assumes this role over a R&D group. However, since this person will also be responsible for securing funding, managing and monitoring all the research projects, writing research reports, and devising new experimental protocols, a different person should be appointed to be the lead in the laboratory. With the right research coordinator or laboratory manager "running the show" in the lab, the principal investigator (i.e., lead researcher) can focus on the key responsibilities, without which the whole group would crumble. This R&D lead investigator must also serve as the ambassador of the lab and the liaison between the ongoing research and those whom it affects. Probably the leader's most important role is to market the services of the research group to ensure that there is outside interest and funding for future work. For the sake of scientific credibility, it is important that the lead scientist never stray too far away from the substance of the group's findings for the sake of style or effectiveness at wooing grant review panels or would-be clients. Truthful, accurate, clear, and realistic (not overstated) reporting will often win the day.

In many cases, a team will be "inherited" from a previous group. If this is the case, competencies and capabilities of laboratory workers should be assessed to ensure that skills and strengths are deployed where most needed for maximum effectiveness and efficiency. Laboratory workers may or may not have been previously involved in R&D. For example, laboratorians accustomed to the rather-constant pace of routine pathogen

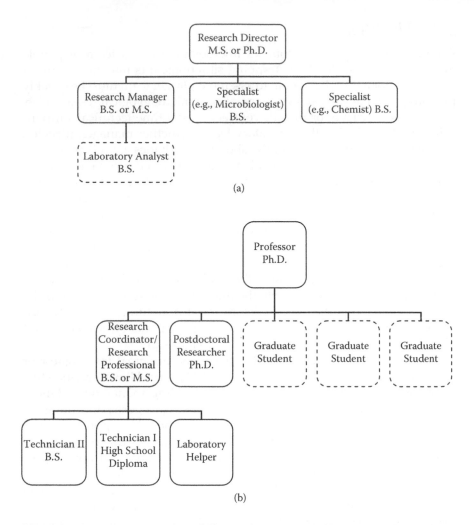

Figure 2.1 Typical organizational charts of food microbiology research groups in industry (a), academia (b), and government (c). Positions outlined with dashed line are typically rotational or filled according to project needs or available funding.

testing may not succeed at the relatively inconsistent (but just as rigor-ous) pace of research microbiology. In such cases, laboratory workers will need to be assessed for skill sets and work styles early on in the transi-tion to research. Technical skill sets can be assessed using a variety of means, such as AOAC proficiency testing (Augustin and Carlier 2002; Edson et al. 2009) and interlaboratory comparisons with reference sam-ples (And and Steneryd 1993). Large laboratories with many personnel will typically develop a hierarchy, with more experienced technicians having more leeway and ability to have the right of first refusal of the

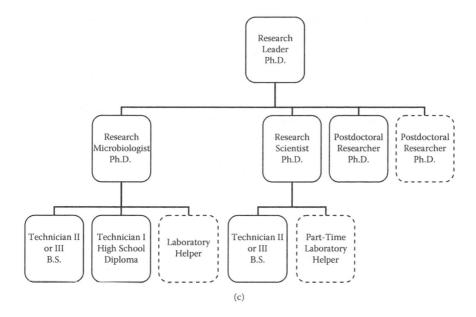

(c)

Figure 2.1 (continued).

choice projects. The laboratory technical lead should take on a professorial mentoring role toward other less-experienced technicians to encourage learning and information sharing, which leads to better repeatability of results and interpersonal harmony within the group. Maintaining strong and healthy interpersonal relationships within a food microbiology laboratory can positively affect productivity and camaraderie but may not necessarily preclude learning and sharing of information about food microbiology (Dykes 2008).

If the purpose of microbial research is to conduct inoculated pack challenge studies on foods and beverages, particularly potentially hazardous foods that support the growth of pathogens, then an expert food microbiologist must design and evaluate the research (Table 2.1) (National Advisory Committee on Microbiological Criteria for Foods 2010). As shown in the table, those overseeing research must have fairly specific education and experience to be qualified to perform certain functions. This does not necessarily mean that these experts must be on staff; outside consultants can accommodate some of the expertise needs.

2.3   The funding

It is difficult to support a team of research microbiologists very long without funding. However, it takes much effort to prime the funding pump with preliminary data or enough proof-of-concept data before

Table 2.1 Recommended Minimum Expertise Needed for Designing, Conducting, and Evaluating Microbiological Studies[a]

Category	Design	Conduct[b]	Evaluate
Knowledge and skills	Knowledge of food products and pathogens likely to be encountered in different foods. Knowledge of the fundamental microbial ecology of foods, factors that influence microbial behavior in foods, and quantitative aspects of microbiology. Knowledge of processing conditions and parameters. Knowledge of statistical design of experiments.[c]	Knowledge of basic microbiological techniques. Ability to work using aseptic technique, to perform serial dilutions, and to work at biosafety level 2.	Knowledge of food products and pathogens likely to be encountered in different foods. Knowledge of the fundamental microbial ecology of foods, factors that influence microbial behavior in foods, and quantitative aspects of microbiology. Knowledge of statistical analysis.[c]
Education and training	PhD in food science or microbiology or a related field or an equivalent combination of education and experience.	BS in food science, microbiology or a related field or an equivalent combination of education and experience. Appropriate hands-on experience in food microbiology is also recommended.	PhD in food science, microbiology, or a related field or an equivalent combination of education and experience.
Experience	Two years of experience conducting challenge studies independently and experience in design of challenge studies under the guidance of an expert food microbiologist.	Two years of experience conducting challenge studies is useful; however, close supervision by an expert food microbiologist may substitute.	Two years of experience conducting challenge studies independently and experience in evaluation of challenge studies under the guidance of an expert food microbiologist.

Table 2.1 (continued) Recommended Minimum Expertise Needed for Designing, Conducting, and Evaluating Microbiological Studies[a]

Category	Design	Conduct[b]	Evaluate
Abilities	Ability to conduct literature searches. Ability to write an experimental protocol.	Ability to read and carry out an experimental protocol. Ability to perform microbiological techniques safely and aseptically.	Ability to analyze and interpret microbiological data.

Source: Adapted from National Advisory Committee on Microbiological Criteria for Foods. 2010. Parameters for determining inoculated pack/challenge study protocols. *Journal of Food Protection* 73 (1):140–202.

[a] State or local regulatory food programs that are presented an inoculation study in support of a variance request may not have expert food microbiologists on staff to confirm the validity of the study. Options available to them include consulting with expert food microbiologists in their state or local food laboratories or requesting assistance from Food and Drug Administration (FDA) food microbiologists through their regional retail food specialist.

[b] Working independently under the supervision of an expert food microbiologist.

[c] It may be appropriate to consult with a statistician with applicable experience in biological systems.

financial support is flowing sufficiently. The startup phase is perhaps the most challenging aspect of funding research programs. The goal of many research leaders is to get consistent financial support for the research program so that attention can be turned to the more interesting (i.e., scientific pursuits) and the more pressing (i.e., personnel management) matters. In academic settings, seed money can only go so far, and any tangible preliminary results that can be extracted from such startup funds will underpin grant proposals for future work. Commercial research groups created with the purpose of supporting a consumer food or beverage product or industrial food ingredient development usually have research funding allocated by the business that will support several months to a few years of research. Contract research laboratories may rely on a cache of funding accumulated from routine laboratory services or consulting fees to finance the startup process. Microbial methods development laboratories for large organizations would likely be sufficiently funded at startup, whereas smaller startup organizations may have to rely on small-business loans or grants. In the United States, small-business innovation research competitive grants in food science (including food safety) are available annually (U.S. Department of Agriculture, National Institute of Food and Agriculture 2011). First-phase awards in 2011 ranged from $70,000 to $100,000, and second-phase support is also available to entities that successfully deliver on first-phase awards. Finally, incubator companies and offshoots of academic research may have university or private equity funding at the outset.

Whatever the source of funding, the one shared aspect of funding by all research labs is the need to produce tangible results that somehow show a return on the investment. Research laboratories often receive lump sums of funding to execute projects. The challenge for long-term success is to perform the project with no more than the amount of funding allocated. R&D managers who can execute a protocol below budget may be rewarded with the privilege of keeping the budget surplus for future operating budget expenses or at least be allowed to purchase new or replacement items for the laboratory that help future projects stay within or under budget. Quality of research produced must never be compromised for running projects under budget. If the output of research includes a patent or a licensed process or technique, this may lead to additional funds for future research, not to mention funds for personal income.

Sometimes, funds for large, long-term (i.e., 2 years or more) projects can come from more than one source. In the United States and in the European Union, parallel funding for projects may come from public and private sources. Industry trade associations often support research in conjunction with other sources, public and private. In the United Kingdom, research funds from the food industry are usually restricted to relatively short troubleshooting projects or to confidential investigations (Roberts 1997). Also, the Ministry of Agriculture, Fisheries, and Food (MAFF) encouraged industrial support of research, with government contributing up to 50% of total funding for projects that had elements of novelty and a consortium of companies involved. These projects included topics like programs on hygienic food processing, separation and detection of pathogens and their toxins, physiochemical principles underlying microbial growth, growth conditions for pathogens, and programs assessing microbiological hazards and risks managing those hazards.

2.4   The laboratory

2.4.1   Laboratory space

Laboratory space is sometimes a contentious issue among competing or even collaborating researchers or between researchers and other technical groups within an organization. Typically, laboratory space is harder to come by in industry settings than on academic campuses. However, academic campuses are not immune to the challenge of acquiring and securing laboratory space for R&D. In many cases, researchers are required to work with less-than-optimal bench space or to share equipment and bench space with other researchers. This can impose restrictions on the research approach. Many researchers follow a planned schedule approach to experimentation and therefore can easily manage around the schedules of other groups, assuming other groups cooperate and follow agreed-on schedules. Some researchers follow a more impulsive and spontaneous approach to projects. In some organizations with more of a focus on innovation and new method and technology developments, liberal lab space is recommended to enable freedom and spontaneity to conduct many small exploratory experiments at short notice. Ample space can also permit research projects to remain set up in laboratories, which saves time by avoiding the need to set up and tear down experiments.

One of the significant needs for food, beverage, and ingredient research laboratories is space for sample storage. Foods, beverages, and ingredients will require specific storage temperatures for shelf life studies and inoculated-pack challenge studies for relatively long periods of time. If laboratories are conducting multiple studies simultaneously, then space can quickly fill. Space limitation problems can be exacerbated if storage studies at more than a few different temperatures are needed. Researchers should take care to measure temperatures accurately in incubators and refrigerators where samples are stored and avoid overloading with samples as airflow obstruction can lead to poor temperature control. Depending on sample mass and quantities under observation, temperature-controlled storage space can be the limiting factor in terms of research capacity.

Laboratories conducting molecular biology experiments may not require extensive bench space but rather significant monetary and time investment in equipment such as polymerase chain reaction (PCR), reverse transcriptase PCR (RT-PCR) thermocyclers, gel docking stations and software, computers for bioinformatics, and microarray sequencers or readers.

As far as ego is concerned, laboratory space is unfortunately a common battleground between competing researchers. Researchers overseeing more space than their peers tend to benefit from a higher perceived value to an organization (especially as perceived by outsiders) whether they deserve such billing or not. This may seem petty, but outsiders (business executives, university administrators, politicians) are usually the people deciding how funding will be allocated, and they often make their decisions after brief laboratory tours. Therefore, their perception of a researcher can and will be slanted by things as easily perceptible as the relative amount of laboratory space commanded by a given researcher compared to his or her peers. There are two basic views to the issue:

1. Productive research should be rewarded with the space required to continue such research and explore new opportunities: "What have you done for me lately?"
2. Seniority rules. "Hey, I was here first, buddy!"

Flexible laboratory space, which accommodates productivity *and* seniority, can be a solution that squelches neither the ambitious nor the egocentric. This could also be considered shared space or multiuse space. Individual researchers still retain their smaller areas for their own use, but these flexible laboratory spaces become an extension of their space, albeit a shared one. These areas are beneficial because of the following:

- They offer economy of scale for the entire lab to save on equipment and materials cost.
- They can become the hub of activity and a catalyst for collaboration.
- They can become showcase spots for the entire lab, engendering a collective pride among the group and improving the overall impression of the lab on visitors (i.e., funders).
- Rather than being dedicated to a sole research group permanently, flexible laboratory space can accommodate multiple research groups over time as projects come and go.

The biggest advantage of flexible laboratory space is that it can change with the changing conditions within a lab. Research laboratories devoting a portion of laboratory space to shared flexible space should experience a minimal amount of idle time for laboratory space and, conversely, fewer times when laboratory space is too crowded and busy. The caveat is that participants will need to cooperate and make sure they leave the shared space and equipment as good as or better than they found it for when the next research group comes in to execute a protocol.

2.4.2   Laboratory layout

The optimal layout of routine microbiological testing laboratories can be far different from microbiological research laboratories. Routine food microbiology testing laboratories are typically designed to receive, log in, and process large volumes of samples from one or more sources. As such, the location of sample receiving and documentation into laboratory control systems would be well suited to have dedicated space, but not necessarily bench space. Research laboratories might also receive large volumes of samples, but not often at the rate and pace of routine testing labs. The goal of a routine testing lab is generally to process large volumes of food and environmental samples as quickly as possible and to report results as soon as they become available. Generally, foodborne bacterial pathogens such as *Escherichia coli* O157:H7, *Salmonella*, and *Listeria monocytogenes* are the principal targets of assays, but various other pathogens, toxins, molds, and even key spoilage organisms are routinely monitored as part of ongoing verification testing or product test-and-release programs. A R&D microbiology laboratory has different objectives. Such laboratories are engaged in short-term and long-term experiments that investigate the detection and behavior of foodborne microorganisms in food systems, simulated food systems, or simulated processing environments. Microbiological R&D in foods can be viewed in stages. The following seven stages of activities take place in laboratories engaged in food microbiological R&D:

1. Protocol development and agreement
2. Equipment and materials planning and assembly
3. Preparation of materials, including media, reagents, and cultures
4. Execution of experiment, with replication
5. Outcome-driven reassessment of protocols with optional modifications
6. Verification and confirmation of results, including statistics
7. Translation of data to reports useful outside the laboratory

Obviously, some of these stages take place strictly within the laboratory, but some activities should take place in separate office space, meeting rooms, or at desks within laboratories. As such, the ideal laboratory layout would include all three workspaces (i.e., laboratory, in-lab desk space, and office/meeting space).

While routine testing microbiology labs deal with consistent and repetitive protocols, research laboratory activity varies from day to day, week to week, and month to month. A well-managed microbiological research program would be capable of performing multiple experiments or studies simultaneously, with sampling times or laborious steps in protocols staggered by hours, days, or weeks when possible. This can be achieved by staggering the scheduled initiation or conclusion of experiments and by modifying sampling times within the constraints of a protocol. However, even the most efficiently run research laboratories will experience so-called downtimes when all the scheduled sampling activities have been completed; all glassware has been cleaned, dried, and reshelved; and all media have been restocked. Such times are opportunities for researching what other labs are using for methods, performing literature reviews, analyzing data, and writing reports. Conversely, even the most efficiently run research laboratories will experience very busy periods from time to time.

There are obviously similarities between the layout of routine testing and research laboratories. For example, both require the ubiquitous waist-level bench, sinks with cold and hot tap water, deionized water, laminar-flow hood, and autoclave. Often, laboratory space dedicated to routine testing and research microbiology are one and the same. However, shared space is not preferred since the differences in laboratory needs for a routine testing versus a research microbiology laboratory can be extreme. For example, in research laboratories, it may be convenient and even necessary to leave experimental conditions set up and ready for the next treatment or sample. This may conflict with the standard operating procedures (SOPs) of accredited routine testing labs. As with most labs, research microbiology labs should have sources and multiple connection points for natural gas, deionized water, vacuum, and compressed air as well as numerous electrical outlets.

One key difference between routine microbiology testing laboratories and research microbiology laboratories is the inclusion of small-scale processing equipment in the latter. Food and beverage microbiology researchers will inevitably need to study the behavior of foodborne pathogens, spoilage microorganisms, or starter cultures during a simulation of a process on a small scale. Access to such pilot plants with food-processing equipment is essential for many aspects of food and beverage research microbiology, although some processes can be mimicked on a laboratory scale. If process validation of pathogen destruction or control is researched on pilot-scale equipment, the work would need to occur in a setting that is designated with biosafety level 2 (BSL-2) status, and the equipment should never again be used to produce food intended for human consumption. Examples of laboratory layouts are shown in Figure 2.2.

2.4.3   Proximity of laboratory to offices

The location of a R&D microbiological laboratory in relation to office or desk space is of consequence. It is appropriate and helpful for desks of technicians, students, and interns (i.e., those performing bench work) to be located inside BSL-1 and BSL-2 laboratories, but not BSL-3. A BSL-3 laboratory should be exclusively for working with BSL-2 or -3 pathogens (U.S. Department of Health and Human Services et al. 1993). Close proximity of desk space to laboratory bench space facilitates good documentation of experimental observations in laboratory notebooks. If there are significant obstacles between the laboratory and office space, productivity can suffer. Desk space invariably means the presence of computers. The abundance of scientific information on the Internet, such as official microbiological or chemical methods, published research, and material safety data sheets (MSDSs), can increase the speed and efficiency of a research program. As laboratory equipment has become more integrated with computers, the lines between desk space and bench space have become blurred. A common conflict arising from office or desk space located within BSL-2 microbiology laboratories is the prohibition of consumption of food or drink. Workers who have no desk space external to the laboratory would likely benefit from having a designated, indoor break area that is outside the laboratory.

For food microbiology research laboratories focused on developing intellectual property and patents, the proximity of the laboratories to

Proposed Microbiology Laboratory Setup

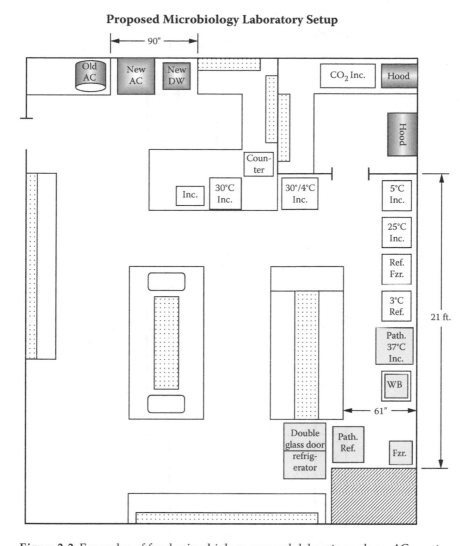

Figure 2.2 Examples of food microbiology research laboratory plans. AC = auto-clave; DW = dishwasher; Fzr = freezer; Inc = incubator; Ref = refrigerator.

business offices could be a factor in productivity. A study of the success of pharmaceutical R&D labs found that productivity, as measured by number of patents, is significantly reduced when research centers are located within 100 miles of the corporate offices (Cardinal and Hatfield 2000). The study suggested that some distance between R&D centers and corporate headquarters could benefit the basic research leading to enhancements of existing drugs but generally decreased new drug discoveries. This can be applied to food microbiology R&D, such as for new antimicrobials,

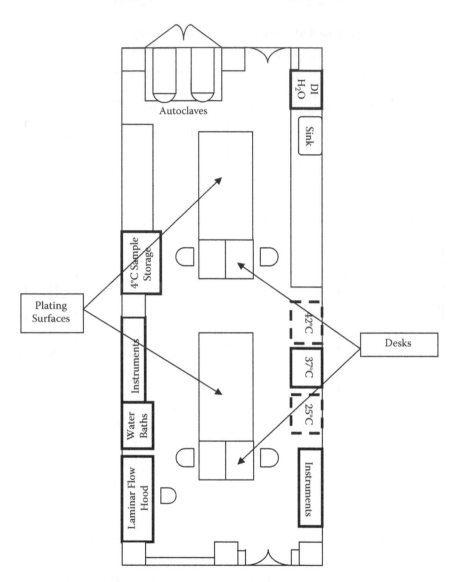

Figure 2.2 (continued).

biocides, and microbial diagnostics. Close proximity of labs to business groups increases interaction, communication, and face-to-face spontaneous information exchanges that can keep R&D focused on strategic goals. Apparently, the trade-off is that close proximity of corporate offices to R&D can stifle the long-term basic research that eventually feeds the pipeline of new developments.

2.4.4 Storage

Most food products require specific storage conditions. As such, food research labs must have capabilities of storing samples at appropriately controlled temperatures at a very minimum. Often, products that are shelf stable (i.e., require no refrigeration) are recommended for storage in "a cool dry place." Many food products are stored in lighted display cases, and many are also susceptible to decomposition due to UV (ultraviolet) light penetration as well. Therefore, fluorescent lighting or even low-UV-emitting fluorescent light may be necessary for research on microorganisms that produce UV reactive metabolites that have an impact on product attributes. An example would be *Lactobacillus viridescens* (now *Weisella viridescens*), which has been shown to grow on meat products, leading to formation of dark gray or green spots (Niven and Evans 1957; Pierson et al. 2003). Also, the heat-resistant mold *Neosartorya fischeri* had enhanced production of the mycotoxin fumitremorgin on media when incubated under light (Nielsen et al. 1988).

These considerations are applicable to both food testing and food research labs. However, a microbiological research lab has the additional obstacle of the need to have restricted access and control of sample and material that may have been inoculated with foodborne pathogens and spoilage organisms. This effectually eliminates the option of shared storage space of food testing and research samples with samples or other foods that are destined for human consumption. The physical location of incubators and coolers that hold samples of foods that have been intentionally inoculated with pathogenic microorganisms must be within the BSL-2 or BSL-3 controlled area (U.S. Department of Health and Human Services et al. 1993).

The variety of food and beverage products and ingredients that may be researched will dictate the storage capabilities for a food microbiology research lab. Flexibility and extra unused storage space would be ideal to meet the ever-changing needs for food microbiology research.

2.5 Equipment, materials, and supplies

The equipment and materials used in food microbiology laboratories are constantly changing with advances in computer technology. The incorporation of computers with most rapid method assays has changed the traditional need for bench space. Instruments such as spiral plating instruments and automated colony counters have markedly reduced the amount of bench space needed for plating large numbers of samples concurrently. Table 2.2 lists some equipment that would be useful in food microbiology R&D laboratories.

Table 2.2 Example of a Research Microbiology Laboratory Equipment Inventory List with Dimensions and Utility Requirements

Instrument	Dimensions (L × W × H)[a]	Electrical requirements	Other requirements	Floor or bench top
Small autoclave	30 × 18¹¹/₁₆ × 28 inches	208 or 230 V, single or three phase	Drain half FPT or ⅝ copper Steam exhaust connections ⅜ IPS	Bench top
Autoclave	51.5 × 41 × 71 inches	120 V, 10 A	Cold (60°F) water 50–80 psig ½-inch NPT minimum Floor drain, 2-in od minimum Steam 50–80 psig ⅜-inch NPT with air/water separator Condensate return, ½-inch NPT minimum (if applicable) Exhaust hood	Floor
Analytical balance	14 × 9 × 15.25 inches	100–120 V AC, 220–240 V AC, 50/60 Hz		Bench top
Top-loading balance	204 × 297 × 81 mm	230 V AC or 115 V AC		Bench top
Top-loading balance	Similar to above	Similar to above		Bench top
Battery power backup	7.6 × 8.5 × 8.5	120 V, 60 Hz, 8.3 A		Bench top
DNA thermocycler	24 × 28 × 23 cm	100–240 V AC rms, 50/60 Hz		Bench top
DNA thermocycler	24 × 28 × 23 cm	100–240 V AC rms, 50/60 Hz		Bench top

Equipment	Dimensions	Electrical	Requirements	Location
Dry bath incubators	12.5 × 11 × 3.5 inches	115 V, 50/60 Hz, 360 W		Bench top
DNA microheating system	5.5 × 11.8 × 7.7 inches	120–240 V AC, 50/60 Hz, 250 W		Bench top
Chemical fume hood	20 × 20 × 30 inches	See manual	Several (see manual)	Bench top or on a new chemical storage cabinet
Biosafety cabinet	50³/₁₆ × 30¼ × 58¹¹/₁₆ inches	115 V AC: 60 Hz, 1 phase, 12 A; or 230 V AC: 50 Hz, 1 phase, 7 A	Main voltage supply not to exceed ± 10%	On a framed mounting stand on the floor
Laminar flow hood	50 × 38 × 46 inches	7 A		Bench top
Refrigerated centrifuge	468 × 695 × 380 mm	110–127 V, 1–60 Hz, 1400 VA, 11.5 A, 1200 W	Room for lid opening	Bench top
Steamscrubber glassware washer	26.71 × 24.5 × 54 inches	115 V (60 Hz), 20 A; 230 V (50/60 Hz)	Minimum water temp 120°F Water 2.5 gallons per fill Minimum water pressure 20 psi Maximum water pressure 120 psi Temporary voltage spikes on AC input line as high as 1500 V for 115-V models and 2500 V for 230-V models	Floor
Hot plate and stirrer	11.875 × 4.5 × 8.625 inches	120 V, 7 A, 840 W, 50/60 Hz		Bench top
Hot plate and stirrer	11.875 × 4.5 × 8.625 inches	120 V, 7 A, 840 W, 50/60 Hz		Bench top

continued

Table 2.2 (continued) Example of a Research Microbiology Laboratory Equipment Inventory List with Dimensions and Utility Requirements

Instrument	Dimensions (L × W × H)[a]	Electrical requirements	Other requirements	Floor or bench top
Incubator	14 × 13 × 16.25 inches	120 V AC, 90 W		Bench top
Bacteriological incubator	16 × 16 × 18 inches	120 V, 50/60 Hz, 120 W, 100 A		Bench top
Dual CO_2 incubator	51.5 × 24 × 37 inches	25 A at 120 V, 60 Hz	Inside pathogen room	Bench top
Digital incubator/shaker	21 × 27.5 × 19.75 inches	100 to 240 V, 50/60 Hz, 600 VA		Bench top
Programmable incubator		115 V, 60 Hz, 7.5 A	Door-opening space	Floor
Programmable incubator	33 × 29 × 76 inches	115 V, 60 Hz, 7.5 A	Door-opening space	Floor
Programmable incubator	33 × 29 × 76 inches	115 V, 60 Hz, 7.5 A	Door-opening space	Floor
Low-temp incubator	24 × 24.5 × 34.5 inches	115 V, 50/60 Hz, 6.6 A, 792 VA, 500 W	Door-opening space	Floor
Refrigerator		115 V, 60 Hz	Door-opening space	Floor
Refrigerator		115 V, 60 Hz	Door-opening space	Floor
Double door 49-ft³ refrigerator	52 × 35 × 64 inches	115 V, 16 A, 60 Hz	Door-opening space	Floor

Equipment	Dimensions	Power	Notes	Location
Compact undercounter refrigerator		115 V, 60 Hz		Floor or bench
Ultralow freezer	28 × 29 × 42 inches	115 V, 60 Hz, 16 A	Hinged-top opening space	Floor
Spectrophotometer	12 × 13 × 7 inches	100 to 240 V, 50 to 60 Hz, 1 A	Inside pathogen room	Bench top
Stomacher	9 × 18 × 13 inches	110 to 120 V, single phase, 60 Hz, 170 W		Bench top
Stomacher	9 × 18 × 13 inches	110 to 120 V, single phase, 60 Hz, 170 W		Bench top
Stomacher	9 × 18 × 13 inches	110 to 120 V, single phase, 60 Hz, 170 W		Bench top
Electrical line conditioner	6.75 × 5.75 × 6.75 inches	120 V, 60 Hz, 1800 W output		Bench top
Programmable liquids dispenser	13 × 9.5 × 6.5 inches	110–120 V AC, 60 Hz, 140 W, 5 A		Bench top
Water bath		120 V AC		Bench top
Spiral plater			Vacuum source	Bench top
Chemical storage cabinet	24 × 19 × 34 inches			Floor
PCR pathogen detection system	14 × 20 × 24 inches	120 V, 50/60 Hz, 10 A, two 6.3-A fuses	Electrical line conditioners	Bench top

continued

Table 2.2 (continued) Example of a Research Microbiology Laboratory Equipment Inventory List with Dimensions and Utility Requirements

Instrument	Dimensions (L × W × H)[a]	Electrical requirements	Other requirements	Floor or bench top
CPU for PCR detection system	Standard tower CPU size	Standard CPU requirements	Line conditioners listed above	Bench top
ELFA pathogen detection system	21 × 32 × 16 inches	120 V, 50/60 Hz, 3 A	Line conditioners listed above	Bench top
CPU for ELFA detection system	Standard tower CPU size	Standard CPU requirements	Line conditioners listed above	Bench top

Note: CPU = central processing unit; ELFA = enzyme-linked fluorescent antibody; FPT = female pipe thread; IPS = iron pipe size; NPT = national pipe thread; od = outside diameter.

[a] Dimensions are shown in either inches or in metric units. In the United States, dimensions of scientific equipment can be provided by suppliers in either unit, and contractors working on the laboratory layout itself may be more comfortable with nonmetric measurements (i.e., inches and feet).

Inventory lists for media, chemicals, reagents, assays, and stock culture are also necessary to maintain a well-functioning research lab. Such inventories should be available in printed form or, better yet, electronically via a networked computer file system or program to ensure that laboratory personnel remain updated on inventories of these important items. Microbiology research is roughly 80% preparation and 20% execution. Hence, the worst nightmare for researchers is to have planned and scheduled an experiment and prepared all the necessary cultures, media, reagents, treatment conditions, and samples only to find that one critical material is absent from the laboratory inventory. In such a case, failure to account for one single, yet critical, material could be the difference between a well-planned and poorly planned experiment. Technicians probably would not mind the "day off," but other colleagues, perhaps those already committed to help that day, might be a bit perturbed. It is obviously better to have used planning resources to the fullest to avoid such uncomfortable situations. Inventories should describe the material; quantity in stock; amount (weight, volume, or number of uses); manufacturer(s); part number; lot number; expiration date; date received; and National Fire Protection Association (NFPA) (safety) rating. Advanced lists could be linked to downloadable MSDSs. Strain inventory lists for stock cultures should indicate the method used to prepare the stock culture (e.g., refrigerated slant, lyophilized, or frozen) and list the genus and species name, quantity in stock, strain number(s), isolation source of strain, source from which strain was obtained, and comments about any special characteristics of the strain. As research laboratories compile data and expertise with certain types of research, these data may best be managed in databases that allow quick access to information and cross referencing of strains with observed responses in different experimental conditions.

Chemical reagents should be stored according to safe practices outlined on MSDSs, supplier documents, or other safety and regulatory bodies. A list of chemicals contained in the laboratory should be kept up to date, as should the expiration dates. Waste containers for spent reagents should be created based on reagent type. For example, waste containers might be needed separately for nonpolar solvents, acids, bases, and so on. Eventually, reagent waste, including spent reagents and expired material, should be discarded, and there are outside services that could handle this for a fee.

By and large, the materials used in microbiology laboratories have become disposable. Gone are the days of reusable glass pipettes and petri dishes. While many materials are made from plastics, even materials still made from glass, like certain test tubes, are designed for single use. Dilution blanks such as phosphate buffer and peptone water are available in prefilled, measured, sterilized plastic bottles. This has decreased the need for glassware washing appliances. A laboratory concerned about water usage may find that plastic disposable materials are a good alternative. Conversely, a laboratory more attentive to the environmental impact of plastic production and eventual disposal of used plastics in landfills may prefer self-preparation of buffers in reusable glass. Of course, cost of labor versus cost of purchasing premade supplies will usually play a large part in the choice of materials.

2.5.1 Environmental impact considerations

Global focus on environmental impact and sustainability has affected operational procedures in many laboratories. The implementation of green chemistry, the design of chemical products and processes that reduce or eliminate the use and generation of hazardous substances, has been promoted as environmentally responsible (Kirchhoff 2005). Microbiology laboratories utilize chemistry in different and varied ways, and green chemistry should be a goal of microbiology laboratories now and in the future. Waste streams for inorganic and organic chemicals, specifically microbiological media supplements such as antibiotics, should be carefully evaluated in laboratories using large quantities. For further reading on waste minimization in laboratories, see the work of Reinhardt et al. (1996).

There are certain specific environmental considerations for food microbiology laboratories. R&D (and routine testing) laboratories commonly use antibiotic supplements for selective enrichment and plating media. Some research laboratories use radioactively labeled cells. Also, new technologies such as nanotechnology have gained use in research microbiology. Each of these (and many other) special circumstances should be evaluated by laboratory directors or biological safety officers.

Populations of microorganisms grown by combining microbiological media with incubation should be contained within the laboratory and disposed of properly after use. This should apply to BSL-1 and above laboratories. Although nonselective enrichment or plating of a food sample, for instance, would result in a mixed population, some of that population could be biologically hazardous. The microbiologist should decontaminate these "unknown" samples to minimize environmental spread of large, concentrated populations of microorganisms that may or may not include pathogens. Laboratories working with BSL-2 pathogens should follow biosafety procedures approved by governmental public health agencies, such as the U.S. Centers for Disease Control and Prevention (CDC) (U.S. Department of Health and Human Services et al. 1993). Autoclaving (i.e., saturated steam under pressure) is considered the most appropriate way to treat laboratory waste properly to destroy microorganisms. Autoclaves are typically operated to achieve a cavity temperature of 121°C for a minimum of 15 min. However, this time and temperature profile was based on direct contact with steam and does not include heat penetration time. Consideration of heat penetration time is often overlooked when setting decontamination cycles for autoclaves. The true time and temperature profiles in waste are affected by distance from the bottom of waste containers, water or solids volume, and composition of waste containers (Lauer et al. 1982). Higher temperatures are generally achieved with greater distance from the bottom of waste containers, greater volume of water, and use of steel rather than plastic containers. While large populations of vegetative bacterial pathogens will probably be destroyed by the standard time and temperature profile of 121°C for a minimum of 15 min in most conceivable waste configurations, a longer process (such as 90 min) or the use of additional water and stainless steel waste containers would be necessary to inactivate *Bacillus stearothermophilus* spores (Rutala et al. 1982). Laboratories working with spore-forming microorganisms or unusually heat-resistant vegetative microorganisms should validate waste decontamination processes with temperature probing or biological indicators.

2.6   External research partnerships

As mentioned in Chapter 1, external R&D capabilities may need to be utilized from time to time. Use of other labs can occur for many reasons, but typically it is done due to the need to use certain instruments, utilize outside expertise with certain techniques, or overcome space limitations or personnel and time limitations. For instance, many researchers wishing to study the behavior of *C. botulinum* in foods or food-related systems must go to one of the few remaining laboratories with the capability to work with select agents. To avoid cumbersome and costly regulatory burdens, a number of public and private labs destroyed microbiological culture collections once the United States imposed regulations associated with the Select Agents and Toxins list (Casadevall and Imperiale 2010). Work with laboratory animals is also becoming more and more regulated and is therefore easier to outsource than to perform "in house."

University and government researchers frequently collaborate on large research projects to share both the funding and the workload. These partnerships can be mutually beneficial. In public/private research alliances, university and government researchers gain the opportunity to work with "real-world" samples of foods and processing environments, while industry researchers gain access to knowledge, resources, and credibility of publicly funded research labs. Private-sector labs may provide some or all of the R&D needs for private companies and institutions. However, unlike food microbiology routine testing, microbiological R&D often does involve defined testing procedures, much less-accredited methods.

Therefore, evaluation and interpretation of results of outside laboratories would still need to be done by a qualified individual (National Advisory Committee on Microbiological Criteria for Foods 2010).

2.7   The reporting mechanism

Every laboratory requires a management and reporting structure. Microbiologically oriented R&D lab personnel are organized in various ways as influenced by research goals, organizational culture, and personal preferences. Previous discussion on the laboratory team can be found in Section 2.2.

Data-reporting systems are essential to making sure results are properly collected, reviewed and approved, and reported. Efficient teams will make sure that data are not unintentionally filed and forgotten. The team of scientists conducting the work should also see the end results, which helps close the knowledge loop and brings better understanding of how day-to-day laboratory activities affect research outcomes. Internal review of reports by scientists can help ensure successful external peer review or favorable acceptance of deliverables by funding sources.

Data should be adequately analyzed with statistics when appropriate. Qualified individuals should perform these statistical analyses, and qualified microbiologists should make interpretations of the results and draw any practical conclusions. Reports should provide details on methods employed and reasoning behind the choices made in conducting the work. However, in many cases, especially in industry, the information beyond the abstract or executive summary will not be read, so refining the outcome of the work into a few salient points is necessary. For more detailed discussion on reporting research outcomes, see Chapter 12.

References

And, M. P., and A. C. Steneryd. 1993. Freeze-dried mixed cultures as reference samples in quantitative and qualitative microbiological examinations of food. *Journal of Applied Microbiology* 74 (2):143–148.

Augustin, J.-C., and V. Carlier. 2002. French laboratory proficiency testing program for food microbiology. *Journal of AOAC International* 85 (4):952–959.

Cardinal, L. B., and D. E. Hatfield. 2000. Internal knowledge generation: the research laboratory and innovative productivity in the pharmaceutical industry. *Journal of Engineering and Technology Management* 17:247–271.

Casadevall, A., and M. J. Imperiale. 2010. Destruction of microbial collections in response to select agent and toxin list regulations. *Biosecurity and Bioterrorism* 8 (2):151–154.

Dykes, G. A. 2008. A technique for enhancing learning about the professional practice of food microbiology and its preliminary evaluation. *British Food Journal* 110 (10):1047–1058.

Edson, D. C., S. U. E. Empson, and L. D. Massey. 2009. Pathogen detection in food microbiology laboratories: analysis of qualitative proficiency test data, 1999–2007. *Journal of Food Safety* 29 (4):521–530.

Kirchhoff, M. M. 2005. Promoting sustainability through green chemistry. *Resources, Conservation and Recycling* 44 (3):237–243.

Lauer, J. L., D. R. Battles, and D. Vesley. 1982. Decontaminating infectious laboratory waste by autoclaving. *Applied and Environmental Microbiology* 44 (3):690–694.

National Advisory Committee on Microbiological Criteria for Foods. 2010. Parameters for determining inoculated pack/challenge study protocols. *Journal of Food Protection* 73 (1):140–202.

Nielsen, P. V., L. R. Beuchat, and J. C. Frisvad. 1988. Growth of and fumitremorgin production by *Neosartorya fischeri* as affected by temperature, light, and water activity. *Applied and Environmental Microbiology* 54 (6):1504–1510.

Niven, C. F., Jr., and J. B. Evans. 1957. *Lactobacillus viridescens* Nov. spec., a heterofermentative species that produces a green discoloration of cured meat pigments. *Journal of Bacteriology* 73:758–759.

Pierson, M. D., T. Y. Guan, and R. A. Holley. 2003. Aerococci and carnobacteria cause discoloration in cooked cured bologna. *Food Microbiology* 20:149–158.

Reinhardt, P. A., K. L. Leonard, and P. C. Ashbrook. 1996. *Pollution prevention and waste minimization in laboratories.* Vol. 3. Boca Raton, FL: CRC Press.

Roberts, T. A. 1997. Maximizing the usefulness of food microbiology research. *Emerging Infectious Diseases* 3 (4):523–528.

Rutala, W. A., M. M. Stiegel, and F. A. Sarubbi, Jr. 1982. Decontamination of laboratory microbiological waste by steam sterilization. *Applied and Environmental Microbiology* 43 (6):1311–1316.

U.S. Department of Agriculture, National Institute of Food and Agriculture. 2011. Small business innovation research: food science and nutrition. Available from http://www.csrees.usda.gov/fo/foodsciencenutritionsbir.cfm; accessed August 24, 2011.

U.S. Department of Health and Human Services, Public Health Service, Centers for Disease Control and Prevention, and National Institutes of Health. 1993. *Biosafety in microbiological and biomedical laboratories.* HHS Publication No. (CDC) 93-8395. Washington, DC: U.S. Government Printing Office.

chapter three

Food process validations

Margaret D. Hardin

Contents

3.1 Introduction

The effective control of any food process begins and ends with validation. The term *validation* can be confusing as it is too often used interchangeably or in concert with the term *verification* with regard to a hazard analysis critical control point (HACCP) program. Although validation and verification are distinctly separate activities, they are not unrelated. Validation involves demonstrating that a process, when operated within specified limits, will consistently produce product meeting predetermined specifications. Verification supports validation and involves ongoing activities, such as auditing, reviewing, inspecting, or testing, to ensure that product and process specifications are correctly implemented by the process. Verification takes for granted that the process has been previously validated. The National Advisory Committee on the Microbiological Criteria for Foods (NACMCF) defined validation and verification (NACMCF 1998) as follows:

> *Validation*: That element of verification focused on collecting and evaluating scientific and technical information to determine whether the HACCP plan, when properly implemented, will effectively control the hazards.

> *Verification*: Those activities, other than monitoring, that determine the validity of the HACCP plan and that the system is operating according to the plan.

Validation is performed for a multitude of systems and for a variety of reasons. In the food industry, validation is often performed for critical control points (CCPs) in HACCP plans (Scott 2005) for processes and procedures. Such validation activities range from chemical and microbiological testing methods, to validation of food and beverage processing equipment, procedures, or interventions.

In May 1987, the U.S. Food and Drug Administration (FDA) published a *Guideline on General Principles of Process Validation*. The purpose of the guideline was to outline general acceptable elements of process validation intended for the preparation of both human and animal drug products and medical devices. Although these guidelines were intended for application to the manufacture of pharmaceutical and medical devices, the general principles and elements of validation described in the document may be applied to food manufacturing processes as well. The document has since been updated, and in January 2011, the FDA published "Guidance for the Industry Process Validation: General Principles and Practices" (FDA 2011c). In the 1987 document, the FDA defined process validation as the establishment of "documented evidence which provides a high degree of assurance that a specific process will consistently produce a product meeting its pre-determined quality attributes." In the 2011 FDA document, process validation is defined more thoroughly "as the collection and evaluation of data, from the process design stage through commercial production, which establishes scientific evidence that a process is capable of consistently delivering quality product." Validation is further outlined in the document as a three-stage approach:

- *Stage 1* emphasizes *process design,* including building and capturing process knowledge and understanding as well as establishing a strategy for process control.
- *Stage 2* involves *process qualification,* whereby the design of the facility and qualifications of the utilities and equipment supporting the process are evaluated to determine if the process is capable of reproducible commercial manufacturing.
- *Stage 3* emphasizes the importance of ongoing *process verification* to ensure the process remains in a state of control.

Process validation for medical device regulatory systems is outlined by the Global Harmonization Task Force (GHTF) in six steps beginning with establishment of a validation team (GHTF 2004). These principles could be applied to food processes. Whether a company chooses three stages or five or six steps to validate their process, it must remember that product safety cannot be inspected or tested into the finished product but must be designed and built into the product process. If correctly done, a process validation study ensures that the process will consistently achieve a food safety objective (FSO) as designed. An FSO articulates the overall performance expected of a food production chain to reach a stated or implied public health goal and provide an appropriate level of protection to the consuming public (van Schothorst et al. 2009; International Commission on the Microbiological Specifications for Foods 1998). Performance objectives of a process and performance criteria can be used to complement FSOs regarding control measures and process criteria as part of operational food safety management (Gorris 2005). While focusing on FSOs and protecting human health is the priority, *in the real world*, the parameters to meet food safety must also be balanced with providing a high-quality product that is acceptable to the consumer.

3.2   Process validation procedures

Each step of the manufacturing process must be controlled to maximize the probability that the finished product meets all food safety standards. Validating a process to achieve a food safety goal makes the equipment and process operate more consistently in delivering a product that meets all the safety standards in place. An additional benefit is that the product will be more consistent in delivering first-pass product quality specifications. Based on the results of process validation, procedures for monitoring and controlling the process and associated process parameters are established and maintained for the validated processes to ensure that the specified requirements are continually met. When changes in the process or process deviations occur, the establishment reviews and evaluates the process and performs additional revalidation where appropriate.

The fundamental approach to validation is very similar whether you are validating meat production processes such as cooking or chilling, a roasting process for peanuts, or an intervention process such as an anti-microbial carcass rinse or flash pasteurization for juice. Validation begins with determination of the overall objective of the project and its proper design and planning. This is best achieved by a multidisciplinary team with representatives who have expertise in quality assurance, technical services, research and development, process control, engineering, maintenance, production, and purchasing. Additional experts such as process authorities may also be called in to provide expertise that is not available inside the company.

3.2.1   Identify qualified experts

There are two types of recognized process authorities for food production defined in U.S. federal regulations. In Title 21, *Code of Federal Regulations,* Part 128B, "Thermally-Processed Low-Acid Foods Packaged in Hermetically Sealed Containers" (since recodified as Part 113), it states that low-acid thermal processes shall be established by "qualified persons" with academic and industrial experience related to thermal processing work. Part 114.83 states that scheduled processes for acidified foods shall be established by "a qualified person." According to the U.S. Department of Agriculture (USDA) regulation 9 CFR 318.300, "Definitions," a processing authority is "the person(s) or organization(s) having expert knowledge of thermal processing requirements for foods in hermetically sealed containers, having access to facilities for making such determinations, and designated by the establishment to perform certain functions as indicated in this subpart." In either case, the process authority acts as a liaison between the production plant and the government to ensure safe production of food. Process authority status is not an official certification of an individual, but rather the unofficial recognition that an individual has knowledge and experience in areas such as thermobacteriology, process engineering, equipment design, performance, sanitation, and so on. In the United States, the recognition would come rather informally through peer recognition in the form of processor(s) and government willingness to work with the person to qualify a food process. It should be noted that a process authority for USDA-regulated meat, poultry, and egg products may not be deemed an authority by FDA authorities responsible for regulating low-acid canned foods since the areas of expertise necessary differ considerably.

Once the team is formed, the next step involves asking questions to determine what it is they are trying to achieve, such as "Are all products in the batch oven reaching an internal temperature (IT) of 71.1°C (160°F)?" or "Does the process result in a 5-log reduction in *Salmonella* spp.?" Ask all questions necessary to gain the knowledge and understanding necessary to establish what to measure, how to measure, when to measure, and where to measure to ensure that both the equipment and the processes are designed to achieve the desired end-product specifications consistently and to establish the protocols for measuring and testing to verify process control. Equipment manufacturers and manufacturers of antimicrobial ingredients may be additional resources who can provide recommendations for validating their processes or recommended operating parameters. A process authority or university extension program may also serve as a resource for expertise in designing and evaluating validation studies. Numerous validation studies already exist for a variety of products and processes. Many of these studies can be found in the scientific literature or be obtained from industry associations, private laboratories, universities, and sometimes government agencies such as the USDA Agricultural Research Service and Food Safety and Inspection Service (FSIS) or the U.S. FDA Center for Food Safety and Applied Nutrition. The study will need to be reviewed to determine how closely the publication mimics the product and process to which it is being applied and what critical parameters of the study must be followed and validated in the plant.

3.2.2 Equipment qualification

In a perfect world, the equipment would be designed and purchased with the desired product and process characteristics in mind. However, *in the real world*, and what is more often the case, the equipment has been installed for a number of years, producing a variety of products for myriad processes and is now going to be validated or revalidated. The importance of properly identifying critical equipment features that could affect the process and end product cannot be emphasized enough. Features may include the overall design of the equipment, including installation and utility requirements, ease of operation, preventive and ongoing maintenance, "cleanability," and complexity and frequency of calibration. Critical control parameters that could affect the process and product must be considered, including critical settings and in-process adjustment requirements as well as any additional critical equipment features. Rigorous testing should be performed to demonstrate the effectiveness and reproducibility of the process. It is important to simulate the actual production conditions as well as "worst-case scenario" situations. These are conditions that will be the most challenging to the process.

The more knowledge and understanding that can be gained about the equipment and the equipment parameters, the more the operations personnel will understand how even the most subtle adjustment will affect the process and final product. These insights and understanding are key to controlling the process and ensuring product quality and safety.

Some validation studies to qualify equipment will validate the equipment alone without performing additional studies of the equipment with the product. Although validation of the equipment without product will provide information about the operation of the equipment itself, it will not provide the necessary details relative to how that equipment functions when loaded with product, including products of varying characteristics, such as size, shape, thickness, temperature, and so on. It will also not tell you how the equipment will perform when it is not filled to capacity, such as a partially loaded smokehouse or roaster or when loaded with multiple and different types of product in the same load. Partial loads and load configurations can significantly affect equipment operations and parameters such as airflow, humidity, and heat transfer. This is true for both batch-type and continuous systems.

3.2.3   Designing and planning the validation experiments

As part of the information gathering and study design, additional considerations include determining if this is a study that can be performed in the processing facility or whether there are some preliminary challenge studies that need to be performed by a third-party laboratory or university. If the equipment or process has been initially developed or validated by the manufacturer or by a research laboratory, additional in-plant studies and equipment and process qualification trials will be necessary to validate the process in the facility, on the equipment, and for the products it manufactures.

One question that is asked repeatedly is how many samples are needed for the test: How much is enough? how much is too much? Many tools and statistical programs are available to assist in the design of these studies (FDA 2011c; GHTF 2004). The number and type of data collected should be able to capture the variability in the process and measure the capability of the process to meet microbial food safety or quality specifications consistently. When processes already exist and products are currently in production, product or environmental sampling and microbial testing to validate the process are often, more realistically, a function of the availability of the process, the product, and the processing and monitoring equipment. When such conflicts arise, process validation will need to be scheduled around plant production schedules, especially if affected product must be withheld from commerce pending laboratory test results.

Before settling on the final design of the study, it is often necessary to perform preliminary trials to capture data that will provide an intimate knowledge and a better understanding of the process and product. For example, variations in product formulation, size, shape, and thickness may affect the process capability and performance; variations in equipment design and operational settings will have an impact on the degree of processing received by the product. Many times, details on how and where to measure variables (including critical limits) such as residence time, product temperature, or pH (product center, external, thickest pieces, etc.) will be learned as this may be the first time the process was truly assessed in detail. While there are optimum final product specifications and desired characteristics, there may be a certain level of acceptance for variability to these specifications, within a certain tolerance, for the company and its customers. It is worth mentioning that this is sometimes referred to as the acceptable quality limit (AQL). Once the data are gathered and analyzed, the process needs to be brought under these acceptable control limits to minimize variability and maximize process and product consistency.

Continuous improvement is built on the goal of first developing a process that is reliable. A reliable process is one that produces the desired output each time with minimal variation (i.e., within the AQL), and reliability is the cornerstone of continuous process improvement. The goal of process control is to minimize this variability to the greatest extent possible within physical, economic, or other constraints and to produce product that is even better able to meet customer requirements. The maximum variability allowed and expected in the process and in the final product will help define the tolerances for validation. Some necessary questions include:

- What are the within-batch/-lot variations, such as for product formulation, fat or moisture content, or the temperature of in-going materials or product? What are the *potential* process variations, and to what degree should these potential variations be addressed now?
- How consistent are these variables between product batches and lots? How do they vary day to day and shift to shift? How do they change when suppliers of raw materials or ingredients change?
- What equipment changes routinely occur or could occur? When are equipment parameters adjusted or readjusted: throughout the day, week, or year?*
- What are the capital expense and ongoing maintenance costs needed to maximize equipment efficiency and minimize this variability?

---

* Editor's note: Efforts to define a process and set its parameters prior to validation are akin to a scientist controlling variables in an experiment prior to conducting it. However, such effort may be opposed by operations personnel who are experts at their own equipment operation. Equipment operators may be accustomed to a high degree of flexibility in making adjustments "on the fly" to processing equipment parameters, such as conveyor belt speed, temperature and humidity settings, raw material loads, and so on. The challenge is to hold them to defined limits of the process and to control against multiple, simultaneous adjustments that can interact to affect the process performance significantly and thereby confound validation efforts.

- How do we verify that the process and equipment are still under control? How often does this need to be done? Who will maintain this control?
- Do we have the necessary expertise in-house?

Following the completion of the preliminary objectives and trials, a *written master plan* should be developed specifying the procedures, tests to be conducted, and data to be collected. This document is required reading for scientists in charge of performing validation of the process. This is where the team asks the who, what, when, where, why, and how questions for many issues and, hopefully, receives real answers. The plan should include a sufficient number of replicate process runs to demonstrate reproducibility and to account for process and product variability. The master plan should also outline the measures for success, including acceptance/rejection criteria. In addition, specifics such as the test conditions for each run, including test parameters, product characteristics, and production equipment, must be outlined. It is essential that these parameters be designed to target the worst-case scenario conditions for the product and process, such as the largest pieces of product, the oldest age of product and ingredients, and the slowest-cooking product formulation or the coldest temperature for product if a heat process is being evaluated. Identifying and testing the worst-case scenario conditions for the product and process will provide a high degree of assurance that the process will deliver product meeting the desired level of quality and safety for all products produced in that process every time the product is manufactured.

The written plan should also document the performance and reliability of the equipment used in the process, including the equipment used to monitor the process, such as thermometers, pH meters, pressure sensors, flow meters, timers, water activity meters, belt speed controller and recorder, and processing equipment such as ovens, roaster, dryers, chillers, retorts, fillers, sterilizers, and freezers.

The plan will identify how key process variables will be monitored and documented, the finished product and in-process test data to be collected as part of the validation process, and the decision points that define what constitutes acceptable test results. If measurements are taken and samples are pulled for objective or subjective analysis, such as microbiological, chemical, or physical analyses, the methods used to acquire data should be published and recognized methods. If available, accredited methods are most preferable. If neither accredited nor published methods are available for a particular circumstance, credible food microbiologists, chemists, and process authorities can design modified methods to suit the need and interpret the findings with appropriate caveats. If there are legitimate concerns related to the validity of the methods used, there will be questions about the results (data) and consequently about the process they are intended to validate.

3.2.4   Perform initial tests

Tests should be repeated enough times to ensure that the results are both meaningful and repeatable until consistent patterns are established. To account for variability in the process such as daily and weekly variations, tests should be conducted at different times and shifts, on different days of the week, at startup and end of each shift of production, and after breaks. In some cases, there may be additional seasonal variation that may affect the process, in which case additional data may need to be collected and compared for summer and winter months. It must be remembered that proper preparation and planning up front will prevent poor performance in the end.

3.2.5   Process validation: Next steps

Once the data from preliminary trials and testing have been generated, they must be analyzed to look for patterns and trends. Statistical process control (SPC) techniques and the application of control charts are useful for this type of analysis. Based on the results of this analysis, the process may need to be adjusted, or there may be a need to establish new process controls. The testing and adjusting should be repeated until repeatability and predictability for the process and worst-case scenario can be established. The bottom line with validation is to control the process to produce safe, high-quality product that meets desired expectations every time. To successfully and effectively control the process, you must know the process, understand the limitations of the process, and remove or at the very least reduce to control process (and product) variability. The data must be complete and reproducible. They must mimic the process and account for all known sources of variability.

The frequency with which validation testing and revalidation need to be performed is established by the results of the initial validation, by the complexity of the process, and by the frequency with which changes to the products and processes occur. The same rule of thumb applies for validation of equipment as well as any recommendations from the manufacturer related to the equipment. Some key factors to consider during revalidation are any changes to the process or product that may have occurred since the last validation, including changes in the flow of the process, new equipment, changes to existing equipment, changes in product formulation or raw materials/ingredients, or new outbreaks, illnesses, or recalls of related products or processes.

To assess the success of the validation, the data should be reviewed to ensure that they have accounted for product variability and equipment variability. Does the validation meet pertinent regulatory requirements? These would have been used to help define the FSO, unless an exposure assessment was done to define the FSO irrespective of the regulations.

A third-party review or an audit may be of value to evaluate success of the validation and resulting changes to improve and control the process. Have the validations included the appropriate and necessary challenge testing and in-plant testing to ensure that the validated process (technology) will result in the same level of safety everywhere, every time, and for every product run through the system?

Once the process has been developed, all of the procedures used in validating the process should be documented for future reference to aid in future validation studies and for employee training purposes. The information must be readily available for review during auditing by governments, customers, or third-party auditors. These documents will also provide useful information for designing any additional in-plant studies or challenge studies.

3.3   What is the role of microbiological data?

In most process validation work, microbiological data are the key metric. Microbiological testing in the form of challenge studies or in-plant studies will often provide much-needed and often-necessary data to support the validation of that process to control or reduce a microbiological hazard associated with certain products and processes. Challenge studies can be useful tools for determining the ability of a food to support the growth of spoilage organisms (shelf life) or pathogens under optimal or less-than-optimal (e.g., temperature abuse) conditions. They are also important in the validation of processes that are intended to deliver some degree of lethality against a target organism, such as a 7-log reduction of *Salmonella* spp. in cooked chicken or a 5-log reduction of *Escherichia coli* O157:H7 during the pasteurization of juice. As is the case with any study or validation, the objectives of the study and methods must be determined and outlined before the study begins.

3.3.1   Microbial inocula

When conducting a challenge study, the appropriate pathogens or surrogates must be selected such that they represent organisms likely to be associated with that food or food formulation and may be organisms or better yet a cocktail of several species of the organism that has previously been isolated from similar foods or isolated from known foodborne outbreaks (Table 3.1).

Normally, the most resistant pathogen of concern that is reasonably likely to occur is a sufficient target organism. In some instances, multiple organisms must be validated. In the validation of some products and processes, vegetative cells of bacteria may be most appropriate, and for other

Table 3.1 Some Pathogens that May Be Considered for Use in Process Validation
Challenge Studies for Various Food Products

Food system	Pathogens
Packaged water	*Cryptosporidium parvum, Giardia lamblia,* hepatitis C
Fresh produce	*Salmonella* spp., *Shigella* spp., enterohemorrhagic *E. coli, Cryptosporidium parvum*
Salad dressings	*Salmonella* spp., *Staphylococcus aureus*
Modified atmosphere packaged products (vegetables, meats, poultry, fish)	*Clostridium botulinum* (proteolytic and nonproteolytic strains), *Salmonella* spp., *Listeria monocytogenes,* enterohemorrhagic *E. coli*
Bakery items (fillings, icings, nonfruit pies)	*Salmonella* spp., *S. aureus*
Sauces and salsas stored at ambient temperature	*Salmonella* spp., *S. aureus, Shigella* spp.
Dairy products	Enterohemorrhagic *E. coli, L. monocytogenes*
Confectionery products	*Salmonella* spp.
Formula with new preservatives	*Salmonella* spp., *S. aureus, C. botulinum,* enterohemorrhagic *E. coli, L. monocytogenes*
Dry-cured pork	*Trichinella spiralis, Salmonella* spp., *Staphylococcus aureus*

Source: Adapted from Institute of Food Technologists. 2001. Chapter VII. The use of indicators and surrogate microorganisms for the evaluation of pathogens in fresh and fresh-cut produce (Task Order No. 3), September 30, 2001. Available from http://www.fda.gov/Food/ScienceResearch/ResearchAreas/SafePracticesforFoodProcesses/ucm091372.htm; accessed July 18, 2011.

products and processes, bacterial or fungal spores may be necessary. Viruses and parasites are less frequently included in validations, but certain processes may call for evidence of destruction of these pathogens as well (e.g., produce decontamination steps; packaged water disinfection). Some research and challenge studies use antibiotic-resistant (streptomycin, rifampicin, etc.) strains of bacteria to differentiate the inoculated strains from naturally occurring strains of an organism in untreated raw material or ingredients. This also permits the optimal recovery of stressed (injured) cells for enumeration on nonselective media (e.g., tryptic soy agar) incorporated with the antibiotics. The method employed to grow inocula, especially bacteria, can have an impact on their response in various food systems. For instance, bacterial populations in logarithmic phase of growth tend to be more resistant to external stresses than stationary-phase populations.

The level of the challenge inoculum must be considered. While a lower inoculum level (10^2–10^3) may be appropriate for a study related to the stability of a product formulation or shelf life, a higher inoculum level (10^5–10^7) may be required to demonstrate the extent of reduction, such as in thermal processing or antimicrobial intervention-related research. The inoculum preparation and method of inoculation must be appropriate for the process and product to be evaluated.

When designing and performing challenge studies, the key is to mimic, as closely as possible, the product and process so that the results are applicable to the actual in-plant processes. This includes consideration for the preparation of the inoculum, such as growth time; temperature; adaptations (cold, heat, acid shocked or adapted); concentration; and substrate (buffer, sterile water, fecal matter). Food microbiologists would need to determine the necessity for adaptation of inocula to stresses associated with the food process. The peer-reviewed literature can be surveyed to determine the most appropriate adaptation/shock/stress procedures to prepare inocula and interpret the impact on ultimate results of the validation. For example, measured heat resistance of cells of *Escherichia coli* O157:H7, *Salmonella enteritidis,* and *Listeria monocytogenes* was up to eight-fold greater when they were grown, heated, and recovered anaerobically rather than aerobically (George et al. 1998). Multiple studies have demonstrated elevated heat resistance in bacterial pathogens as a consequence of prior heat shock, such as in *Clostridium perfringens* (Heredia et al. 1997), *Listeria monocytogenes* (Linton et al. 1990), and *Salmonella* (Mackey and Derrick 1986; Bunning et al. 1990). Although numerous other examples could be given, the point here is that food microbiologists must determine the need for selection, preparation, and recovery of the inoculum.

The duration of the study becomes important particularly when validating survival or recovery of injured organisms or for shelf life and frequency of sampling. The product formulation factors (matrix, pH, water activity, percentage moisture, etc.) and storage conditions (temperature, packaging, atmosphere, etc.) can affect the validation study and must be based on worst-case scenario expectation for the product and process. The types of tests (physical, chemical, microbiological) and methods of analyses of samples (culture, convenient, rapid, sublethally injured/stressed cell recovery) must be appropriate for the product and process evaluated. The measurement of success will depend on the objectives of the study and the results of the data analysis. Results such as average reductions in levels of bacteria are not appropriate in most instances. The focus should be on the worst-case conditions and defining the limitations of the process.

When the use of pathogenic organisms is not appropriate (e.g., production plant, pilot plant, growing field), indicator organisms or surrogate organisms may be used (Busta et al. 2003). Surrogate microorganisms are those microorganisms, usually bacteria, that are used in challenge or process validation studies if a target pathogen would not be used. Mossel et al. (1998) referred to these as marker organisms and defined them as "types of bacteria* whose response in foods processed-for-safety reflects the microbicidal or microbistatic goals of the treatment process." Whether referred to as markers or surrogates, these organisms offer an alternative for validating in-plant processes to control pathogens. Surrogates are nonpathogenic microorganisms that are stable and possess similar growth and inactivation characteristics and respond to a particular treatment or to processing parameters (i.e., pH, temperature, oxygen, heat, cold, antimicrobials, sterilization, etc.) in a manner that is equivalent to or more resistant than that of a target pathogen. The use of surrogates can allow the researcher to quantify the effect of processes and interventions on a nonpathogenic organism in commercial food processing environments where pathogens cannot be utilized because of safety concerns. This may include when it is undesirable to inoculate the actual in-process food product(s) or to compromise processing equipment or the production environment with high levels of pathogens when trying to assess the efficacy of various new or experimental intervention strategies or to perform in-plant validation. These nonpathogenic organisms should be easily prepared, easily enumerated, and easily differentiated from other microflora. In addition, some of these surrogate (index) organisms may be transformed to express fluorescent proteins (Cabrera-Diaz et al. 2009) (red, green, or yellow) or antibiotic resistance (Fairchild and Foegeding 1993) (streptomycin, ampicillin, rifampicin). This aids in differentiation of the index organism from background microflora during microbiological examination. Examples of some surrogates include indicator organisms, such as *E. coli*, *Enterobacteriaceae*, and coliforms, or the use of *Clostridium sporogenes* PA 3679 for *Clostridium botulinum*, *Listeria innocua* for *Listeria monocytogenes*, and *Escherichia coli* K12 for *E. coli* O157:H7. For details on the use of and definitions of surrogate microorganisms and index and indicator organisms, see an expert report done for the FDA (Institute of Food Technologists 2001) or other sources (Kornacki 2010).

3.4   Considerations for specific processes

3.4.1   Interventions: Chemical, physical, biological treatments of foods

Interventions that focus on eliminating, reducing, or controlling microbial populations to reduce contamination or extend product shelf life include a multitude of processes and ingredients. They may be applied at one or more steps in a food production process. Physical methods

---

* Yeast, mold, viruses, and parasites might also be utilized as "markers."

include various heat treatments (pasteurization, hot water, dry heat, or steam), dehydration and drying, refrigeration, freezing, wash and rinse systems, vacuum systems, irradiation, pulsed electric fields, ultrasound, high pressure, and controlled atmosphere packaging. The list of chemical interventions that have been developed to be applied to a food product or included as ingredients in product formulations is rather extensive. Traditional and commonly used antimicrobials applied to fresh and ready-to-eat food products include organic acids (lactic, acetic, and citric); chemicals and sanitizers (covered in detail in Chapter 6) such as chlorine compounds, ozone, electrolyzed water, acidified calcium sulfate, potassium lactate, sodium diacetate, sodium citrate, trisodium phosphate, and various salt compounds; naturally occurring spices and oils; and a variety of antioxidant compounds that exhibit antimicrobial properties. A combination of more than one intervention treatment may be used in a hurdle approach to microbial control in which the synergistic effects of the combined treatments is more effective than either treatment used alone. Biological approaches such as bacteriocins and bacteriophages have been more recently developed. Some of these interventions are applied to food contact surfaces or directly to the product itself, some during packaging, or to in-package product depending on the regulatory approvals, labeling requirements, and the overall desired effect on product quality and safety.

During the validation and implementation of an antimicrobial intervention, ingredient, system, or process, all pertinent equipment and product characteristics should be considered. Thermal processes such as hot water or steam pasteurization will require equipment qualification and the development of a temperature profile by mapping the process to better understand the process and the impact of the product on the process and equipment performance. For any and all processes involving real-time monitoring of critical process variables such as temperature, it is important to include in the design precise instructions of how and where to take the temperature. All monitoring devices must be used as intended and calibrated to ensure that they accurately reflect the process and process variations. The calibration and expected variability of the monitoring equipment must also be accounted for during equipment and process qualification and validation and during ongoing monitoring of the process.

As previously mentioned, during the information-gathering and qualification stages of validation, questions related to the product(s), process, and equipment will assist in gaining the insight needed to validate and control the process properly. In the case of a postpasteurization treatment such as heat or steam, it is important to understand how equipment performs to be certain that all product surfaces that have been exposed to possible recontamination may also be reached by the pasteurization treatment. Questions related to the process and product may include:

1. Have hot spots, cold spots, and partial loads been accounted for?
2. How does the in-going product temperature effect the process and equipment performance? What dwell time (exposure to hot H_2O) is needed?
3. Do all products receive equal heat treatment/exposure?
4. Is there adequate spacing between products to ensure all product surfaces receive equal heat treatment/exposure?
5. What about chilling after heating (e.g., product surface time/temperature profile)?
6. How does the product shape affect the dwell time, and are there cold spots in the process that will effect/dictate the dwell time?

Product shape is an important consideration as the smoother the surface, the more efficient the pasteurization process will be. Crevices or cuts in product, net marks, or wrinkles can become harborage sites for spoilage organisms or pathogens such as *Listeria monocytogenes* and may be more impervious to heat and steam treatments. Spacing of product during the process, whether hanging in a batch house or continuous process or loaded on a carrier designed to move product through a pasteurization system, must allow for all sides (surfaces) of the product to be treated evenly. To obtain effective coverage, product overlap should be avoided so that cold spots in the product are easier to reach, resulting in more uniform process lethality.

As with all validations, the objective of the process must be kept in mind during validation of an antimicrobial treatment. For example, is the post-lethality treatment sufficient to eliminate the levels of *L. monocytogenes* contamination likely to occur? In cases where potential contamination levels are unknown, risk assessment can be used to define the FSO. This approach may be used in the validation of most post-pasteurization lethality treatments, including exposed product and in package hot water and steam pasteurization processes. Once again, if published literature is used to validate the log reduction, products, treatments, or other variables that are used in the establishment's process, the product, process, and equipment used by the establishment should be the same as those used in the published literature, or the treatment will need to be validated for the plant's specific conditions and product characteristics.

While the limitations in use of physical lethality treatments such as hot water or steam are based primarily on their cost and effect on product quality, the use of chemical-based antimicrobials is based not only on cost and product quality but also on regulatory approvals (USDA FSIS and FDA) and any special labeling requirements. USDA FSIS approvals and associated requirements for labeling are fairly straightforward and outlined in the USDA FSIS Directive 7120.1 Safe and Suitable Ingredients Used in the Production of Meat, Poultry and Egg Products (United States

Department of Agriculture, FSIS 2011). According to the FDA, any substance that is reasonably expected to become a component of food is a food additive that is subject to premarket approval by the FDA unless the substance is generally recognized as safe (GRAS) (FDA 2011b). Additional resources are available for determining if a substance is approved by the FDA for food contact or use, such as 21 CFR 174-179 (appropriately regulated indirect additive), 21 CFR 182-186 ("Generally Recognized as Safe—GRAS"), or 21 CFR 181 ("Prior Sanctioned"), the list of Threshold of Regulation Exemptions or the Effective Food Contact Substance Notifications (FDA 2011b). A company considering using an approved antimicrobial such as one listed in the USDA FSIS Directive 7120.1 must follow the stated approvals as written for their particular product. These vary from one product to another, and a particular antimicrobial carcass rinse approved for use at a specified level on poultry carcasses with no special labeling requirements may not be approved for use with pork or beef carcasses unless declared on the label. In another example, one antimicrobial may be applied to beef bologna as a spray at a certain level (ppm) or for a specified amount of time but may not be approved for use as a dip for the same product and may not be approved for any other deli meats. These approvals are based on the data and documentation submitted by a company intending to use the product or by the manufacturer of the particular ingredient or process during the approval process. The more specific the data and research submitted, the more specific the resulting published approval will be. Therefore, it is recommended that one of the first steps taken in the evaluation of an antimicrobial is a search of the current regulatory approvals related to that particular ingredient or process (USDA 2011; FDA 2011a, 2011b).

When assessing or validating an antimicrobial agent or process, a part of the validation documentation may include a demonstration that the antimicrobial agent or process, as used, is effective in suppressing or limiting the growth of the target organism. This is the point at which a company must determine if a challenge study or in-plant study is necessary. For example, the validation may need to support an initial reduction in levels of a pathogen that the antimicrobial agent or process can achieve (e.g., 2-log reduction). For some antimicrobial ingredients and processes, the length of time (days) that the antimicrobial agent or process is effective throughout the shelf life of the product (45 days, 60 days, 90 days, or longer) must also be determined. In this case, additional studies at optimal and suboptimal (worst-case) shelf life times and temperatures may be warranted.

As an example, in validating an antimicrobial carcass rinse for a processing plant, the company may first investigate the regulatory approvals for the antimicrobial. If no prior approvals exist, the company or the manufacturer of the antimicrobial must seek and obtain approval for use from FSIS to ensure that the product or process does not adversely affect product safety or government inspection activities. This will likely entail conducting trials showing the effectiveness and safety of the antimicrobial. The company may also evaluate the current research on the particular antimicrobial or on a similar antimicrobial and on the effectiveness, feasibility, limitations, and costs associated with the particular delivery system (rinse cabinet, dip, spray, etc.). Evaluating the research can become overwhelming as the research related to antimicrobials and systems to deliver or apply antimicrobials is extensive and varied. So, what does the research say about organic acids, for instance? There have been numerous and varied types of organic acids published in the literature, from citric acid to lactic acid to acetic acid used alone or in combination with each other or with other treatments such as steam and hot water. They may have been sprayed at pressures from 20 to 40 psi in amounts from 100 to 1000 mL per surface, at concentrations ranging from 1% to 5%, and at temperatures ranging from 25 to 60°C at varying distances from the carcass surface. Some research has evaluated the effects of antimicrobials applied to a "hot" carcass surface, while other studies have evaluated the effects of antimicrobials applied to a carcass surface postchill. Different types of spray cabinets with different nozzle types (i.e., flat or fan-type nozzle) and with varying oscillation and spray patterns have been studied. A company must weed through the myriad of research studies to determine which is the best fit and most closely mimics their process and product. An equipment manufacturer may recommend a specific nozzle type, pressure, exposure or dwell time, and patterns to achieve maximum coverage and effectiveness. The equipment (i.e., cabinet system) must be validated in the plant on a quarterly, semiannual, or annual basis. For example, in validating the spray pattern and total coverage of a carcass with an organic acid, the carcass can be covered with large pieces of pH paper, which will change color when exposed to the acid. After sending the carcass through the spray cabinet, the paper is removed and evaluated for any gaps in coverage, and the nozzles or process can then be adjusted as needed. In the case of hot water cabinets, the temperatures and dwell times applied in a plant situation may vary with the facility's in-house hot water capability (and sustainability) and worker safety, as well as with desired line speed and available space on the line for a prefabricated final wash cabinet. When validating hot water cabinets or steam and evaluating the maximum (and minimum) temperatures reached on different areas of the carcass, several temperature indicator labels can be placed in different areas of the carcass, or type K thermocouples can be threaded just under the carcass surface and back out so they lay on the meat surface and are connected to a thermometer. Following the evaluation of temperatures in different areas of the carcass, the process may be adjusted by adding nozzles or changing the nozzle pattern, increasing the water temperature, or slowing the line or extending the length of the cabinet to increase the time the carcass is exposed to the hot water. Lean and fat color may become an overriding factor when using hot water as prolonged exposure to hot water temperatures can result in irreversible denaturation of meat proteins. The cost of removing the "cooked" meat and loss in yield may be more than the benefits for this type of antimicrobial process.

On the practical side, the company will most likely evaluate factors such as the cost and ease of application for both the antimicrobial and the delivery system; where it will be applied in the process; whether a spray cabinet will keep up with line speeds or if a dip will work better and take up less square footage on the production floor; what the operating parameters (pressure, flow rate, temperature) and utility needs are, such as exhaust, filtration, electrical, water usage, and expected maintenance of the equipment; if there are any potential worker safety issues, such as chemicals or temperatures from steam or hot temperatures that necessitate additional safeguards be put in place, such as containment and exhaust for heat and or chemicals used in the cabinet; how it will fit into the current hazard analysis and critical control point (HACCP) supporting and prerequisite programs; what the expectations and requirements are for monitoring and documentation and if they will require additional employees; and how the antimicrobial will affect product quality, safety, and shelf life.

Additional in-plant validations can also become overwhelming as the plant designs protocols for validating the process in the plant while limiting exposure or contamination of other parts of the process and carcasses. With this in mind, validations of the microbial effectiveness of antimicrobial carcass rinses and interventions are often performed on the lower region of the carcass, such as the leg, neck, or jowl, using a fecal slurry and documenting the reductions in indicator organisms such as generic *E. coli*, coliforms, or surrogate organisms (Cabrera-Diaz et al. 2009; Eggenberger-Solorzano et al. 2002). These types of in-plant validations may require prior approval from in-plant regulatory inspection personnel and are often carried out at the end of the production day. Documentation should also include qualification of the equipment and parameters such as overall cabinet operation, including but not limited to nozzle pattern, dwell time, flow rate, pressure, line speed, and temperature. Different types of equipment and systems are available to measure and monitor these parameters, such as nonreversible temperature labels and wireless datalogging systems for time, temperature, humidity, and pressure. The more data the company can accumulate about the process and product, the better it will understand the process and equipment and how to manage the process for optimum effectiveness and to maintain control of the process.

3.4.2   Thermal processing

Thermal processes such as cooking play a significant role in food quality and safety. Thermal processes may preserve a food, alter the texture, improve tenderness, lower the water activity, stabilize the color, and intensify the flavor of a food. Most thermal processes are designed around the principles of heat transfer and microbial death (Scott and Weddig 1998). The basic premise in thermal processing (and in chilling) is the transfer of heat and the passage of thermal energy (heat) from a hot to a colder body. The driving force for heat transfer is the surface-to-core temperature difference. The rate with which energy is transferred to and disbursed throughout a product is affected by the surface-to-core temperature difference, as well as by product characteristics, and by the method of heat transfer (conduction, free convection, forced air convection, or radiation). Thermal processes range from dry heat to moist heat and radiant heat or combinations thereof and may occur in a batch or as a continuous process. The thermal processes most often encountered in the processing of food involve the use of hot air, water, oil, steam, or pressure. Thermal processing of some products, such as in the cooking of meat products, involves both the transfer of heat from the air into the product and transport of heat throughout (within) the product. At the same time, moisture migrates from the product interior to the surface, and evaporation occurs at the product surface. The rate of migration, in this instance, depends on product temperature, composition, moisture concentration, and water-holding capacity of the product and is influenced by factors such as the temperature, airflow, and relative humidity of the surrounding cooking medium (air, water, oil, steam). Most cooking loss is then the result of evaporation from the surface, while the product's interior may show little change. Thermal conductivity or the "speed" at which heat travels through a product to its surface may be affected by product composition. For example, products with higher moisture content generally exhibit higher thermal conductivity, while higher fat content may have an adverse effect on conductivity due to its insulating ability. Is it a ground product (pork, beef, mixed species, or soy), whole muscle, or chunked and formed product cooked in a bag or mold? Is it emulsified and stuffed in casing (natural, collagen fibrous) or a roast beef or ham in netting? Is it tuna in a retort pouch or chicken noodle soup in a can or a jar? Different product characteristics will affect heat transfer and the distribution of heat throughout the product.

Several factors need to be considered during the equipment qualification for thermal processes to maximize heat transfer and microbial death. Significant variations can occur depending on the load configuration, the initial temperature of the product, and product load variations. These factors should be monitored to determine the zones with slowest heating in the oven, water tank, sterilizer, or fryer or across a cook belt. For example, for USDA inspected meat and poultry products, the objective of validation may be to ensure all products meet the USDA FSIS (1999) "Compliance Guidelines for Meeting Lethality Performance Standards for Certain Meat and Poultry Products" (e.g., Appendix A). The performance standards, as outlined in the guidance material, recommend that the cooking time, product temperature, and relative humidity of the process be monitored and documented. To qualify the equipment, validate the process, and ensure uniform thermal lethality, the company must understand and recognize all possible sources of variability in their process(es). This includes validating the equipment and taking into consideration equipment design (airflow, heat flow); product loading, variations in product shape, size, and composition; and variations in the thermal process itself from initial heating through final cooling. Variables that may influence product heating include cold spots (in product and equipment); product size, weight, shape, fat or moisture content; raw materials and ingredients; density; initial product temperature; cook-in packaging, molds, and casing materials; and air spaces in the product. Variables to consider in the lethality process that may influence heat transfer and resulting lethality include the temperature and humidity of the process, proper spacing of product to ensure all products and all areas of the product are exposed to the heat equally for maximum treatment and lethality, circulation and airflow, and the heat transfer coefficient of the heating medium (water, air, oil, steam, etc.) for every type of product. The goal is to control these sources of variability and prove (i.e., validate) that the process is under control and verify that the process is staying in control.

3.4.2.1 Temperature measurement and control

The 18th century mathematical physicist and engineer Lord Kelvin (Sir William Thomson) is credited with stating that "to measure is to know"; "if you cannot measure it, you cannot improve it." Keeping in mind that the primary goal of any thermal process is to kill bacteria and the secondary goal is to develop quality characteristics, proper temperature measurement and control are critical to the success of the process. Therefore, when validating and monitoring critical process parameters such as temperature, it is important to understand how the monitoring equipment works and to know the expected variability and accuracy of the equipment. For temperature, this includes taking into account the variability of any in-process as well as handheld monitoring devices and determining if the accuracy of a thermometer is to within $\pm1°C$ or $\pm2°F$. Temperature variation will also need to be accounted for in any measurements taken, as will worst-case scenario conditions of product and process temperature. Whether it is wireless, infrared, or a probe-type monitoring device, it is important to know how to use the device properly. For instance, if the temperature sensor is located a quarter inch from the end (tip) of the thermometer, then the device may need to be inserted into the geometric center of the product to the sensor point rather than to the tip end. In addition, the thermometer or sensor may need to be left in the product for a certain period of time to equilibrate (stabilize) and record an accurate product temperature. This time may vary from several seconds to minutes depending on the type of sensor and product being monitored. It is equally, if not more, important to know how and when to calibrate the device properly. Probes should be calibrated against a certified standard thermometer such as an NIST (National Institute of Standards and Technology) reference thermometer. If using a dry well calibrator, be sure the manufacturer provides a certificate of accuracy and traceability. It is also important to calibrate the thermometer to temperatures within the range of expected use. An ice water method is generally employed for calibrating thermometers used to monitor colder temperatures and boiling water for hot temperature calibrations. Since microbial death is a function of both time and temperature, devices used to monitor and record the time should also be properly functioning and accurate. Additional humidity and airflow sensors may also be necessary to document the distribution of moisture and proper patterns and fluctuations of the flow of air throughout the system. Appropriate sensor and temperature monitoring equipment (i.e., thermocouples) should be placed throughout the load (in the air as well as in product) to measure process variation and to determine the cold spots and the areas in the process that are slowest to heat.

The design of the equipment will influence the heat transfer and resulting lethality of a process; therefore, when validating processes such as batch ovens and continuous ovens, factors to be considered include air temperature (dry bulb), moisture in the air (wet bulb), fluid velocity (air, water), and time. All of these factors combined will define the processing (cooking) schedule for the process. During qualification of the equipment, sources of variability include differences in temperature and variation from equipment top to bottom, side to side, and front to back. Whether it is a batch house, continuous house, flow process, cook belt, water cook, fryer, steam cooker, roaster, or retort, all sources of variation inherent in the process must be accounted for and controlled since the goal is to establish uniform and balanced temperature, humidity, air, and liquid flow rates. Based on the results of the initial equipment qualification and information-gathering phase in validation, it may be necessary to solicit the assistance of the equipment manufacturer and maintenance technician to calibrate or balance processing equipment properly and to set up a schedule for regular preventive maintenance and recalibration to maintain equipment (i.e., cook tanks, fans, air ducts, heating elements, dampers, etc.). Some changes may be necessary to control inherent equipment variability more effectively. For example, in some water cook systems, the injection of air into the cook tank during the cook cycle agitates the water, which in turn reduces thermal layering and improves heat exchange, resulting in a more balanced and uniform cook process.

Without a well-established process control program, product consistency (quality and food safety) can drift. Proper product loading into any cooking equipment, whether it is a smokehouse, fryer, or retort, is necessary to ensure the product is evenly dispersed with adequate spacing between product for even and adequate heating. Validation must also be performed on partial loads and loads of mixed products, such as products of differing sizes, shapes, and composition. All of these factors can influence the airflow and overall heat transfer for the product and process. Some configurations of loading racks or trucks in house allow for better air flow for partial loads. The product carrier design (closed, open) can also affect airflow and lethality as a closed framework can block oven airflow. Paying attention to spacing out product uniformly on a smoke stick and minimizing "touchers" will ensure more uniform cooking of link-type products such as hot dogs, franks, and sausages. For some continuous systems, a sheet of stainless steel placed behind the last row of product entering a continuous system may contain the heat and air and provide more even cooking for the end-of-batch or end-of-shift product in the system. Product parameters such as the shape (geometry) of the product affect heat transfer; heat transfer is greatest along the shortest dimension of the product. Uneven, naturally shaped products like bacon and bone-in hams will inherently have more temperature variation than a more uniformly shaped hot dog or molded ham.

Other process consistency issues arise with balancing the airflow in the process and understanding where the hot and cold spots occur, how they move, and how to manage them. Process variability will be easier to manage for product of uniform shape, similar composition, and uniform load, allowing for adequate spacing.

Additional considerations for a thermal process include understanding the process and how it relates to the concepts of integrated time-temperature lethality and the D, F, and z values involved in microbial death. Integrated time-temperature considers the entire heat process and takes into account all of the heat to which the product is exposed from the time the product is first exposed to heat until the product begins to cool. Products do not reach their final internal temperature immediately, and the temperature may even increase after a cook cycle is completed and the heat is removed. This is why it is so important that the food processor understands and validates the heat transfer into its products under the most challenging conditions possible for the product and process. This is best achieved by mapping the process (equipment and product) from initial in-going product temperature through final chill, usually with the aid of data-logging temperature and humidity probes loaded into representative products at various spots in the processing equipment. It is also important to realize that the heating and chilling parameters will likely vary with products of differing size, shape, and composition and in different areas of a cook system. Therefore, validation of processes generally begins using the most challenging product produced in that equipment (system), such as the coldest temperature for in-going product, the largest products that will heat and chill more slowly, or the area of the product where heating (and chilling) is expected to be the slowest, using the product formulation that is expected to take longer to heat and chill.

Microbial life and death can have an impact on both the safety and the shelf life of the product. Therefore, it is important to know the microbiological quality of the raw materials included in the product. Process validation procedures may include some element of microbial testing of raw materials for indicator or index microorganisms to help establish the requirements of the process.

The death of bacteria may begin at temperatures slightly above the maximum, allowing multiplication. While bacterial spores may survive temperatures much higher than the maximum, allowing for the potential multiplication of vegetative cells, the number of viable cells will be reduced exponentially with the time of exposure to a lethal temperature. The D value is the time in minutes, at a constant temperature, necessary to destroy 90% or 1 log of the organisms present (e.g., for *Salmonella*, $D_{145F} = 0.7$ min). The z value is the number of degrees Fahrenheit required to cause a 10-fold (1 log) reduction change in the D value. The z values account for the relative resistance of an organism exposed to different temperatures. For instance, if the z value for *Salmonella* is 10°F and D value at 145°F is 0.7 min, then $D(155) = 0.07$ min and $D(135) = 7.00$ min. The z value is used for products that may need to be processed at a lower temperature for a longer period of time or at a higher temperature for a shorter time to maintain product quality characteristics while achieving the same lethality. The F value measures the process lethality as the time, in minutes, at a specified temperature required to destroy a specific number of viable cells having a specific z value.

If a processor is unsure of the impact a product formulation or the ingredient is likely to have on the process lethality, then it is best to conduct the necessary tests to be certain rather than to wait until the process fails. Even the slightest change in a supplier or to ingredients of lower microbiological quality or a failure to thaw frozen raw ingredients completely, resulting in colder in-going product, may result in a failure in the process, a recall, or a public health issue. Particular attention should be paid to product formulation changes affecting fat, salt, moisture, and acidity as these are known to have an impact on microbial resistance to various processes.

3.4.3 Cooling and stabilization

The proper cooling and stabilization of food products, particularly following lethality, involves the same principles of heat transfer discussed previously. The goal of cooling and stabilization is to chill product rapidly to below temperatures that will preclude microbial growth and product deterioration. While this temperature may vary with the product composition, microorganisms of concern, packaging, atmosphere, and the time the product is exposed to the higher temperature, generally temperatures above 10–15°C (50–60°F) are considered excessive (Tompkin 1996), with temperatures between 5 and 54.4°C (41 and 130°F) identified as the temperature danger zone (high risk) for microbial growth. The rate and effectiveness of chilling depends on the chemical (composition) and physical (weight, thickness, density) characteristics of the product and the method (cooling medium) and efficiency of heat transfer. In contrast to lethality, during cooling there is a passage of thermal energy (heat) from a hotter body (product) to a colder body (surrounding medium) in such a way that the product and surroundings reach a thermal equilibrium. The transfer of heat from the product to the cooling medium may be of a more passive nature, such as occurs in a conventional cooler, or there may be a more active transfer of heat to the cooling medium, such as in agitated chilling water or fan-driven air.

Conventional air chilling is the most predominant method of chilling used by the food industry. Air cooler and blast air chill are also widely used, combining air as a cooling medium with highly efficient air-handling systems. In these situations, the combined efficiency of air temperature and air velocity determine cooling rate. However, they may not be appropriate for all products. Liquid chilling methods such as chilled water (0 to 4°C) are frequently used for chilling products such as fresh poultry. The addition of countercurrent flow (agitation) in these systems improves cooling uniformity and efficiency. Some cooling systems chill products in a stepwise manner so they are not chilled too fast, which may have a negative impact on product quality, such as in the chilling of beef and lamb carcasses. The type of liquid used in the chiller will also have an impact on the rate of cooling. Some chilling systems employ highly effective liquid cooling media of either sodium chloride brine or propylene glycol. The addition of salt to water lowers the freezing temperature of the solution, and the heat transfer efficiency can be greatly enhanced for a comparatively low cost. Glycol works in the same manner, allowing the liquid to chill product at a much lower temperature without freezing the cooling medium. However, as glycol is not approved for direct contact with food products, it is therefore used more frequently as a liquid chilling method for packaged products. Liquid chilling systems may be applied to food processes as a spray or as an immersion-type chilling process. Both types of systems must be validated based on the challenges and requirements of the equipment and of the products. Another type of chilling is evaporative or intensive chilling; the chill effect is achieved by wetting the product's surface with water while directing air movement over the product. There are also chemical chilling methods, such as the use of nitrogen.

Each chilling system presents different challenges when it comes to validating the equipment and the products for the process. The equipment must be validated as appropriate to the chill tank, cooler, tunnel, and freezer. As previously discussed in the qualification of equipment used in thermal processing, the equipment used in cooling and stabilization must be qualified to ascertain areas of uneven cooling. Approaches to cooling equipment qualification will likely vary with the chosen chilling method: chill tank, cooler, tunnel, freezer, or a combination of methods. During the qualification of all cooling processes, temperature (and time) is a major factor to monitor. Understanding how the temperature varies throughout the equipment (water tank, cooler, tunnel, etc.) and during the chilling process for all products is essential for controlling the process. Some additional factors to consider when qualifying and validating chilling equipment and processes vary with the type of chill system and may include flow rates in countercurrent flow systems, air velocity, brine concentration, or the spray system in brine or glycol or in an evaporative chilling system.

Just as with thermal processes, food products are chilled more effectively when there is adequate spacing between products. Proper spacing between hot products allows for the chilled air or water, whether it is moving or not, to pass across all surfaces of the product. This may include providing several inches of space between carcasses, alternate stacking of boxes in patterns on pallets, or using slip sheets to allow adequate airflow between products during chilling. For example, in coolers the proximity of hot product in a cooler can have a significant impact on heat transfer and cooling rate. For instance, product placed near or under cooler fans tends to chill faster, while product placed near cooler doors that are opened and closed several times a shift may chill more slowly. The type of packaging (covered combos of product, individual packages, plastic tubs, boxes of hot product, boxes of packages, no packaging, vacuum or modified air packaging, polyvinylchloride [PVC] overwrap of a foam tray) will also affect the rate and effectiveness of chill.

The thermal properties and thermal conductivity of the food product affect the rate at which heat travels through a product to its surface. High moisture content usually indicates high thermal conductivity, while fat content may have an adverse effect on conductivity due to its insulating ability (Singh and Heldman 1984; Woodams and Nowrey 1968). Product geometry, including configuration, thickness, and diameter, affect the rate of heat transfer (Singh and Heldman 1984; Mangalassary et al. 2004). The product chills more rapidly across the narrowest portion of the product. The temperature differential, the relationship between product core temperature, and the temperature of the cooling differential will determine the cooling rate.

Validation of chilling processes, as with other processes, must be based on worst-case conditions. It is important to understand the limitations of the chilling process and how factors such as the warmest, largest, or heaviest product will affect the process and to measure how the process will effectively control this process variability. If more than one product type is processed through the chiller, then separate validation specifications may be required for each product. For the worst-case conditions, slowest-cooling products are generally considered (validated) first. For example, larger carcasses can take twice as long to chill as smaller carcasses chilled under the same conditions, so it is important to be realistic in deciding on the weight and size range to ensure the system will chill all carcasses adequately. A throughput profile is needed to be sure that the chiller capacity is designed and validated to manage the maximum daily or weekly throughput of product through the chiller. The company (validation team) must first identify the objectives of the chilling process, which will most likely be to chill the product to below a certain temperature within a certain amount of time. In this case, and to validate the process, the company must identify the current flow and parameters, such as types of products run through the system (whole carcasses, pieces, species, etc.), throughput, initial and final product temperatures, and any change in yield or temperature that is desired in the process. The next step is to measure what is currently being achieved in the current process under worst-case conditions (highest prechill temperatures, larger/ heavier carcasses/pieces, closer spacing, etc.). When a chiller is already filled to capacity and additional space is needed, an almost-automatic response is to push products (carcasses) closer together on the rail or move pallets of product closer together. This practice of overloading and improper spacing of product can have a significant negative impact on product chilling and push the cooler beyond its chilling capacity. The next step is to identify what limitations or constraints there are with the current process, which may include inadequate chiller space available, lack of time available to hold carcasses prior to fabrication, or limitations on the amount of power or chilling capacity available. Compare the desired outcomes with what is currently being achieved, adjust the process such that it will achieve the desired results, and measure again to verify the process is doing what the master plan states.

The failure to remove the required heat can be due to a number of reasons, such as insufficient time allowed for chilling, a refrigeration capacity that does not adequately address the high initial product load, overloading the system, variability in size of products, or incorrect environmental conditions (temperature, humidity, air velocity). Therefore, some final words when measuring and validating chilling processes are to understand the product characteristics and how they are affected by cooling, evaluate the cooling system performance and monitor it closely, and consider needed improvements, the need for revalidation, or the use of alternate chilling methods as production increases. Chilling needs to be designed to give uniform air distribution and sufficient refrigeration to reduce product temperatures in a fully loaded chiller.

3.4.4   Refrigeration and freezing

Keeping perishable food cold is important to maximize product quality throughout shelf life. Storing these products within a controlled range of refrigeration (0 to 4.4°C; 32 to 40°F) or freezing (<0°C; <32°F) temperatures will prevent or at least limit the growth of pathogenic and spoilage microorganisms and maintain product quality (yield, taste, texture, appearance, etc.) until product preparation or consumption. Temperature is generally regarded as the most important environmental factor affecting the shelf life of perishable foods and the growth of microorganisms. Reducing the temperature of the product extends the lag phase of microorganisms and reduces oxidative and enzymatic changes that may negatively affect product quality. Even moderate temperature abuse, generally defined as above 10–15°C (50–60°F) may drastically shorten shelf life and shorten the time required for microorganisms to grow.

In validating refrigeration and freezing, it is important to measure and manage the chill chain from raw material to consumption and to understand not only the chilling and storage at each stage of the cold chain but also the interaction between the stages (Figure 3.1). Temperature monitors can be placed in boxes, on pallets, or in trucks and refrigerated display cases, and time and temperature data can be downloaded using data-logging software.

Temperature and time monitoring is the first step (Figures 3.2 and 3.3). The dynamic temperature information can be input into predictive microbial models (see Chapter 10) to perform an assessment for the potential of microbial growth. From that information, the expected behavior of microorganisms within specific food systems can be determined using predictive models, inoculated-pack challenge studies, or both.

Freezing, a process whereby a liquid turns into a solid, also slows food decay (slows enzymatic reactions) and slows (or completely stops) the growth of microorganisms, depending on the temperature at which an intrinsic food solutes. Lower temperatures slow metabolism and reaction rates and sequester water necessary for bacterial growth (as it freezes). Some of the major influencing factors concerning refrigerated

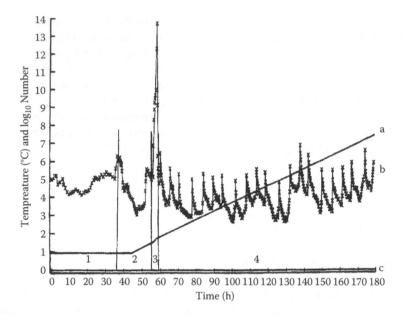

Figure 3.1 Temperature profile during food product distribution and microbial growth curves stages 1, 2, 3, and 4 depict the harvesting, packing, transport, and retail display of food, respectively. Curve a is predicted microbial growth at temperature profile b. (From Baird-Parker and Kilsby, 1987. *Journal of Applied Bacteriology*, 63:43s–49s. With permission from John Wiley & Sons, Ltd.)

and frozen storage are the storage temperature, the degree of fluctuation in the storage temperature (controlled and uncontrolled), chiller capacity, and the type of wrapping/packaging in which the product is stored. While minor temperature fluctuations in a stored product are generally considered unimportant, the extent and effect larger temperature fluctuations may have on a particular product vary with the temperature of the fluctuation, the amount of time the product is exposed to the temperature variation, and product type (composition). The effect of additional factors, such as the potential additive effect of fluctuations in temperature, may also need to be considered. Therefore, it is important to consider and measure equipment defrost cycles during equipment and process validation. The thermal properties of the food are a function of both its composition and its temperature.

Refrigeration equipment (coolers and freezers) is always used to control temperature, However, for some processes the cooler or freezer may serve one or two purposes: either to maintain product at its initial temperature or to reduce and store product at a specified temperature. It is important to understand that these two functions may require different equipment and different process capabilities such that one cooler may not

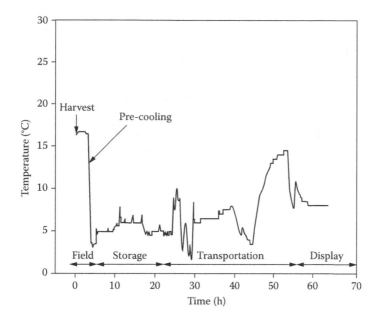

Figure 3.2 Temperature history of iceberg lettuce. (From Koseki and Isobe, 2005. *International Journal of Food Microbiology,* 104:239–248. Wth permission from Elsevier.)

be adequate to perform both functions. When a process, such as a cooler or freezer, is required to serve several functions, then each function must be clearly identified and validated to ensure that the process performs all functions effectively for all product types. The maximum time product is stored at refrigeration or frozen temperatures is generally based on a measure of product quality and may be determined through organoleptic and chemical and microbiological analysis.

One area that does not receive the attention it should either in the published literature or in commercial processing is the process of thawing (defrosting, tempering). There are several methods for thawing currently in use in commercial operations, including microwave technology, air, and water thawing. The major issue with the process of taking product from a frozen state to a refrigerated or warm state is that all too often it is uncontrolled. As previously mentioned, the process validation procedure may be the first time that the process was truly defined by nonoperational personnel. Often, it will be discovered that thawing or slacking-out procedures have not been defined as part of the master plan.

Companies will often purchase raw ingredients, which can be stored frozen, when they are available or when prices are low, to be used later for further processing such as grinding, chopping, injecting, or cooking.

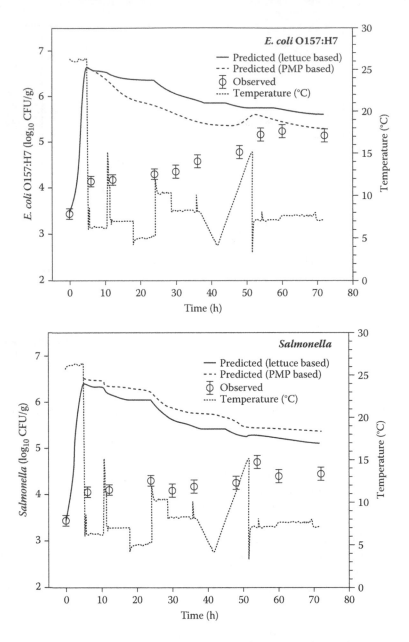

Figure 3.3 Growth of bacterial pathogens on lettuce under real temperature history. PMP = pathogen modeling program. (From Koseki and Isobe, 2005. *International Journal of Food Microbiology*, 104:239–248. With permission from Elsevier.)

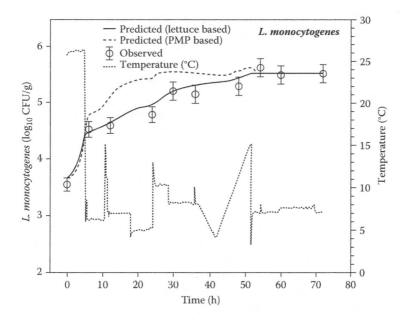

Figure 3.3 (continued).

Most thawing systems (water and air) apply heat to the product surface and rely on conduction to transfer that heat to the center of the product. The main concern with this manner of thawing when it is not controlled is related to the majority of bacteria (pathogenic and spoilage), which are located on the surface of the product. During thawing when the product surface temperature begins to rise, microbial growth begins (where it left off before freezing), and if a product is particularly large, the surface can reach fairly high temperatures and high levels of microorganisms before the center of the product is thawed and ready for use. Uncontrolled methods include putting a water hose in a bin of frozen meat to thaw overnight or storing product in a warm staging area of the plant or on a loading dock to thaw. While some of these methods are used for last-minute thawing to meet production demands, they are highly variable and may negatively impact product quality and safety. These methods are difficult to measure and to validate with regard to time and temperature and may pose additional risks with regard to water quality and sanitation when thawing occurs in uncontrolled areas of the plant or outside production areas on loading docks. At the very least, the worst-case conditions should be subjected to a validation study. If results are not favorable, the worst-case time/temperature limits can be modified, and the new process can be validated.

Controlled methods for thawing include controlled air, controlled water, and microwave technology. Microwave technology uses electromagnetic radiation to generate heat within the meat. It is primarily used to temper product such as meats destined for immediate further processing (i.e., cooking). Controlled air can be used for thawing any type of product. Products need to be separated to allow airflow, and air circulation needs to be constant such that all products are thawing at the same rate. If different products are being thawed simultaneously, validation should take into consideration the worst-case conditions and which products thawed under which conditions thaw first. For temperature-controlled cooler thawing, the cooler temperature is often kept at about 5°F above the desired final core temperature. Thawing under these conditions may take a few days depending on the product; therefore, proper planning to meet the needs of production is necessary when using an air thaw method. Controlled water uses either running water or water that is changed frequently. Water temperature should not exceed 50°F, and both surface and core product temperatures as well as water temperatures should be measured during validation of this process. In addition, product may need to be protected (packaged) from water damage, water absorption, and cross contamination. Proper validation of the thawing process is a must as uncontrolled and inadequately thawed product can also jeopardize further processing. When further processing, such as a lethality process, is based on a specific in-going product temperature, the inclusion of inadequately thawed raw materials in the blend can result in some product not reaching the final cook temperature. Depending on the frequency and type of monitoring for the process, the product can be missed during normal process monitoring. This is why it is so very important to build safety into the process and validate up front each and every step in the process flow. When validating commercial thawing practices, a key to success is remembering that the time-temperature pattern during thawing can affect microbial survival and negatively impact both product safety and quality.

3.4.5   Drying and desiccation

The processes of drying and desiccation do not often receive the same attention as other lethality processes and have been described as much an art as it is a science. For many of these processes, the underlying, preliminary steps in the process may be as important, if not more important, to the validation process as the more obvious lethality (drying and desiccation) steps. This is where a team that includes individuals with expertise in the "art" of making the product may be just as important as the scientists. Validation of these processes often includes many steps that occur early in the process, before the product enters the processing facility, and as far back as the proper chilling of carcasses, storage of raw materials, and product pH, as is the case for country-cured hams and prosciutto. The uptake of salt/cure is not as complete in colder hams with a higher pH. Therefore, validating refrigerated storage of raw materials and even thawing or tempering of product to ensure that all product is at the correct temperature before curing play a significant role in validating the total process. Temperature, time, and relative humidity must also be controlled throughout the curing, drying, and ripening processes. The size and weight of the product can have a significant impact on the curing and drying process as heavier product and thicker pieces of product will not uptake salt and salt/cure as fast as smaller, lighter pieces. During commercial production, hundreds of hams may be processed together through the same system and under the same conditions; therefore, understanding, measuring, and controlling any variability that may occur in product, size, weight, temperature, and pH can significantly reduce variability in final product quality and safety. In addition, it is important to understand and reduce, or at least account for, the variability in different areas of the coolers and drying rooms to ensure all products are curing and drying at the same rate. Validation of these process control factors may involve systematic collection of a_w samples throughout multiple lots of production in different areas of drying rooms or chambers. If all other factors are the same, larger (thicker) pieces generally will cure and dry more slowly and smaller pieces more quickly. Validation and monitoring the process can help to address any deviation or variability that may occur in the product or process before it can cause undesirable quality or safety concerns.

Similarly, the production of dry, semidry, and fermented products is both artisinal and scientific. These processes rely on acid production during fermentation as well as drying to develop the proper organoleptic results and reduce microbial concerns. The quality of the in-going raw materials, including product temperature, pH, composition (moisture, fat, protein), and ingredients (starter culture, acid, sufficient fermentable carbohydrate) are significant factors in achieving safe and uniform fermentation and drying. For fermented products, proper temperature and humidity control is required in all areas of the fermentation house or room to ensure the proper growth of starter cultures and associated acid production. Therefore, proper validation of the equipment, including the fermentation chamber (house or room), drying rooms, ripening rooms, and measuring devices (humidity, pH, air velocity, water activity, and temperature), is important. Fermentation, drying, and ripening rooms, much like smokehouses and coolers, may have areas in the room (or house) where product may ferment, dry, or ripen more slowly. Also, overloading rooms, improper spacing of product, or inclusion of multiple products of varying size, weight, shape, and composition can change the rate and completeness of the fermentation or drying process. Additional factors include the type of casing the product is stuffed into and orientation of the product, such as whether the product is hanging on trees versus product placed on shelves. Any or all of these factors can have an impact on airflow, which is critical for proper drying. As with other processes, proper identification of all the key factors necessary to produce product meeting the desired final product specifications is necessary to validate the process properly. Validation of these products may involve demonstrating control of growth of *Staphylococcus aureus* during fermentation and adequate levels of destruction of *Salmonella* and *E. coli* O157:H7 throughout the fermentation, heating (if any), and drying stages. Processes can be mimicked in properly equipped biosafety level 2 pilot plants or laboratories.

Dried products, which are generally defined as products that are microbiologically stable at ambient temperature, often require validation from in-coming raw materials as mentioned, through packaging. Proper identification of key steps may include in-coming product specifications such as raw material temperature and composition and in-process parameters such as product pH, temperature, relative humidity, airflow in dryers, and product spacing. In addition, microbial stability and product quality throughout shelf life depend on maintaining low water activity during distribution, storage, and display until consumption or use. Therefore, validation of these processes may necessitate validation of storage and packaging systems to ensure the packaging system (material and atmosphere) is suitable to limit oxygen transmission and moisture reabsorption throughout shelf life.

3.4.6   Packaging and packaging systems, aseptic or otherwise

Packaging plays a key role in delivering high-quality and safe products to the consumer and end user. Primary packaging provides a barrier to outside contamination as well as maintains the desired product quality specifications. As important as packaging is to product integrity and safety, not much attention is paid to validating packaging systems or materials. Often, a packaging film manufacturer will assist with initial validation steps, such as product package compatibility, sealing, seal strength, and package integrity tests, particularly when existing packaging equipment will be used with their product (film). This is often where commercial food processors begin and end their validation along with some additional quality assurance monitoring of leakers for troubleshooting. However, the validation of packaging materials and systems should include a written protocol, including worst-case conditions that may compromise package integrity and concurrent product quality and safety. Leakers, caused by a loss of seal, pinholes, slits, cuts, and tears in plastic films are a major packaging defect. Many of these defects can be limited by validating the process for worst-case conditions prior to implementing a packaging system. Test packaging with product that may leave a fat smear on the seal area or product that has sharp edges such as bone, stems, or cracked spices that can cause pinholes in packaging materials, particularly during rough handling and transportation. Overpacking individual packages into boxes, cartons, or combos can cause tears, cuts, and pinholes. Overpacking products in containers or packing large product into small containers or bags can result in leakers, cuts, pinholes, or broken containers. Packing smaller product into larger packaging can cause excess moisture to accumulate in corners of the package and reduce product shelf life. Validation testing should also include worst-case conditions for storage of packaging materials (bags, film, cups, jars, lids, etc.) and include humidity, time, light, and temperature. High temperatures or humidity during storage may cause the materials to warp, melt, or become misshapen and reduce their integrity.

Aseptic processes present a unique and specialized set of challenges in validation, usually related to demonstrating proper destruction of pathogenic and spoilage Gram-positive spore-forming bacteria. Developing and validating aseptic processing and packaging systems generally require the assistance of process authorities and engineers who specialize in this area to oversee equipment selection, process development, and validation. There are specific regulatory requirements that must be considered particularly for low-acid foods, versus acid or acidified foods, when validating an aseptic processing or packaging system. In addition, aseptic processes are typically a continuous flow process in which one system may be used to produce multiple and diverse product lines (Chandarana et al. 2010). As with any validation, the team must identify the objective of the validations, identify the products and characteristics of the products to be produced, and identify variables to be monitored during initial trials and validation of the equipment and the process. While the process of identifying key factors in an aseptic process is similar to other processes, the complexity of the processes and the challenges and interactions of multiple processes in a continuous system are unique. There may be as many as 50 critical factors that must be monitored in any one of these processes from the temperature of incoming raw materials, to the particle size, flow rate, diameter of the tube, pressure, heating, sterilization agents, and sterility of packaging.

Validation of an aseptic process and packaging system must also include validation of the sterilant used for the machine, packages, and closures. Sterilants used in aseptic packaging range from saturated steam to hydrogen peroxide, and ionizing radiation. Some questions to consider may include how much sterilant to apply, how it will be applied, and whether it is compatible with the packaging machine, packaging material, and product. The packaging material, whether it is preformed rigid or semirigid containers, laminated paper, thermoform-fill-seal container bags, or pouches, must also be validated. Each type of material presents a different set of challenges to be considered during validation of the process. Since these are aseptic processes, it is imperative that the equipment, monitoring and recording devices, and controls are functioning as intended (Chandarana et al. 2010), and proper documentation is critical.

3.5   Validation creates better understanding of food-processing capability

Process validation incorporates various technical (i.e., engineering, statistics, microbiological testing) and nontechnical (i.e., supply chain distribution, material purchasing, sales channels) aspects of food production. The steps to food and beverage process validation are important for a number of reasons. During the information-gathering phase, a process receives a thorough review by microbiologists and other qualified persons, perhaps for the first time. This review is supplementary to the HACCP reassessment or process control monitoring activities. It is at this point that the critical parameters of the process, including any assumptions, are challenged and perhaps redefined in terms of meeting specific criteria, which could be FSOs. As the master plan and critical parameters are defined, the specific experimental needs regarding the process validation experiments should become clear. The proper performance of a process validation reveals the actual achievement of the process and either verifies or refutes the expected outcomes. Any flaws in the process or insufficiency in achieving microbial control should be uncovered during this activity and should become the focus of process improvement. As an added side benefit of these efforts, efficiencies in the process may be developed, leading to improved quality and cost savings. Even if processes are found through validation to be insufficient and an equipment, energy, or raw material cost must be added to achieve sufficient microbial control, a long-term benefit of improved product performance and public health would likely pay for that cost over time. Process validation not only is a good idea but also is becoming more of a focus of regulators and an expectation within the global food trade.

References

Baird-Parker, A. C., and D. C. Kilsby. (1987). Principles of predictive food microbiology. *Journal of Applied Bacteriology Symposium Supplement* 63:43s–49s.

Bunning, V. K., R. G. Crawford, J. T. Tierney, and J. T. Peeler. 1990. Thermotolerance of *Listeria monocytogenes* and *Salmonella typhimurium* after sublethal heat shock. *Applied and Environmental Microbiology* 56:3216–3219.

Busta, F. F., T. V. Suslow, M. E. Parish, L. R. Beuchat, J. N. Farber, E. H. Garrett, and L. J. Harris. 2003. The use of indicators and surrogate microorganisms for the evaluation of pathogens in fresh and fresh-cut produce. *Comprehensive Reviews in Food Science and Food Safety* 2:179–185.

Cabrera-Diaz, E., T. M. Moseley, L. M. Lucia, J. S. Dickson, A. Castillo, and G. R. Acuff. 2009. Fluorescent protein-marked *Escherichia coli* biotype I strains as surrogates for enteric pathogens in validation of beef carcass interventions. *Journal of Food Protection* 72 (2):295–303.

Chandarana, D. I., J. A. Unverferth, R. P. Knap, M. F. Deniston, and K. L. Wiese. 2010. Establishing the aseptic processing and packaging operation. In *Principles of aseptic processing and packaging*, ed. P. E. Nelson. West Lafayette, IN: Purdue University Press.

Eggenberger-Solorzano, L., S. E. Niebuhr, G. R. Acuff, and J. S. Dickson. 2002. Hot water and organic acid interventions to control microbiological contamination on hog carcasses during processing. *Journal of Food Protection* 65 (8):1248–1252.

Fairchild, T. M., and P. M. Foegeding. 1993. A proposed nonpathogenic biological indicator for thermal inactivation of *Listeria monocytogenes*. *Applied and Environmental Microbiology* 59 (4):1247–1250.

Food and Drug Administration (FDA). 1987. *Guideline on general principles of process validation*, ed. U.S. Department of Health and Human Services and Food and Drug Administration. Washington, DC: Federal Register.

Food and Drug Administration (FDA). 2011a. Determining the regulatory status of components of a food contact material. Available from http://www.fda.gov/Food/FoodIngredientsPackaging/FoodContactSubstancesFCS/ucm120771.htm#iiidrs; accessed May 1, 2011.

Food and Drug Administration (FDA). 2011b. Determining the regulatory status of a food ingredient. Available from http://www.fda.gov/Food/FoodIngredientsPackaging/FoodAdditives/ucm228269.htm; accessed May 1, 2011.

Food and Drug Administration (FDA). 2011c. Guidance for industry, process validation: general principles and practices. Available from http://www.fda.gov/downloads/Drugs/GuidanceComplianceRegulatoryInformation/Guidances/UCM070336.pdf.

George, S. M., L. C. C. Richardson, I. E. Pol, and M. W. Peck. 1998. Effect of oxygen concentration and redox potential on recovery of sublethally heat-damaged cells of *Escherichia coli* O157:H7, *Salmonella enteritidis* and *Listeria monocytogenes*. *Journal of Applied Microbiology* 84 (5):903–909.

Global Harmonization Task Force. 2004. Quality management systems—process validation guidance, SG3/N099-10:2004 (2) 2004. Available from http://www.ghtf.org/documents/sg3/sg3_fd_n99–10_edition2.pdf; accessed March 10, 2011.

Gorris, L. G. M. 2005. Food safety objective: an integral part of food chain management. *Food Control* 16 (9):801–809.

Heredia, N. L., G. A. García, R. Luévanos, R. G. Labbé, and J. S. García-Alvarado. 1997. Elevation of the heat resistance of vegetative cells and spores of *Clostridium perfringens* type A by sublethal heat shock. *Journal of Food Protection* 60 (8):998–1000.

Institute of Food Technologists. 2001. Chapter VII. The use of indicators and surrogate microorganisms for the evaluation of pathogens in fresh and fresh-cut produce (Task Order No. 3), September 30, 2001. Available from http://www.fda.gov/Food/ScienceResearch/ResearchAreas/SafePracticesforFoodProcesses/ucm091372.htm; accessed July 18, 2011.

International Commission on the Microbiological Specifications for Foods. 1998. Principles for the establishment of microbiological food safety objectives and related control measures. *Food Control* 9 (6):379–384.

Kornacki, J. L. 2010. How do I sample the environment and equipment? In *Principles of microbiological troubleshooting in the industrial food processing environment*, ed. J. L. Kornacki. New York: Springer Science+Business Media.

Koseki, S., and S. Isobe. 2005. Prediction of pathogen growth on iceberg lettuce under real temperature history during distribution from farm to table. *International Journal of Food Microbiology* 104:239–248.

Linton, R. H., M.D. Pierson, and J. R. Bishop. 1990. Increase in heat resistance of *Listeria monocytogenes* Scott A by sublethal heat shock. *Journal of Food Protection* 53:924–927.

Mackey, B. M., and C. M. Derrick. 1986. Elevation of the heat resistance of *Salmonella typhimurium* b sublethal heat shock. *Journal of Applied Bacteriology* 61:389–393.

Mangalassary, S., P. L. Dawson, J. Rieck, and I. Y. Han. 2004. Thickness and compositional effects on surface heating rate of bologna during in-package pasteurization. *Poultry Science* 83:1456–1461.

Mossel, D. A. A., G. H. Weenk, G. P. Morris, and C. B. Struijk. 1998. Identification, assessment and management of food-related microbiological hazards: historical, fundamental and psycho-social essentials. *International Journal of Food Microbiology* 39 (1–2):19–51.

National Advisory Committee on Microbiological Criteria for Foods. 1998. Hazard analysis and critical control point principles and application guidelines. *Journal of Food Protection* 61 (9):1246–1259.

Scott, J., and L. Weddig. 1998. Principles of integrated time-temperature processing. Processed Meat Industry Research Conference, Philadelphia, September 17–19.

Scott, V. N. 2005. How does industry validate elements of HACCP plans? *Food Control* 16:497–503.

Singh, R. P., and D. R. Heldman. 1984. Heat transfer in food processing. In *Introduction to Food Engineering*, ed. B. S. Schweigert and J. Hawthorn. San Diego, CA: Academic Press.

Tompkin, R. B. 1996. The significance of time-temperature to growth of foodborne pathogens during refrigeration at 40–50°F. Joint FSIS/FDA Conference on Time/Temperature, Washington, DC, November 18.

United States Department of Agriculture, Food Safety and Inspection Service. 1999. *Appendix A: Compliance guidelines for meeting lethality performance standards for certain meat and poultry products*. Available from http://www.fsis.usda.gov/Frame/FrameRedirect.asp?main=http://www.fsis.usda.gov/OPPDE/rdad/FRPubs/95-033F/95-033F_Appendix_A.htm; accessed April 17, 2012.

United States Department of Agriculture, Food Safety and Inspection Service. Directive 7120.1 Revision 6. 4/5/11. *Safe and Suitable Ingredients Used in the Production of Meat, Poultry and Egg Products*. Available from http://www.fsis.usda.gov/oppde/rdad/FSISDirectives/7120.1.pdf; accessed 5/1/11.

van Schothorst, M., M. H. Zwietering, T. Ross, R. L. Buchanan, and M. B. Cole. 2009. Relating microbiological criteria to food safety objectives and performance objectives. *Food Control* 20 (11):967–979.

Woodams, E. E., and E. J. Nowrey. 1968. Literature values of thermal conductivities of food. *Food Technology* 22:150–158.

chapter four

Food product validations

Peter J. Taormina

Contents

4.1   Introduction

In many ways, food product validation means the same thing as "challenge study" or "challenge testing." In such an experiment, a product is "challenged" by being inoculated with a known level of specific microorganisms, and the behavior of the microbial inoculum is monitored in or on the product under a set of predetermined conditions. If the microorganism is adequately controlled within the food system according to design, the product passes the challenge. Often, this amounts to validating the lack of growth of a microorganism or toxin formation in a food or beverage product during the desired shelf life, but the purpose of product validation can vary. For example, a preservative system may itself be challenged with microorganisms in various matrices, food and beverage products, or even microbiological media. Also, a microbiological diagnostic product may be validated for adequate recovery of microorganisms from various foods, beverages, ingredients, and environments. A sanitizer product may also be validated against pathogens or surrogates on various produce surfaces.

Since food products are constantly being developed and redesigned in response to market demands, it has been said that a food microbiologist could make a career out of simply conducting these types of challenge studies. Regardless of studies having been conducted on similar products or even published in peer-reviewed journals, a study of the actual product in question with the actual pathogen (or surrogate) or spoilage organism of concern subjected to the exact processing and storage conditions is much easier to explain and offers more peace of mind about approving products based on the final results. While there is something to be said for "job security" for microbiologists engendered by this constant need for challenge studies, some results-oriented researchers seeking resolution to these types of studies may resort to modeling as a means to assess product performance or microbial control on a more inclusive scale. The use of modeling for microbial risk analysis rather than product-by-product validation can be more efficient and rapid for determining microbial control for large groups of products or for multiple product variations. Such models themselves are constructed from data derived from inoculation studies* and are constrained to predictions that are within or close to the parameters of those studies. More can be found on modeling in Chapter 8.

---

* Microbiological models are derived from inoculation studies in broth media, in simulated food matrices (such as autoclaved ground meat), or in actual food matrices.

4.2 Background on challenge testing of foods

Chapter 3 of this book discussed process validation in detail, with some insight into finished product formulation. It also laid the groundwork for this chapter by defining and discussing just what validation is and what it means for food safety and quality. Process and product validation may be conducted simultaneously in instances when it is apparent that the process and the product interact to affect microorganisms. For example, a new sanitizer used in vegetable flume water may have an impact on the microbial flora of bagged salad at the point of use and during the resultant shelf life. Another example would be a dairy fermentation process expected to achieve acidification within a certain number of hours that might be impacted by product reformulation with δ-gluconolactone, which could affect time to acidification and resultant microflora development during shelf life. When process and product interact, Chapters 3 and 4 should both be consulted for insight into both process and product validation, respectively.

In many ways, a distinction can be made between food process validation and food product validation; in fact, such distinction should be made for certain objectives. Granted, there are many areas of crossover between the two types of studies, but there are just as many instances when an ingredient, food, or beverage product must be validated in such a way that is irrespective of the process from which it was derived or the process to which it is going. This chapter covers the aspects of food product validation that are independent of the process.

There are two key reference documents that are useful for determining the need for product validation and for aiding in the design and conductance of these types of studies. A comprehensive report by the Institute of Food Technologists (IFT) for the Food and Drug Administration (FDA) of the U.S. Department of Health and Human Services on the evaluation and definition of potentially hazardous foods includes many important considerations for validating the microbial safety of food products (IFT 2001). This document was assembled by an expert panel of industry, academic, and government scientists and was reviewed by multiple experts in the field of food microbiology. Several chapters of the report provide background regarding which microorganisms are of concern based on intrinsic and extrinsic properties of foods. Chapters 6 and 7 of the IFT document deal specifically with challenge studies and provide details on protocols. Much of the information in the report is also discussed on the FDA Web page concerning evaluation and definition of potentially hazardous foods (FDA 2010). A more recent report by the National Advisory Committee on Microbiological Criteria for Foods (NACMCF) deals solely with the best practices for inoculated pack/challenge studies for food products (NACMCF 2010). This report was also produced by an expert panel and is indispensible for those involved in food product validation. Both references (IFT and NACMCF) discuss challenge studies for bacterial growth inhibition, inactivation, or combination studies of both.

These and other documents on the subject (Vestergaard 2001; Scott et al. 2005; Uyttendaele et al. 2004; NACMCF 2005) give detailed guidance on the selection of challenge organisms, inoculum level, inoculum preparation and method of product inoculation, duration of studies, formulation factors, storage considerations, sample analysis, data interpretation, and determination of pass/fail criteria. A researcher involved with food and beverage product validation for bacterial control would be able to find all the information necessary to manage the entire product validation process from these two documents. Neither reference considers growth and survival of toxigenic fungi or survival and inactivation of viruses. Not much is written concerning spoilage microorganisms since spoilage is not viewed as a public health issue. However, many of the principles outlined in these references might also be inferred for product validation with spoilage bacteria as well as fungal, viral, or parasitic inocula. Rather than regurgitate the details from the two reports mentioned, this chapter provides practical insight into product validation with the assumption that the reader can find specific details on best practices in the referenced material. Further, this chapter also discusses considerations for validating methods for detection of microorganisms in food products, which is not the focus of either report.

Researchers may be called on to conduct food and beverage product validation against microorganisms for several reasons. Some studies are undertaken to demonstrate compliance to a regulatory standard, food safety objective (FSO), to a hazard analysis and critical control point (HACCP) prerequisite program, or to satisfy customer requirements for food safety or quality standards. Other reasons might be to assess the impact of product reformulation on shelf life. Basically, any time something changes about the way a product is produced, handled (packaged, stored, and distributed), or consumed, an inoculated pack/challenge study may be warranted. While published studies and predictive models cover a wide array of products and microorganism interactions, if there are differences in critical factors of products from those in publications, a product validation study is warranted. Product validation is a necessary part of research and new product development or redevelopment.

4.3   *Product validation procedures*

Chapter 2 of this book outlined seven key steps in microbiological research and development. A researcher in charge of performing food product validation can follow these seven general steps and produce excellent information, especially when consulting and using the very specific criteria

---

**SEVEN STEPS OF MICROBIOLOGICAL
RESEARCH AND DEVELOPMENT**

1. Protocol development and agreement
2. Planning and assembling equipment and materials
3. Preparation of materials, including media, reagents, and cultures
4. Execution of experiment, with replication
5. Outcome-driven reassessment of protocols with optional modifications
6. Verification and confirmation of results, including statistics
7. Translation of data to reports useful outside the laboratory

---

and best practices outlined in the previously mentioned references by IFT and NACMCF. In addition, published peer-reviewed literature should be researched for the most up-to-date or specific best practices for inoculating and sampling certain product types.

For microbiological research and development, steps 1–3 involve developing objectives and preparing and planning. Steps 4 and 5 involve conducting the experiments and collecting data. Steps 6 and 7 involve analyzing data and arranging it into a presentable format. At each step, the researcher will be faced with decisions about how much detail is appropriate: one or five strains; use of preconditioned cells or log-phase cells; two or three replicate samples; single or duplicate assays per sample; selective media or use of recovery media for stressed cells; number of samples per time over the duration of the experiment; level of statistics applied to the data; length and detail of the report. Each of these decisions will depend on why the study is being performed, who requested it, who is performing it, and to whom the report is directed. For example, a validation of a simple formulation change to a product may warrant only two replicates inoculated with one strain and sampled at the beginning, middle, and end of shelf life plus 1.25 times the shelf life. In such cases, there is no need to overdo it and waste precious resources on product validations that are extremely likely to succeed. Instead, it is advisable to save the triplicate samples, multiple strains, multiple sample times, and multiple plating media for validating the lesser-known products that appear to be potentially hazardous on paper.

4.3.1   Define the purpose and objective of the validation

Generally, this chapter focuses on food, beverage, or ingredient product validations that demonstrate microbial control in food systems as part of qualifying products as safe and wholesome for consumption.

However, the product validation concept from a microbiological per-spective can be applied to microbial diagnostic products, antimicro-bial products, and biocide and sanitizer products. After all, these also are products and must be challenged with microorganisms to validate performance. In some cases, a food product validation will be related to the performance of these associated products within a specific food matrix. The following are some examples of objectives for different product validation studies:

- *A food ingredient* such as a pasteurized neutral pH sauce that is sold refrigerated to other food businesses to use in making finished food products
  - Objective: Validate that the ingredient will not spoil during refrigerated storage within the specified shelf life
- *A raw material product* such as brined seafood that is sold refrigerated to other food businesses to use in making finished food products
  - Objective: Validate that brine concentration is sufficient to pro-vide stasis or cause death of pathogenic bacteria at the specified storage temperature
- *A consumer product* such as sparkling soft drink that is cold filled into aluminum cans
  - Objective: Validate that the beverage will not permit growth of yeasts
- *A microbial diagnostic test kit* for presumptive identification of *Listeria monocytogenes* being used on a food matrix that was not covered in the collaborative study for certification
  - Objective: Validate the expected recovery of inoculated *L. mono-cytogenes* and natural contamination from representative food matrices
- *A novel antimicrobial* (*preservative*) that shows efficacy against micro-organisms *in vitro*
  - Objective: Validate the control of specific target microorganisms in different food or beverage matrices
- *A sanitizer* proposed for use in a fresh produce processing plant for the first time
  - Objective: Validate the log reduction of target indicator microor-ganisms on representative fresh produce surfaces

The criteria of success for each of these examples will differ. Some product validation criteria would be vague, as in the first example: to validate that the ingredient "will not spoil." Other objectives will need to be very specific as the success would depend on adherence to a regu-latory or customer-driven requirement like less than 1-log growth of a pathogen or greater than 2-log reduction of a pathogen. A rapid microbiological method may be required to detect greater than or equal to 10^1 CFU (colony-forming units) of the target pathogen per gram of semisoft cheese, for instance.

4.3.2   Define the product

The next step in designing product validation studies is to define the nature of the product. This can be a difficult task depending on the purpose of the validation study. In the case of proving the ability of an assay to recover target microorganisms from various product matrices, the goal is to pare down the number of product types from a potentially vast selection to keep the study manageable. One strategy might be to segregate products as is done in various comprehensive food microbiology textbooks that discuss relevant microorganisms by food commodity (Lund, Baird-Parker, and Gould 2000; Doyle and Beuchat 2007; International Commission for the Microbiological Specifications of Foods 2005). Since there are multiple food product types and ingredients within these commodities, some further selection must be made of representative food or ingredient matrices. Another approach may be to identify the target microorganisms and then decide which food commodities are relevant. For instance, validation of a *Salmonella* detection methodology would likely involve validations in a variety of products that have been recalled due to contamination or were vehicles of foodborne illness, like eggs, raw meat and poultry, dry milk powder, spices, nuts, and chocolate. Conversely, validation of specific food products may only require a specific complement of pathogens. The standard microbiological challenge test method by the public health and safety organization, NSF International, for validation that a food product is not potentially hazardous requires the use of specific strains of *Bacillus cereus, Escherichia coli* O157:H7, *Listeria monocytogenes, Salmonella* spp., *Staphylococcus aureus*, and *Clostridium perfringens* (NSF International 2000).

    If the purpose of the study is to validate the performance of a certain food formulation to control the growth of and survival of a microorganism, then the difficulty is not in selecting which food system to test, but rather in ensuring that proper samples are obtained of the actual product or ingredient in the form in which it is to be produced. This usually requires detective work on the part of the food microbiologist to truly identify the possible product range under the scope of the validation. The microbiologist will have to understand some detail about process, the formulation, and the expected performance of the product in the marketplace under normal conditions of handling, distribution, storage, and display. It can be difficult to ensure that test samples are produced in the final version and are not altered in an extraordinary way that would influence results. For instance, product validations might require the preparation of plant test runs or test plots for growing and harvesting produce. The microbiologist should determine to the best of his or her ability that samples are truly representative of the actual product being validated. The best way to confirm this is to analyze the critical parameters of the test samples and compare to those normally seen in product. These critical parameters include intrinsic properties of the food and external factors, which are known to affect microbial behavior.

4.3.2.1   Intrinsic food properties: Attributes that affect microorganisms

Intrinsic properties of foods, beverages, and ingredients must be well defined as they influence microbial growth, survival, and detection. Typically, the main factors used to determine microbial behavior are pH, moisture content, oxidation reduction potential (Eh), nutritional content, antimicrobial content, and physical structures such as biological barriers (Jay 2000). The following are some of the food properties that are typically measured:

- Moisture (% moisture, a_w, brine concentration [water-phase salt], moisture-to-protein ratio)
- Other proximates: fat, protein, carbohydrate, ash (wet chemistry)
- Acidity (titratable acidity and pH)
- Inhibitors, preservatives, preservative enhancers (analytical chemistry techniques such as high-performance liquid chromatography [HPLC], gas chromatography mass spectrometry [GC-MS], liquid chromatography mass spectrometry [LC-MS], etc.)
- Naturally occurring microflora: autochthonous (microbiological plating, multiplex polymerase chain reaction [PCR], etc.)

Naturally occurring microflora of the product are also very important in determining behavior of an inoculum. For instance, challenge studies with *L. monocytogenes* may be affected by lactic acid bacteria. Depending on the type of lactic acid bacteria (e.g., bacteriocin producers vs. non-producers) and starting population, a *L. monocytogenes* inoculum can be inhibited on control meat samples that do not contain any antimicrobial agents (Bredholt, Nesbakken, and Holck 1999). This phenomenon causing lack of growth on the positive-growth control can call into question whether the inhibition seen on test variables with antimicrobial agents was in fact due to the agents inhibiting growth or due to the lactic acid bacteria competitively excluding the inoculum. This makes it important to ensure that samples are produced in an environment roughly equivalent to that of actual, full-scale production. Also, researchers may want to analyze test variables for levels and type of spoilage flora prior to initiation the inoculated pack study.

The factors of pH and a_w are the primary determinants of the potential for pathogen growth in foods. Ranges of these two factors within foods were used to indicate which pathogens would be targeted for challenge testing (NACMCF 2010) (Table 4.1).

4.3.2.2 Commingling and interfaces

Commingling of food matrices is often overlooked as a common practice that creates new microenvironments within assembled food products. Often, food components are assembled in a food-processing plant into sandwiches, snack packs, party trays, and so on. Commingled products consist of numerous individual products (e.g., meats, cheeses, fruit, crackers, breads, spreads, and sauces) that in and of themselves have been validated to control microorganisms. However, when these components are brought together, the possibility of cross contamination of one food component to another (e.g., from meat to produce) may cause issues. Also, the interfaces between different food components can create new microenvironments that need validation. Rajkowski et al. (1994) demonstrated *S. aureus* growth at interfaces between salami and cheese where a_w became elevated due to moisture migration. Individually, the salami and cheese were sufficiently dry to restrict growth, but when brought together, the interface became suitable for *S. aureus* growth. Five points of inoculation were recommended to validate that chocolate meringue pie is not potentially hazardous: (1) meringue, (2) filling and meringue interface, (3) filling, (4) filling and crust interface, and (5) crust (NSF International 2000). Similar points of inoculation (components and interfaces) were recommended for validation of vegetable and cheese bread. Microbiologists must be astute in identifying potential points of moisture migration, pH change, and other changes at interfaces that would potentially become hazardous during a product's shelf life.

4.3.2.3 Extrinsic properties

Extrinsic factors of time, temperature, humidity, and atmosphere can have profound effects on microorganisms. In many foods, beverages, and ingredients, the intrinsic properties are of little impact until and unless extrinsic factors (like temperature changes) force the microorganisms to deal with them. Obviously, cold temperatures restrict the growth of microorganisms in food systems that are otherwise conducive to growth. Other food systems that are unsuitable for microbial growth may permit survival of microorganisms depending on the storage temperature. For instance, decimal reduction times of *L. monocytogenes*, *Salmonella*, and *E. coli* O157:H7 decreased with increasing temperatures from 4 to 10 to 21°C in fermented and dried sausage, a food system with low pH and a_w (Porto-Fett et al. 2008). While pathogen growth rates increase in neutral pH and products with high a_w with increasing temperature, sometimes

Table 4.1 Potential Pathogens[a] of Concern for Growth Studies Based on Interaction of Product pH and a_w[b]

a_w values	pH values					
	<3.9	3.9 to < 4.2	4.2–4.6	>4.6–5.0	>5.0–5.4	>5.4
<0.88	NG[c]	NG	NG	NG	NG	NG
0.88–0.90	NG	NG	NG	NG	Staphylococcus aureus	S. aureus
>0.90–0.92	NG	NG	NG	S. aureus	S. aureus	L. monocytogenes, S. aureus
	NG	NG	L. monocytogenes, Salmonella	Bacillus cereus, Clostridium botulinum, L. monocytogenes, Salmonella, S. aureus	B. cereus, C. botulinum, L. monocytogenes, Salmonella, S. aureus	B. cereus, C. botulinum, L. monocytogenes, Salmonella, S. aureus
>0.94–0.96	NG	NG	L. monocytogenes, Pathogenic E. coli, Salmonella, S. aureus	B. cereus, C. botulinum, L. monocytogenes, Pathogenic E. coli, Salmonella, S. aureus, Vibrio parahaemolyticus	B. cereus, C. botulinum, L. monocytogenes, Pathogenic E. coli, Salmonella, S. aureus, V. parahaemolyticus	B. cereus, C. botulinum, C. perfringens, L. monocytogenes, Pathogenic E. coli Salmonella, S. aureus, V. parahaemolyticus

| >0.96 | NG | Salmonella | Pathogenic E. coli Salmonella S. aureus | B. cereus C. botulinum L. monocytogenes Pathogenic E. coli Salmonella S. aureus V. parahaemolyticus | B. cereus C. botulinum L. monocytogenes Pathogenic E. coli Salmonella S. aureus V. parahaemolyticus V. vulnificus | B. cereus C. botulinum C. perfringens L. monocytogenes Pathogenic E. coli Salmonella S. aureus V. parahaemolyticus V. vulnificus |

Source: Adapted from National Advisory Committee on Microbiological Criteria for Foods. 2010. Parameters for determining inoculated pack/challenge study protocols. *Journal of Food Protection* 73 (1):140–202.

[a] *Campylobacter* spp., *Shigella*, and *Yersinia enterocolitica* do not appear here because they are typically controlled when the pathogens listed are addressed.

[b] Data are based on the PMP, ComBase predictor, ComBase database, or peer-reviewed publications (see Chapter 8 for descriptions of these programs).

[c] NG, no growth; when no pathogen growth is expected but formulation or process inactivation studies may still be needed.

the opposite occurs in products with low pH or low a_w. Examples of slower death rates of pathogens in low-pH or low-a_w food systems at lower temperatures have been documented in kippered beef (Jacob et al. 2009), precooked bacon (Taormina and Dorsa 2010), and peanut butter spreads (Park, Oh, and Kang 2008), to name a few foods.

Food packaging influences the extrinsic properties affecting foodborne microorganisms. Packaging provides barriers against air and moisture penetration. Virtually all flexible packaging materials permit some degree of oxygen and moisture penetration over time, so the interior is often a dynamic environment. The degree and rate of penetration will depend on the film thickness, quality of sealed film junctures, and the food product itself. Vacuum packaging (VP) is frequently used to remove air and thereby provide an additional hurdle against microbial growth by reducing O_2. The use of modified atmosphere packaging (MAP), by which packages are flushed with N_2, or N_2 and CO_2 mixtures, is also used in many products to inhibit microorganisms. Both VP and MAP can extend the shelf life of refrigerated foods by shifting the balance of the microflora to favor innocuous microorganisms. Cans and retort pouches provide a very protective and stable environment in terms of atmosphere. These materials permit very minute changes in atmosphere relative to other food packaging and essentially establish a static environment for the enclosed food. However, in various packaging types, metabolic activity by surviving microorganisms growing in the product can alter the internal environment. Microbial metabolism can cause changes in the composition of dissolved gasses, lead to production of acidic or basic by-products in the food, and induce changes in headspace atmosphere. Depending on the product and microflora involved, these changes can create an opportunity for other microorganisms to grow or, conversely, may create unfavorable conditions for other competing microorganisms. Interaction of competing microflora and packaging atmosphere in fresh produce has been the focus of much research (Devon 1999). Microflora/atmosphere interactions should be investigated since there are specific examples of foodborne pathogens becoming more hazardous as a result of the behavior of spoilage microorganisms. Growth of proteolytic molds on tomatoes raised pH such that they became more conducive to growth by *C. botulinum* (Draughon, Chen, and Mundt 1988) and *Salmonella* (Wade and Beuchat 2003). As another example, growth of *Pseudomonas fragii* but not *L. monocytogenes* was inhibited in cooked pork tenderloin when packaged under MAP with 100% and 80% CO_2 (Fang and Lin 1994). The last example demonstrates how a product might not spoil but permit pathogen growth.

The temperature of storage of inoculated samples is very important, especially for determining the growth potential of psychrotrophic microorganisms. The foodborne pathogens *L. monocytogenes*, *Yersinia enterocolitica*, psychrotrophic *C. botulinum*, and *B. cereus* are potential hazards in refrigerated ready-to-eat foods and necessary hazards to address in establishing safety-based consume-by date labels (NACMCF 2005). Intended conditions of storage and use must be considered in designing challenge studies for refrigerated foods in particular. The storage temperature used in microbial challenge studies should include the typical temperature range at which the product is to be held and distributed (IFT 2001). This might include anticipated degrees of temperature abuse or humidity changes for exposed products.* Some studies may incorporate temperature cycling in the way inoculated samples are stored to simulate the relatively colder temperature of distribution centers and the relatively warmer temperatures of retail display and consumer use (Audits International 1999; Pouillot et al. 2010). This type of study conducted on frankfurters against *L. monocytogenes* during freezing, thawing, and home storage led to recommendations for consumers to freeze frankfurters immediately after purchase and to discard frankfurters formulated without antimicrobials within 3 days of thawing or opening (Simpson Beauchamp et al. 2010).

4.3.2.4   Associated products

The researcher should determine what other related products will be covered by the study. Usually, it is best to select the worst-case product to represent other similar products (e.g., highest moisture, highest storage temperature, etc.). Modeling can be helpful here to establish which products are most at risk for lack of microbial control. This can help eliminate variables and reduce the burden of the challenge testing project.

4.4   Identify pathogens or spoilage microorganisms of concern

Once the focus of the validation is identified and defined and products have been selected, the microorganisms of interest can be chosen. The selection of types of microorganisms, number of strains, and culture conditions for the product validation will depend on the objectives. Some of the questions to ask at this stage are as follows:

- Is the formulation hostile to these microorganisms?
- Should the food formulation necessitate a decrease of the microorganisms of concern? If so, how much of a decrease is necessary?

---

* Temperature abuse studies are a valid part of microbial challenge testing for risk assessment. However, the use of "accelerated shelf life testing," by which product is stored at temperatures higher than intended for use in commerce for the purpose of speeding up the development of results, is generally not recommended. Elevated temperatures will shift the advantage away from psychrotrophs and toward mesophiles, so extrapolation of accelerated shelf life testing data to a refrigerated product is invalid.

- Is the formulation required to control growth of these microorganisms during shelf life?
- If microbial growth is possible, how much of an increase is tolerable?
- What are the time constraints of these microbial growth limits?

The researcher should build an approach that considers food safety or quality objectives for each product and then design a study to demonstrate achievement of those objectives. Table 4.2 describes products and relevant microorganisms for use as inocula for survival and death studies.

These would typically be useful when the presence of the microorganisms in the food, beverage, or ingredient is not acceptable from a scientific, public health, or regulatory standpoint. For instance, any metabolic growth of yeast in sparkling beverages would likely result in spoilage. Therefore, a challenge study could be set up to determine rates of death of yeasts from various starting populations to understand the tolerable contamination limits of that product. Another example could be a shelf-stable product that has potential for postprocessing contamination by *L. monocytogenes* (a pathogen with zero tolerance in ready-to-eat foods in the United States and regulatory limits at the point of consumption in Canada and Europe). In such a product, it may be useful to know how long the pathogen would survive after a potential contamination event during packaging.

Many challenge studies are undertaken to validate the lack of growth of microorganisms during the shelf life of the product. Such growth studies are largely for refrigerated, ready-to-eat foods, but not all. Some examples are given in Table 4.3 of food, beverage, and ingredient products and pertinent microorganisms for growth challenge testing.

Cultures can be obtained by either purchase or through culture sharing with other researchers. A list of various culture collections throughout the world can be found on the Internet (World Federation for Culture Collections 2011). Cultures can be purchased from some of these collections. Some university or government researchers may be willing to share strains. University researchers might request a "gift in kind" to defray the cost of packaging and shipping the cultures.

4.5 Obtain representative product samples

One of the key steps in designing a product validation is in the food sample matrix itself. As mentioned previously, the research microbiologist must take into consideration the intrinsic and extrinsic properties that would be factors in microbial behavior during the study and prepare representative samples that mimic these same conditions. Larger whole pieces of solid products or large volumes of fluid products may be difficult to manage.

Table 4.2 Examples of Food Products and Relevant Pathogenic[a]
and Spoilage Microorganisms Suggested for Inoculated
Product Survival and Death Rate Challenge Studies

Food product or ingredient type	Major exclusionary attributes after processing	Surviving or recontaminating microorganisms[b]
Beverage base	Acidity, antimicrobial flavor oils, low nutrient	Mold: *Thamnidium, Cladosporium, Fusarium*
Concentrated pasteurized juice	Acidity	Bacteria: *Sporolactobacillus* Mold: *Neosartorya fischeri*
Refined fats and oils	Low moisture, absence of nutrients	Bacteria: *Salmonella* Mold: *Aspergillus, Cladosporium*
Dry spice, dry powder ingredients (e.g., soy protein isolate, maltodextrin)	Low moisture	Bacteria: *Salmonella, Cronobacter*
Nuts, nut spreads and pastes	Low moisture	Bacteria: *Salmonella*
Acidic carbonated beverage	Acidity, carbonation	Yeast: *Dekkera (Brettanomyces), Saccharomyces cerevisiae, Zygosaccharomyces bailii* Virus: Norovirus (or viral surrogate)
Acidic noncarbonated cold-filled beverage	Acidity	Bacteria: *Gluconobacter, Acetobacter, Leuconostoc* Yeast: *Saccharomyces cerevisiae, Zygosaccharomyces bailii, Candida, Issatchenkia* Mold: *Penicillium, Fusarium* Virus: Norovirus (or viral surrogate)
Acidic noncarbonated hot-filled beverage	Acidity	Mold: *Byssochlamys, Neosartorya fischeri* Bacteria: *Alicyclobacillus*
Canned thermally processed acid food	Acidity, water activity	Bacteria: *Bacillus coagulans, B. polymyxa, Clostridium pasteurianum, C. butyricum, C. thermosaccharolyticum*
Canned thermally processed low-acid canned food	Water activity	Bacteria: *Bacillus stearothermophilus, B. coagulans, Clostridium bifermentans, C. thermosaccharolyticum, Desulfotomaculum nigrificans*

continued

Table 4.2 *(continued)* Examples of Food Products and Relevant Pathogenic[a]
and Spoilage Microorganisms Suggested for Inoculated
Product Survival and Death Rate Challenge Studies

Food product or ingredient type	Major exclusionary attributes after processing	Surviving or recontaminating microorganisms[b]
Shelf-stable meat, fermented or dried meats	Acidity, low moisture, possible presence of curing agents, possible presence of starter culture or bacteriocins	Bacteria: *L. monocytogenes*, *Salmonella*, enterohemorrhagic *E. coli* Yeast: *Debaryomyces hansenii*

[a] Viruses and parasites might also be considered for product lethality validation because presence in finished food, beverage, or ingredient products would be unacceptable.

[b] Presence of these microorganisms would be unacceptable in the product; therefore, a rate of death would be of interest.

Therefore, samples can be segmented into smaller representative quantities as necessary to save space, reduce the amount of materials needed for the study, and reduce biohazardous waste. However, there are limits to scaling down as accuracy and relevance can be lost if the product samples are scaled down too much. For instance, the researcher should ask, "Are 1 × 1 × 1 cm pieces of a ready-to-eat turkey roast product adequate to validate an 8-kg deli-style product?" or "Is 1 g of dried infant formula truly representative of an entire 2.5-kg canister?" Consistent ratios of microorganism to food are essential and should mimic real contamination scenarios. Care should be taken to account for the scaled-down sample size by adjusting inocula and any pertinent treatment dosage, packaging, and so on. Sampling methods also will have to account for the difference in sample size or quantity of validation samples versus normal product. Techniques like excision sampling, package rinsing, filtration, and whole-sample enrichment will need to be performed in such a way that resulting data can be extrapolated to actual full-scale product.

It is helpful and even necessary to retain uninoculated samples and analyze for spoilage microflora and any other relevant analytical chemistry or proximate tests. If absence of pathogens needs to be established, care should be taken to perform such testing only if any associated product is fully under control and not released into commerce. Performing these analyses can help researchers interpret the findings from the inoculated samples, verify that the samples were truly representative of the actual product, and prove they were within specifications at the time of inoculation. Without such information, results could be questionable.

Table 4.3 Examples of Food Products and Relevant Pathogenic
and Spoilage Microorganisms Suggested for Evaluating
Growth Inhibition for Product Shelf Life Determination

Food type	Microorganisms of shelf life concern
Pasteurized refrigerated milk	Bacteria: *Pseudomonas, Paenibacillus, Bacillus, Microbacterium*
Processed meat and poultry	Bacteria: *Clostridium perfringens, Listeria monocytogenes, Lactobacillus, Leuconostoc*
Refrigerated raw ground beef, pork, and poultry	Bacteria: *Salmonella, E. coli (biotype I), Pseudomonas, Hafnia, Shewanella, Brocothrix*
Bagged salad, sprouts	Bacteria: *Clostridium, Salmonella*, enterohemorrhagic *E. coli, Listeria monocytogenes, Pseudomonas, Serratia marscescens, Erwinia* Yeast: *Rhodotorula* Mold: *Penicillium*
Cut fruit	Bacteria: *Salmonella, Listeria monocytogenes, Erwinia, Pseudomonas* Yeast: *Pichia, Candida, Rhodotorula* Mold: *Penicillium, Alternaria, Geotrichum*
Prepared deli salad	Bacteria: *Staphylococcus aureus, Salmonella, Listeria monocytogenes* Yeast: *Hansenulla* Mold: *Penicillium* Virus: Norovirus (or surrogate)
Bakery product	Bacteria: *Staphylococcus aureus, Serratia marscescens, Bacillus subtilis* Yeast: *Zygosaccharomyces rouxii* Mold: *Rhizopus stolonifer, Neurospora intermedia, Mucor*
Pasteurized refrigerated juice	Bacteria: *Acetobacter, Gluconobacter* Yeast: *Saccharomyces, Candida* Mold: *Neosartoria, Byssochlamys, Cladosporium*
Pickled refrigerated produce	Bacteria: *Lactobacillus casei, L. paracasei, Propionibacterium, Gluconobacter* Yeast: *Saccharomyces, Pichia, Hansenula*
Fruit puree	Yeast: *Zygosaccharomyces rouxii, Hanseniaspora, Candida*

4.5.1   Multiple lots of production

If a challenge study is to be conducted on an existing product (already being produced), it is best to acquire multiple lots of the product produced at different times and accounting for any known or suspected process variation. However, in practice the production of multiple test samples can be problematic. Some samples can only be made in a processing plant, harvesting environment, field, orchard, and the like; therefore, only a short window of opportunity may exist to acquire the test samples.

4.5.2 Packaging and other extraneous materials intimately associated with product

Any packaging material used for the inoculated pack/challenge study should be similar to that of the actual product in attributes that could affect microbial growth. These factors may include moisture, oxygen, and light transmission rates; heat conductivity of packaging film; and use of oxygen absorbers. If product packages are boxed together, it should be established whether this would have an impact on the results. If the product is exposed to light during retail display, light simulation may be necessary during the challenge study.

4.5.3 Confirm that the producer is following normal procedures

Confirming that the producer is following normal procedures can be a difficult task without good understanding of the process. Communication with those preparing samples is important and should be followed with verification by records review. The person(s) responsible for producing test samples should be asked ahead of time to describe how the samples will be made. The researcher should instruct how samples are to be treated from the point of production (whether in an agricultural setting, a food plant, or a pilot plant) to the point they are ready for inoculation, making sure that samples are not exposed to improper temperature and other external factors. The researcher should note differences (if any) between the process of making test samples and the real production process and compensate as necessary. If a difference is found between the two products but is deemed not to be a critical parameter that would affect results, this should be explained and scientifically supported in the report.

4.5.4 Verify absence of processing aids unless part of the validation

Processing aids are commonly used for production of a variety of food products. Many of these are designed to be antimicrobial treatments. Others may not be related to control of microorganisms but may interact with treatments that are being tested as part of the product validation. Examples of antimicrobial treatments could be post-lethality treatments applied to ready-to-eat meats that would retain antimicrobial properties several days after production or an antimicrobial food contact surface sanitizer that is permitted to carry over into product at a low level. Other processing aids such as surfactants in liquid products or anticaking agents in dry materials might affect the dispersion of the inoculum or its recovery during microbiological enumeration. In many cases, processing aids will have lost the technical effect by the time of inoculation (generally a few days after production). If necessary, samples can be made without a processing aid, but it can be difficult to verify absence of the material.

4.6 Examples of other forms of food product validation

4.6.1 Validating the performance of a novel unproven antimicrobial ingredient in various food systems

An extract of a plant shows promise in terms of *in vitro* efficacy against *L. monocytogenes*, especially when combined with a certain surfactant at a certain ratio. This novel antimicrobial system may only work in the test tube and would need to be validated in various food systems. A series of food product validations would be necessary to determine efficacy in specific food products.

4.6.2 Validating the performance of a known antimicrobial system in a specific food product

Perhaps a known robust preservative system, such as sodium benzoate, potassium sorbate, and EDTA (ethylenediamine tetraacetic acid) is being proposed to protect a new shelf-stable, high-sugar sauce product. The sauce is pasteurized but could be exposed to postprocessing contaminants during packaging. This new formulation is less acidic than normal such that it approaches the dissociation constant (pKa) of the weak acid preservatives. In this case, the known preservative system would need to be validated in a reformulated food product for control of yeast, mold, and acidophilic bacteria.

4.6.3 Validating a sanitizer or antimicrobial processing aid

A proprietary acid peroxide-based sanitizer mixture shows greater efficacy *in vitro* against *Salmonella*, *L. monocytogenes*, and enterohemorrhagic *E. coli* than existing sanitizers. With such broad potential use, the sanitizer could have application for removal and reduction of pathogens in fresh produce and in fresh meat and poultry carcasses. A series of laboratory experiments is designed to measure log reductions of the pathogens on produce and meat surfaces treated with the proprietary mixture and compare with log reductions achieved by water (control) and existing sanitizers on the market. Perhaps the proprietary sanitizer is also efficacious against certain viral and parasitic foodborne pathogens. An additional series of experiments would be set up to measure efficacy against viral and parasitic pathogens on fresh produce.

4.6.4   Validating performance of a known and valid process in different food or beverage formulations

In-plant process data reveal that a process is providing sufficient inactivation of target microorganisms, as evidenced by process time and temperature data and microbial counts before and after the process. However, a new food formulation with different critical parameters is being proposed for that process. A laboratory product validation study may be sufficient, rather than a full-blown, in-plant, process validation study. The product validation may simply be a matter of performing laboratory-scale heating studies with the target microorganisms using small samples of the normal and new formulation. A successful outcome would be equal or greater reductions of the target microorganisms in the new formulation compared to the normal formulation.

4.6.5   Validating shelf life of a food system

It should be mentioned that product validation may also involve testing the microbiology of food, beverage, or ingredient products during the course of the shelf life without adding additional microorganisms to the products. Microbiological assays of the native microflora of products include aerobic plate counts (APCs), anaerobic plate counts (AnPCs), lactic acid bacteria, acetic acid bacteria, yeast, and mold, among others. A shelf life "study" in which product samples from a production run are retained and analyzed over the course of the product shelf life are used to validate the initial shelf life of a product. Ongoing shelf life testing using microbiology is part of verifying that the product is continually achieving that shelf life. Shelf life studies can be performed relatively easily and without much of the equipment and expertise needed for inoculated product studies. They are a research and development function and provide useful information concerning microbiological performance of the product. Even though products are not inoculated with known, specific microorganisms, the microbiological assays used for shelf life studies can be indicative of potential behavior of particular microorganisms known to be a food safety or spoilage hazard for the product. In that sense, shelf life studies do provide value toward microbiological product validation.

4.6.6   Validating performance of a microbiological method for recovery of analyte from various food or environmental matrices

4.6.6.1   Qualitative

Certain presence/absence tests would need to be validated in different food matrices. While the Association of Official Agricultural Chemists (AOAC) or Association Française de Normalisation (AFNOR) (or other method certifications) may have performance tested or validated the method, microbiologists will often run across food, beverage, ingredient, or environmental samples that were not even remotely included in the validation and possess some properties that warrant a product-specific validation.

4.6.6.2   Quantitative

Quantitative testing such as plate count methods and impedance may require validation in different food matrices. Quantitative growth-based methods are discussed in Chapters 9 and 10. Quantitative real-time PCR (qPCR) can also be used in foodborne pathogen diagnostics. While qualitative (i.e., presence/absence) real-time PCR relies on enrichment steps to about 10^4 CFU/mL preceding target DNA detection, qPCR must be performed without an enrichment to amplify only existing DNA in the sample. Therefore, qPCR must distinguish between live and dead cell DNA, as dead cell DNA can be falsely amplified, leading to inaccurate quantification. It was shown that dead cell DNA degraded the fastest in chicken homogenate (1 log unit per 0.5 h) and slowest in pork rinse (1 log unit per 120.5 h) (Wolffs, Norling, and Radstrom 2005), which could lead to overestimation of cell counts by up to 10 times the amount of live cells in the sample.

4.7   Concluding thoughts

Food product validation is an important part of research and development and a critical role of the research microbiologist. Food product validation is used in research and development of new food, beverage, and ingredient products to determine the microbiological safety and quality of the product and may be necessary to qualify it for commercial production. There are fairly detailed procedures for conducting food product validations, especially for assessing microbial food safety by challenge testing foods. Procedures for validating control of spoilage microorganisms in foods, beverages, and ingredients is less defined, but also critical.

Other nonfood products like microbial diagnostics, antimicrobials, and sanitizers must be validated in regard to effective use in and around foods and beverages. Validation of these nonfood (but food-related) products will inevitably involve introducing microorganisms to food systems and will require some of the same principles of food product challenge testing.

4.8   Frequently asked questions

Q. How many lots of product (i.e., production runs) should I validate?

A. Ideally, three, but realistically two or even one can suffice when data already exist for similar products. Regardless, collecting product properties like pH, a_w, moisture, protein, fat, salt, preservative level, and so on is necessary to verify that the samples were within specifications. It may be difficult to get multiple lots of product for research and development test samples, and if there are numerous variables, it may make the study too large to perform anyway. Whatever research and development test samples go into full-scale production, they should be revalidated at that point. Highly sensitive, borderline products should be validated in three separate production lots. In the event of unfavorable or unexpected results, any repeats of those studies should be done on three separate lots.

Q. Do I have to validate every product?

A. This is not likely. If you establish a product grouping system based on intrinsic and extrinsic factors that affect microbial growth, then you can select the worst-case product (most microbiologically sensitive) from each group and validate that product. All the other less-sensitive products would be validated by default. There may be other nonscientific reasons to validate every product or every brand.

Q. What strains should we use and how many?

A. This depends on the objectives. Five strains is the standard practice for food microbiology product validation. Select at least one strain that has been frequently used in published literature (like *L. monocytogenes* Scott A, for instance) so that there is a body of work for comparison of results. Select a few related environmental isolates if they are available. Include isolates from product, especially for spoilage microorganisms. If validating against pathogens, try to use strains that were associated with outbreaks in similar products. Once all these sources for strains have been exhausted, resort to laboratory stock strains if necessary.

Q. I have conflicting results from two separate replicates from separate lots. Which one do I use as validation support for my HACCP prerequisite program?

A. This could be due to process variability affecting the samples, lab error, or both. If you collected pH, a_w, and proximates on the product samples and they showed no marked differences, repeat on two separate lots—going to two separate labs if possible. The worst thing you can do is repeat studies until you get a favorable outcome. Every time a study is repeated, there should be a defensible reason why it is more (or less) accurate than the one that preceded it. If a product failed a validation but it was later reformulated or the process changed prompting a new study, be sure to document what changed and when.

Q. My routine testing third-party lab says it can do product validation studies for me. Should I go with it or keep looking for another lab?

A. My advice is *caveat emptor* (buyer beware): While many labs do have the specialized expertise to do these studies, just as many do not. Private microbiology testing labs tend to make their money by handling large sample volumes for routine tests and turning around results as quickly as possible. Product validation is research, and because the pace and duration of research are so different, it can be difficult for some labs to be good at both routine testing and research. I have personally seen some product validation data coming from private microbiology testing labs that contradict the published, peer-reviewed literature as well as years of my own work on similar products. This could be due to several factors: strain selection, inoculum preparation, inoculation method, product storage temperature, sampling method, plating media, and incubation conditions. This is why the NACMCF report explicitly describes what sort of education and expertise is needed to manage product validation studies. Make sure whatever lab you use (even if it is your own) meets these criteria. I cannot endorse any particular labs in this chapter, but I do advise that you look for labs with specific expertise and experience with food research microbiology (they are out there). Experience with clinical, pharmaceutical, and cosmetic microbiology research is good for those types of products, but I recommend food labs for food research.

References

Audits International. 1999. Audits/FDA temperature databases 1999. Available from http://www.foodrisk.umd.edu/exclusives/audits/index.cfm; accessed March 11, 2010.

Bredholt, S., T. Nesbakken, and A. Holck. 1999. Protective cultures inhibit growth of *Listeria monocytogenes* and *Escherichia coli* O157:H7 in cooked, sliced, vacuum- and gas-packaged meat. *International Journal of Food Microbiology* 53 (1):43–52.

Devon, Z. 1999. Effects of post-processing handling and packaging on microbial populations. *Postharvest Biology and Technology* 15 (3):313–321.

Doyle, M. P., and L. R. Beuchat. 2007. *Food microbiology: fundamentals and frontiers.* 3rd ed. Washington, DC: ASM Press.

Draughon, F. A., S. Chen, and J. O. Mundt. 1988. Metabiotic association of *Fusarium*, *Alternaria*, and *Rhizoctonia* with *Clostridium botulinum* in fresh tomatoes. *Journal of Food Science* 53 (1):120–123.

Fang, T. J., and L.-W. Lin. 1994. Growth of *Listeria monocytogenes* and *Pseudomonas fragi* on cooked pork in a modified atmosphere packaging/nisin combination system. *Journal of Food Protection* 57 (6):479–485.

Food and Drug Administration. 2010. Evaluation and definition of potentially hazardous foods 2010. Available from http://www.fda.gov/Food/ScienceResearch/ResearchAreas/SafePracticesforFoodProcesses/ucm094141.htm; accessed December 5, 2011.

Institute of Food Technologists. 2001. Evaluation and definition of potentially hazardous foods *Comprehensive Reviews in Food Science and Food Safety* 2 (Supplement).

International Commission for the Microbiological Specifications of Foods (ICMSF). 2005. *Microorganisms in foods 6: Microbial ecology of food commodities*, Vol. 6. New York: Kluwer Academic/Plenum.

Jacob, R., A. C. Porto-Fett, J. E. Call, and J. B. Luchansky. 2009. Fate of surface-inoculated *Escherichia coli* O157:H7, *Listeria monocytogenes*, and *Salmonella typhimurium* on kippered beef during extended storage at refrigeration and abusive temperatures. *Journal of Food Protection* 72 (2):403–407.

Jay, J. M. 2000. *Modern food microbiology*. 6th ed. Gaithersburg, MD: Aspen.

Lund, B. M., T. C. Baird-Parker, and G. W. Gould. 2000. *Microbiological safety and quality of food*, Vols. 1–2. 2 vols. Vol. 2. New York: Springer-Verlag.

National Advisory Committee on Microbiological Criteria for Foods. 2005. Considerations for establishing safety-based consume-by date labels for refrigerated ready-to-eat foods. *Journal of Food Protection* 68 (8):1761–1775.

National Advisory Committee on Microbiological Criteria for Foods. 2010. Parameters for determining inoculated pack/challenge study protocols. *Journal of Food Protection* 73 (1):140–202.

NSF International. 2000. *American national standard for non-potentially hazardous foods*. Ann Arbor, MI: NSF International.

Park, E. J., S. W. Oh, and D. H. Kang. 2008. Fate of *Salmonella* Tennessee in peanut butter at 4 and 22 degrees C. *Journal of Food Science* 73 (2):M82–M86.

Porto-Fett, A. C., C. A. Hwang, J. E. Call, V. K. Juneja, S. C. Ingham, B. H. Ingham, and J. B. Luchansky. 2008. Viability of multi-strain mixtures of *Listeria monocytogenes*, *Salmonella typhimurium*, or *Escherichia coli* O157:H7 inoculated into the batter or onto the surface of a soudjouk-style fermented semi-dry sausage. *Food Microbiology* 25 (6):793–801.

Pouillot, R., M. B. Lubran, S. C. Cates, and S. Dennis. 2010. Estimating parametric distributions of storage time and temperature of ready-to-eat foods for U.S. households. *Journal of Food Protection* 73 (2):312–321.

Rajkowski, K. T., R. Schultz, F. Negron, and A. Dicello. 1994. Efficacy of water activity on growth of *Staphylococcus aureus* at meat-cheese interfaces. *Journal of Food Safety* 14:219–227.

Scott, V. N., K. M. J. Swanson, T. A. Freier, W. P. Pruett, W. H. Sveum, P. A. Hall, L. A. Smoot, and D. G. Brown. 2005. Guidelines for conducting *Listeria monocytogenes* challenge testing of foods. *Food Protection Trends* 25 (11):818–825.

Simpson Beauchamp, C., O. A. Byelashov, I. Geornaras, P. A. Kendall, J. A. Scanga, K. E. Belk, G. C. Smith, and J. N. Sofos. 2010. Fate of *Listeria monocytogenes* during freezing, thawing and home storage of frankfurters. *Food Microbiology* 27 (1):144–149.

Taormina, P. J., and W. J. Dorsa. 2010. Survival and death of *Listeria monocytogenes* on cooked bacon at three storage temperatures. *Food Microbiology* 27 (5):667–671.

Uyttendaele, M., A. Rajkovic, G. Benos, K. Francois, F. Devlieghere, and J. Debevere. 2004. Evaluation of a challenge testing protocol to assess the stability of ready-to-eat cooked meat products against growth of *Listeria monocytogenes*. *International Journal of Food Microbiology* 90 (2):219–236.

Vestergaard, E. M. 2001. Building product confidence with challenge studies. *Dairy, Food and Environmental Sanitation* 21 (3):206–209.

Wade, W. N., and L. R. Beuchat. 2003. Metabiosis of proteolytic moulds and *Salmonella* in raw, ripe tomatoes. *Journal of Applied Microbiology* 95 (3):437–450.

Wolffs, P., B. Norling, and P. Radstrom. 2005. Risk assessment of false-positive quantitative real-time PCR results in food, due to detection of DNA originating from dead cells. *Journal of Microbiological Methods* 60 (3):315–323.

World Federation for Culture Collections. 2011. World Data-Centre For Microorganisms. Available from http://www.wfcc.info/ccinfo/collection/; accessed November 9, 2011.

chapter five

Competitive research and development on antimicrobials and food preservatives

Keila L. Perez, T. Matthew Taylor, and Peter J. Taormina

Contents

5.1   Introduction

Food products will begin to deteriorate and lose their inherent quality almost from the point of harvest, primarily a result of (1) growth of indigenous or cross-contaminating microorganisms and the metabolic products of their growth and proliferation and (2) enzymatic and chemical reactions producing undesirable olfactory, visual, or physical changes in the food. The application of one or multiple processing or preservation technologies can slow, inhibit, or temporarily arrest the onset of product deterioration, producing a product capable of extended shelf stability and safety. Many methods have been developed and applied to the processing or preservation of foods, including thermal processing, reduced-temperature storage, ionizing and nonionizing irradiation, dehydration, fermentation and biopreservation, and the addition of chemical food antimicrobials and preservatives.

Davidson and Taylor (2007) identified food antimicrobials as chemical compounds added to, or naturally occurring in, foods that act to inhibit or inactivate naturally occurring or cross-contaminating microorganisms. Most food antimicrobials utilized in the preservation of foods exert inhibitory bacteriostatic or fungistatic effects in the food product rather than lethal (i.e., bactericidal, fungicidal) effects at usage concentrations (Davidson and Branen 2005). Food antimicrobials are often applied in conjunction with other processing measures, commonly referred to as hurdle processing, for the control of contaminating pathogenic and spoilage microorganisms (Leistner 2000; Leistner and Gorris 1995). The U.S. *Code of Federal Regulations* (CFR) defines food antimicrobials as "preservatives," although this identifier allows the inclusion of chemical additives that serve to preserve the food's quality (e.g., butylated hydroxytoluene [BHT]) [21 CFR 101.22(a)(5); 21 CFR 70.3(o)(2)].

Food antimicrobials may be identified as either natural or traditional (or possibly both) based primarily on the source of the antimicrobial compound, history, and regulatory approvals of use across differing countries (Davidson and Taylor 2007). Many food antimicrobials occur naturally but bear specific regulatory allowances for their use in foods (e.g., hen egg white lysozyme); such compounds would be primarily identified as traditional antimicrobials. Examples of natural antimicrobials include some weak organic acids (e.g., citric, malic acids) and bacterially synthesized antimicrobial polypeptides (i.e., bacteriocins) (e.g., pediocin from *Pediococcus* spp.). Some naturally occurring antimicrobials, like lysozyme, are also considered traditional antimicrobials due to availability of synthetic manufacturing or processing systems and federal regulatory limits on their use (e.g., nisin, lactic acid). Food antimicrobials are distinguished from various therapeutic antibiotics (e.g., penicillin, ciprofloxacin, etc.) for the purposes of this discussion, based on (1) federal regulations disallowing therapeutic antibiotics to be added to foods during processing and (2) differing sources and mechanisms of action of food antimicrobials versus antibiotics.

This chapter reviews the various issues and concerns associated with food antimicrobial research and development. Attention will be paid to the processes required for the development of a commercial food antimicrobial, in particular its creation, assessment of antimicrobial efficacy, and essential data to gather for the submission of petitions for patent and federal approval for use and for the marketing of a developed antimicrobial. The combination of active antimicrobials and assay of antimicrobial combinations is reviewed in light of the development of food antimicrobial products that contain multiple inhibitory compounds with similar or differing mechanisms of activity. Finally, a review of novel trends in food antimicrobial delivery systems, specifically via encapsulation or incorporation into packaging films, is given. While the chapter provides a brief historical survey of the differing food antimicrobials approved for use in foods, the reader is directed to other well-written reviews of food antimicrobial use and methods of assay (Davidson, Sofos, and Branen 2005; Davidson and Harrison 2002; Gould 1989; López-Malo Vigil et al. 2005; López-Malo Vigil, Palou, and Alzamora 2005; Roller 2003; Davidson, Taylor, and Santiago 2005).

5.2   Food antimicrobial applications in foods

5.2.1   Key factors affecting antimicrobial activity

A variety of factors will have an impact on the activity and utility of a food antimicrobial added to a food for the inhibition of microbial growth/replication. These include the processing of the food product and the point of antimicrobial application within the process continuum, the purpose for the antimicrobial's usage, the antimicrobial's chemistry and activity (i.e., spectrum of activity, mechanisms of action, concentration at which inhibitory activity is observed *in vitro* and in the food product), and the microorganisms targeted for inhibition (Davidson and Taylor 2007). Factors that affect the efficacy of a food antimicrobial in a food product may be classified as intrinsic, extrinsic, process related, or microbial related (Gould 1989).

Davidson and Branen (2005) differentiated the microbial factors affecting antimicrobial efficacy into those associated with the microorganisms targeted for inhibition and those associated with the antimicrobial's chemistry. Factors that may have an impact on the activity of a food antimicrobial include the inherent resistances of contaminating microorganisms to the antimicrobial (vegetative cells, bacterial spores, strain-specific antimicrobial tolerances); initial number of microbes and growth rates; interactions with other indigenous or contaminating microorganisms; cellular physiologies (Gram reaction, growth requirements); life cycle and status (logarithmic-phase growth, degree of injury); and ability to secrete capsular material in the food (Davidson and Branen 2005). Intrinsic factors are those associated with the food product's physicochemistry, including physical structure and presence of physical barriers to invasion, availability of water and essential nutrients, pH, oxidation/reduction potential (E_H), and presence of other antimicrobials or compounds that may interact with antimicrobials, resulting in synergistic or antagonistic impacts on the added antimicrobial with regard to microbial inhibition. The food's postprocessing storage temperature and duration of storage, the gaseous atmosphere and relative humidity of the storage, and packaging environments of a particular food are grouped into the extrinsic factors affecting antimicrobial activity (Davidson and Taylor 2007). Changes in the food product-specific microbiota, moisture content, nutrient availability, and structure are identified as process-related factors (Davidson and Taylor 2007).

5.2.2   Traditional food antimicrobials

5.2.2.1   Organic acids, fatty acids, and fatty acid esters

The pH of a food product is one of the most significant intrinsic factors that will affect the growth of microorganisms in that food product and will have a significant impact on the activity of an added antimicrobial. Weak organic acids (e.g., lactic acid, acetic acid, citric acid) and the fatty acids (e.g., caprylic acid, lauric acid) are most effective as food antimicrobials when applied in a reduced pH system due to an increased content of protonated acid molecules as compared with the content of deprotonated anion molecules (Doores 2005; Gould 1964). Thus, acid-specific pK_as and food product pH conditions must be known by the antimicrobial researcher or food processor to predict the likelihood of successful application of an organic acid antimicrobial in a food. Eklund (1983, 1985a, 1985b) reported that organic acids inhibit microbial growth through the disruption of the proton motive force (PMF) of microbial cells, disrupting the electrochemical gradient across the microbial membrane, leading to breakdown of multiple cellular processes, often referred to as uncoupling. Russell (1991, 1992) suggested rather that the previous model was incomplete and did not explain the lack of observed microbial inhibition in some research and suggested rather that the accumulation of the anion in the cellular cytoplasm following the acid's penetration into the cell was essential for microbial inhibition. Cherrington et al. (1990, 1991) determined that intracellular accumulation of deprotonated acid anions resulted in damage to microbial DNA, proteins, and other cellular components. Other reports have described degradative reactions of organic acids and their associated salts on noncytoplasmic portions of microbial cells (Alakomi et al. 1999). Organic acids or fatty acids are widely applied in the food industry and have been widely researched for their utility in the inhibition of contaminating microorganisms. An exhaustive review of the organic and fatty acids, their mechanisms of action, foods to which they have been applied, and the concentrations observed as being inhibitory to foodborne microorganisms is beyond the scope of this chapter. Rather, other texts provide further discussion on these and other relevant topics (Chipley 2005; Doores 2005; Ricke 2003).

5.2.2.1.1 Acetic acid and acetates Acetic acid (pK$_a$ 4.75; molecular weight [MW] 60.05 Da) and its sodium, potassium, and calcium salts; sodium and calcium diacetate; and dehydroacetic acid are some of the oldest known food antimicrobials. Bacteria inhibited by acetic acid include differing species of *Bacillus* and *Clostridium*, *Campylobacter jejuni*, *Escherichia coli*, *Listeria monocytogenes*, *Arcobacter butzleri*, *Yersinia enterocolitica*, *Salmonella enterica*, *Staphylococcus aureus*, as well as various fungal organisms, including toxigenic molds and some yeasts (Červenka et al. 2004; Karapinar and Gonul 1992; Siragusa and Dickson 1993; Theron and Lues 2007; Doores 2005). Acetic acid and its salts have shown some success as antimicrobials in foods. Acetic acid increased poultry shelf life when added to chicken parts in cold water at pH 2.5 (Mountney and O'Malley 1965). Over et al. (2009) reported application of 150 mM acetic acid to raw chicken tissue resulted in about 2.0 log$_{10}$ reduction in *S.* Typhimurium and *E. coli*. Acetic acid has variable effectiveness for decontamination of meat. Use of acetic acid resulted in reduction of viable *E. coli* O157:H7 on beef held at 5°C (Siragusa and Dickson 1993). *Escherichia coli* O157:H7 and *Salmonella* on beef trim exposed to 2% or 4% acetic acid were reduced by 2.0–2.5 log$_{10}$ versus on untreated controls (Harris et al. 2006).

Sodium acetate and other salts of acetic acid have been investigated for antimicrobial functionality in differing types of food products. Sodium acetate (1%) extended the shelf life of catfish muscle by 6 days when stored at 4°C (Kim et al. 1995). Application of 2.5% (w/v) sodium acetate to rainbow trout fillets resulted in a 1.4 log$_{10}$ reduction of mesophilic aerobes (Kilinc et al. 2009). Sodium diacetate (pK$_a$ 4.75) is approved for use in processed meat and poultry products by the U.S. Department of Agriculture (USDA) Food Safety and Inspection Service (FSIS) (9 CFR 424.21) and the Canadian Department of Health (2008) at a level not exceeding 0.25% of the product formulation. It is frequently combined with sodium or potassium lactate or other antimicrobials to inhibit *L. monocytogenes* following postprocess cross contamination in fully cooked meats or poultry. Barmpalia et al. (2004) reported that inclusion of 0.25% sodium diacetate inhibited *L. monocytogenes* on frankfurter surfaces over 40 days of storage at 10°C; levels of the pathogen were 2.5 log$_{10}$ less than controls following storage. Sodium diacetate and sodium lactate were described to work synergistically for pathogen inhibition in media and food. Incorporation of a 3% sodium lactate/sodium diacetate blend into chicken meat extended the lag phase of *Salmonella* Typhimurium cells by approximately 28 h (Jung, Min, and Yoon 2009).

5.2.2.1.2 Benzoic acid and benzoates Benzoic acid (pK_a 4.19; MW 122.12 g/mole) and sodium benzoate were the first antimicrobials permitted by the Bureau of Chemistry, the predecessor of the U.S. Food and Drug Administration (FDA) (Chipley 2005). Benzoic acid occurs naturally in various plants, including plums, prunes, cinnamon, cloves, and most berries. Sodium benzoate is highly soluble in water (66.0 g/100 mL at 20°C), whereas benzoic acid is much less so (0.27% at 18°C). Benzoic acid and the benzoates are effective antifungal agents. The inhibitory concentrations for benzoic acid at pH less than 5.0 ranges from 20 to 2000 µg/mL acid for inhibition of various foodborne yeasts and molds (Chipley 2005). Yousef et al. (1989) observed inhibitory activity of benzoic acid in broth at 1000–3000 ppm; bactericidal activity was not demonstrated. Benzoic acid at 0.1% reduced *E. coli* O157:H7 in apple cider by 3.0–5.0 log_{10} colony-forming units (CFU)/mL after 7 days of refrigerated storage (Ceylan, Fung, and Sabah 2004). Potassium benzoate (5.0%) applied to bologna or ham slices inhibited *L. monocytogenes* growth over 48 days of vacuum storage at 10°C (Geornaras et al. 2005).

5.2.2.1.3 Lactic acid and lactates Lactic acid (pK_a 3.79; MW 90.08 g/mole) is synthesized during anoxic fermentation by members of the lactic acid bacteria (LAB). Lactic acid and lactates are added to foods as antimicrobials, acidulants, or flavorants in differing food products. Lactates also serve to aid in retarding onset of lipid oxidation and off-flavor development in processed meats (Papadopoulos et al. 1991). Significant work has been completed describing the antimicrobial potential of lactic acid and the lactates for meat and poultry carcass decontamination (Toldrá 2009). The USDA allows lactic acid application on carcasses prior to or after chilling (<5.0% acid solution), subprimal cuts and trimmings (2–3% acid, <55°C), and beef heads/tongues (2–2.8% in wash systems) (USDA-FSIS 2011). Sodium lactate (2.5–5.0%) inhibits *Clostridium* spp., *L. monocytogenes*, *Salmonella*, *S. aureus*, *Yersinia* spp., and multiple genera of spoilage bacteria in meat and poultry products (Jung, Min, and Yoon 2009; Nuñez de Gonzalez et al. 2004; O'Bryan, Crandall, and Ricke 2008; Shelef 1994). Blending of lactate and diacetate, alone or in the presence of other antimicrobials such as lauric arginate or pediocin, has proven effective for inhibiting growth of *L. monocytogenes* and *Salmonella* Typhimurium on multiple further-processed meat and seafood products (Barmpalia et al. 2004; Porto-Fett et al. 2010; Maks et al. 2010). Stopforth et al. (2010) observed that lauric arginate ester (LAE; 0.07%) applied to hams containing potassium lactate

(1.68%) effectively inhibited *L. monocytogenes* growth for 90 days at 4°C under vacuum.

As sodium or potassium lactate has relatively little pH-lowering impact in some meat and poultry systems, the mechanism of inhibition via these compounds has not been fully elucidated. Lactates have been reported to depress the water activity (a_w) of a food effectively, thereby resulting in increased inhibition of contaminating microorganisms (Červenka et al. 2004; Shelef 1994). Others have contradicted that a_w reduction by lactates is not the primary mechanism of microbial inhibition as observed in various studies, but rather the observed microbial inhibition by lactate is a function of the combined effects of all mechanisms discussed here (Weaver and Shelef 1993).

5.2.2.1.4   Propionic acid and propionates   Propionic acid and sodium, potassium, and calcium propionates are very effective antifungal agents, although some bacteria may also be inhibited by the addition of propionic acid (Doores 2005). Combination of 0.4% propionate with starter cultures including multiple species of *Lactobacillus* was observed to lengthen the shelf life of a bread by 15 days (Gerez et al. 2010). Dipping of apple slices, however, in 0.5–2.0% calcium propionate solutions failed to produce significant reduction of *E. coli* populations (Guan and Fan 2010).

5.2.2.1.5   Sorbic acid and the sorbates   Sorbates are the best characterized of all food antimicrobials regarding their spectrum of action; sorbic acid was first identified in 1859 from the berry of the mountain ash tree (rowanberry) (Stopforth, Sofos, and Busta 2005). Sorbic acid is a *trans-trans*, unsaturated monoprotic fatty acid with a pK_a of 4.75, indicating activity is maximal in foods with a pH less than 6.0 (Eklund 1983). The sorbates are inhibitory toward both fungal and bacterial microbes. Foodborne yeasts inhibited by sorbates include *Brettanomyces, Byssochlamys, Candida, Cryptococcus, Debaryomyces, Hansenula, Pichia, Rhodotorula, Saccharomyces, Sporobolomyces, Torulaspora,* and *Zygosaccharomyces* (Stopforth, Sofos, and Busta 2005; Fujita and Kubo 2005). Molds are also inhibited by the sorbates; genera of sensitive molds include *Alternaria, Aspergillus, Botrytis, Cephalosporium, Fusarium, Geotrichum, Helminthosporium, Mucor, Penicillium, Pullularia, Sporotrichum,* and *Trichoderma* (Stopforth, Sofos, and Busta 2005). The sorbates have been shown to inhibit the synthesis of mycotoxins from species of the toxigenic molds *Aspergillus, Alternaria,* and *Penicillium* (Stopforth, Sofos, and Busta 2005; Bullerman 1983, 1984). Bacteria inhibited by sorbic acid and the sorbates include *Acinetobacter, Aeromonas, Alicyclobacillus acidoterrestris, Bacillus, Campylobacter, E. coli* O157:H7, *Lactobacillus, Listeria innocua, Pseudomonas, Salmonella, Staphylococcus, Vibrio,* and *Yersinia* (Santiesteban-López, Palou, and López-Malo 2007; Stopforth, Sofos, and Busta 2005). Sorbic acid inhibits the catalase-producing bacteria, while many catalase-negative organisms are not inhibited. This permits the use of sorbates in various products fermented by the LAB. Sorbate inhibits the growth of many spoilage and pathogenic bacteria in or on foods (Walker and Phillips 2008; Tassou, Lambropoulou, and Nychas 2004; Ananou et al. 2010; Ukuku and Fett 2004). Sorbates are effective anticlostridial and anti-listerial antimicrobials in cured meats and seafood (Drosinos, Skandamis, and Mataragas 2009; Fernández-Segovia et al. 2007). Sorbates prevent *Clostridium* spore germination and *C. botulinum* neurotoxin synthesis in cured meats and poultry; substantial research has been completed on the replacement of nitrites in cured meats with sorbic acid (Stopforth, Sofos, and Busta 2005). Combinations of 0.1% (w/w) benzoate, 0.2% propionate, and 0.3% sorbate worked well in cured meat products, preventing growth of *L. monocytogenes* on ham stored at 4°C for 12 weeks (Glass, McDonnell et al. 2007). Similar formulations were effective in cured pork/beef bologna, but not in uncured turkey (Glass, Preston, and Veesenmeyer 2007).

One of the primary cellular targets of sorbic acid in vegetative cells appears to be the cytoplasmic membrane. Sorbic acid inhibits cellular uptake of amino acid, which in turn has been theorized responsible for PMF depletion (Freese 1978; Freese, Sheu, and Galliers 1973; Sheu and Freese 1972; Sheu, Konings, and Freese 1972; Sheu et al. 1975). Eklund (1983) reported much higher concentrations are required for complete loss of transmembrane electrical potential ($\Delta\Psi$) in cells. Researchers have demonstrated that sorbic acid does not function exclusively as a weak organic acid, exhibiting membrane permeabilization efficacy in addition to inducing dissipation of cellular PMF (Stratford and Anslow 1998).

5.2.2.1.6   Miscellaneous organic acids, fatty acids, and fatty acid esters   In addition to the weak acids already discussed, multiple other acids are effective antimicrobials useful for inhibition of microbes in differing food products, including citric, malic, tartaric, fumaric acid, and their salts. Addition of 50 mM fumaric acid to lettuce surfaces produced reductions of *E. coli* O157:H7, *S.* Typhimurium, and *S. aureus* by 1.2–1.5 $\log_{10}$, although browning was observed on treated tissues. Combination of fumaric acid (0.15%) and sodium benzoate (0.05%) in raw apple cider (pH 3.3) produced a 5.0 $\log_{10}$ CFU/mL reduction in *E. coli* O157:H7 numbers (Comes and Beelman 2002). Kim, Kim, and Song (2009a, 2009b) reported 0.5 g/100 mL fumaric acid applied to surfaces of sprouts reduced *E. coli* O157:H7, *Salmonella*, and *L. monocytogenes* by 2.8–3.4 $\log_{10}$ CFU/g.

Del Río et al. (2007, 2008) reported that citric acid extended the lag phase of foodborne microbes, effectively lengthening the time required for bacteria to achieve stationary phase. Citric acid inhibits *Salmonella* and *E. coli* O157:H7 on beef and poultry (Cutter and Siragusa 1994; Laury et al. 2009). Citric acid inhibits bacterial pathogens on fresh and fresh-cut produce (Bari, Ukuku et al. 2005). As a triprotic organic acid, citric acid has been reported to function not only as a weak organic acid but also its salts are described to function as chelating agents (Miller, Call, and Whiting 1993). Levulinic acid has recently been investigated for antimicrobial efficacy against bacterial pathogens on multiple foods. Vasavada, Carpenter, and Cornforth (2003) reported that aerobic plate counts in sausages were 1.5–2.0 $\log_{10}$ CFU/g lower than controls after 14 days of refrigerated (2°C) storage when levulinic acid was added at between 1% and 3%. *Salmonella* Typhimurium and *E. coli* O157:H7 were reduced by 4.8–5.0 $\log_{10}$ CFU/g on lettuce and sprouts when 0.5% levulinic acid and 0.05% sodium dodecyl sulfate (SDS) were applied (Zhao, Zhao, and Doyle 2009, 2010).

Increased attention has been paid to the utility of various medium-chain fatty acids with regard to antimicrobial utility. Nakai and Siebert (2004) reported that caprylic and caproic acids effectively inhibited *Listeria* spp. Minimum inhibitory concentrations (MICs) of lauric and caprylic acid against *Listeria* ranged from 0.63 to 5.0 mM (Nobmann et al. 2009). Brandt et al. (2011) demonstrated 25.0 µg/mL octanoic (caprylic) acid suppressed growth of *L. monocytogenes* at pH 5.0 *in vitro*; a combination of octanoic acid and acidified calcium sulfate enhanced pathogen inhibition. Caprylic acid combined with SDS resulted in significant inactivation of *S. Enteriditis* in produce wash water; the pathogen was suppressed to nondetectable levels (Zhao, Zhao, and Doyle 2009).

Monoesters of fatty acid esters have been repeatedly reported to inhibit the growth of multiple bacterial genera in model and food systems; one of the most effective is glycerol monolaurate (monolaurin) (Branen and Davidson 2004). Monolaurin is inhibitory to Gram-positive bacteria, including *Bacillus* spp. and *L. monocytogenes* (Bala and Marshall 1996a, 1996b; Branen and Davidson 2004; Mansour and Milliere 2001). Monoester of caprylate (monocaprylin) exerts inhibitory activity toward both Gram-positive and Gram-negative bacterial pathogens (Nobmann et al. 2009). *Escherichia coli* O157:H7 and *Salmonella* exposed to 25 mM monocaprylin for 1 min at room temperature were reduced by 3.0 and 2.0 $\log_{10}$ CFU/mL, respectively (Chang, Redondo-Solano, and Thippareddi 2010). LAE (N^α-lauroyl ethylester) is a cationic surfactant food antimicrobial inhibitory toward multiple Gram-positive and Gram-negative bacterial genera, including *L. monocytogenes*, as well as various fungal microbes (Dai et al. 2010; Brandt et al. 2010; Stopforth et al. 2010). Parameters of LAE treatments necessary to produce 2-log reductions of *L. monocytogenes* consistently on frankfurters and hams have been reported (Taormina and Dorsa 2009a, 2009b).

5.2.2.2   Nitrites

Sodium nitrite ($NaNO_2$) and potassium nitrite (KNO_2) have specialized use in cured meats. The primary antimicrobial application for nitrites in cured meats is the inhibition of clostridial spore germination and toxin synthesis by *C. botulinum* (Duncan and Foster 1968). Roberts and Ingram (1966) reported nitrite addition does not increase spore killing but rather inhibits the outgrowth of vegetative cells postheating.

Antimicrobial effects of nitrite on microorganisms other than the clostridia are variable. Tsai and Chou (1996) reported 200 mg/L nitrite at pH 5.0 resulted in 6.5–7.0 $\log_{10}$ CFU/mL reduction of *E. coli* O157:H7, while more than 400 mg/L nitrite suppressed the pathogen to nondetectable levels. Gill and Holley (2003) described the application of 180 mg/L nitrite in broth (pH 6.0) and the resulting inhibition of *Salmonella* Typhimurium, *Shewanella*, *Serratia*, and *E. coli*, as well as Gram-positive fermentative and pathogenic microbes. Gibson and Roberts (1986a, 1986b) reported limited inhibition of *C. perfringens*, *Salmonella*, enteropathogenic *E. coli*, and the fecal streptococci via nitrites in acidified medium containing up to 6% NaCl. In addition to the various cured meats containing nitrites, these compounds are employed in curing of fish and poultry. Concentrations of nitrite used in these products varies but is generally limited to 156 µg/g for most cured products; levels for bacons may range from 100 to 120 µg/g. Utilization of ascorbate can accelerate product curing and may inhibit formation of nitrosamines (Tompkin 2005). Tompkin (2005) described the history of nitrite usage in the United States, mechanisms of antimicrobial activity, and health and regulatory concerns related to nitrite addition to foods during curing. The reader is directed to Tompkin's well-written chapter for information on nitrites and concerns related to their use, as well as discussion of research seeking to replace nitrites with alternative curing agents (e.g., sorbic acid).

5.2.2.3 Sulfites

Sulfur's use as an antimicrobial dates to at least the ancient Greeks (Gould 1989). Sulfites are effectively used to control growth of spoilage and fermentative fungal microbes in fruit and vegetable products, as well as various members of the acetic acid bacteria (Ough and Were 2005). Sulfites demonstrate antioxidant properties, inhibiting enzymatic and nonenzymatic browning in various foods. Sulfur's antimicrobial activity is most heavily impacted by the environmental pH:

$$SO_2 \cdot H_2O \rightleftarrows HS^{3-} + H^+ \rightleftarrows SO^{3-}_2 + H^+$$

Aqueous solution of sulfur dioxide yields sulfurous acid in theory (H_2SO_3), although evidence points to the formation of the molecule $SO_2 \cdot H_2O$ (Gould and Russell 1991). The pK_a's for sulfur dioxide, depending on temperature, are 1.76–1.90 and 7.18–7.20; the undissociated form of sulfites or $SO_2 \cdot H_2O$ exhibit greatest antimicrobial activity (Gould and Russell 1991). Microbial targets for sulfites include enzymes and their cofactors, ATP (adenosine triphosphate), the cytoplasmic membrane, and membrane-located proteins (Ough and Were 2005). Bacterial microbes are sensitive to sulfite activity. Sulfites are used for inhibiting various spoilage microbes, the acetic acid bacteria, LAB, and some Gram-negative pathogens (Gould 2000). *Salmonella, E. coli,* and *Yersinia* isolates were inhibited in sausage containing 15–200 µg/mL free sulfite, added as metabisulfite (Banks and Board 1982). Tong and Draughon (1985) reported the addition of 0.066% sodium bisulfite effectively halted synthesis of ochratoxin A by aflatoxigenic molds in medium incubated at 28°C.

5.2.2.4 Preservative enhancers

Chemical compounds that serve as chelators or sequestrants of micronutrients can contribute to the efficacy of preservatives used in foods. Some of the more common chelators and sequestrants used in foods are calcium and disodium salts of ethylenediamine tetraacetic acid (EDTA), various chain-length polyphosphates, citric acid, and phytic acid.

Several molds isolated from food were shown to be inhibited by phosphates alone on an agar media assay (Suarez et al. 2005). Although *Aspergillus ochraceus* and *Fusarium proliferatum* were resistant to phosphates at concentrations as high as 1.5% (w/v), *Byssochlamys nivea, Aureobasidium pullulans,* and *Penicillium glabrum* were sensitive. Phosphates with the highest sequestering power and chain lengths greater than 15 phosphate units were found to be most inhibitory. Polyphosphates, primarily sodium hexametaphosphate, have been patented as preservative enhancers for weak acid preservative systems used in acidic beverages (Calderas et al. 1994, 1997). Phosphates in conjunction with organic salts (i.e., lactate and diacetate) also provide shelf life extension of fresh pork organic salts (Ruzek 1996). Sodium pyrophosphate (SPP) has been found to increase the heat sensitivity of *L. monocytogenes* (Juneja and Eblen 1999), although SPP in pork meat protected starved *L. monocytogenes* cells from lethal effects of heating (Lihono et al. 2003). Increasing levels of SPP and salt increased heat resistance of *Salmonella* in ground beef (Juneja, Marks, and Mohr 2003).

In laboratory media, several phosphates were inhibitory to growth of enteropathogenic *E. coli,* and this effect was enhanced by the presence of sodium chloride and sodium nitrite (Hughes and McDermott 1989). Growth rates of *C. perfringens* in broth were restricted by the levels of SPP, suggesting that the chemical could have protective value in meats, especially in conjunction with salt and acidulant (Juneja et al. 1996). SPP was also shown to prevent outgrowth during extended cooling of cooked pork (Singh et al. 2010).

Chelators like EDTA and citric acid potentiate the effects of weak acid preservatives, monolaurin, and H_2O_2 (Brul and Coote 1999). The influence of NaCl and EDTA combinations on growth rates of stressed *L. monocytogenes* was studied (Zaika and Fanelli 2003). It was observed that 0.3 mM

EDTA caused the following changes to cell populations: (1) a sixfold increase in generation time and lag phase duration, (2) loss of flagellae, and (3) cell elongation. Phytic acid and EDTA were used in combination with various other antimicrobials to reduce levels of *L. monocytogenes* on fresh-cut produce (Bari, Ukuku, et al. 2005). Interactive effects of nisin and EDTA against Gram-negative bacteria were disputed by Gill and Holley (2000, 2003), who conducted studies on log-phase cells, not nutrient-starved populations as pre-dated studies had used.

5.2.3　Naturally occurring compounds and systems

Many foods contain compounds that possess antimicrobial potential and can serve to extend the shelf life of the food. Although many such compounds have been investigated and their antimicrobial activity and utility characterized, concerns exist surrounding the use of such compounds as food additives. Naturally occurring antimicrobials are seldom added to a food at effective concentrations without some form of processing or refinement being completed prior to incorporating in the food. For many of the plant-derived compounds with antimicrobial utility, the compound will often impart an undesirable organoleptic characteristic to the food (e.g., essence of garlic is incompatible in processed beverages). Ultimately, the food industry is challenged with the identification of an antimicrobial substance (or substances) that may be effectively added to foods without extensive refinement and without noticeable sensorial/palatability impact (Stopforth, Skandamis, et al. 2005).

5.2.3.1　Animal sources

5.2.3.1.1　Lactoferrin and lactoferricin　Lactoferrin (76.5 kDa) is a glycoprotein that exists in milk primarily as a tetramer with Ca^{+2}, possessing two iron-binding sites per molecule. Its exact biological role is not known, although it may function to prevent infection of a nonlactating mammary gland and aid in protecting the newborn against gastrointestinal infection (Teraguchi et al. 1995a, 1995b). Lactoferrin inhibits the foodborne microorganisms *Carnobacterium*, *L. monocytogenes*, *E. coli*, and *Klebsiella* (Venkitanarayanan, Zhao, and Doyle 1999; Al-Nabulsi and Holley 2005; Murdock et al. 2007). Branen and Davidson (2004) investigated the interactions of lactoferrin with monolaurin, nisin, and lysozyme against *L. monocytogenes*, *E. coli*, *Salmonella*, and *Pseudomonas*. Lactoferrin combined with nisin and monolaurin in microbiological medium inhibited *L. monocytogenes* and *E. coli* O157:H7, respectively. Limitation of microbial access to nutrients via iron chelation represents a likely contributor to the antimicrobial mechanism of lactoferrin (González-Chávez, Arévalo-Gallegos, and Rascón-Cruz 2009). Microorganisms with low iron needs, such as the LAB, are not overly sensitive to lactoferrin (Pandey, Bringel, and Meyer

1994). Lactoferrin induces lipopolysaccharide (LPS) release by chelation of LPS-associated cations (calcium, magnesium, iron). Orsi (2004) observed that lactoferrin degraded microbial adhesion to inanimate surfaces, inhibiting biofilm formation by differing bacteria.

Lactoferricin is a small polypeptide (25 residues) obtained from proteolytic digest of lactoferrin by pepsin (Arseneault et al. 2010). Enrique et al. (2009) reported that lactoferricin was inhibitory to various members of the LAB (e.g., *Lactobacillus, Pediococcus*). Jones et al. (1994) reported bacteriostatic and bactericidal activity against multiple pathogenic genera. Venkitanarayanan et al. (1999) reported that addition of 100 μg/g into ground beef effectively resulted in a reduction of *E. coli* O157:H7 of about 1.0 $\log_{10}$ CFU/g over 5 days of refrigerated (4°C) incubation. Recently, investigators have determined that pepsin hydrolysates of lactoferrin reduced *E. coli* O157:H7 in liquid medium by 2.5 $\log_{10}$ CFU/mL (Del Olmo, Calzada, and Nuñez 2010).

5.2.3.1.2 *Lactoperoxidase* Lactoperoxidase is a glycoprotein enzyme secreted in raw milk, colostrum, saliva, and other biological secretions (Davidson and Taylor 2007). Bovine lactoperoxidase has a molecular weight of approximately 78 kDa; approximately 10–30 mg of native enzyme is contained per liter of milk (Wilkins and Board 1989; Cals et al. 1991). Stopforth, Skandamis, et al. (2005) described the enzyme's antimicrobial activity as resulting from the oxidation of halides and thiocyanate in the presence of excess peroxide (H_2O_2), producing hypothiocyanate ($OSCN^-$) and hypothiocyanous acid (HOSCN). Lactoperoxidase inhibits Gram-negative and Gram-positive bacterial organisms, although the catalase-producing Gram-negative organisms (e.g., *Salmonella, E. coli*, etc.) were reported to be especially sensitive to the enzyme's activity (Kussendrager and Van Hooijdonk 2000).

Lactoperoxidase contributes to the shelf quality/stability of raw milks (Babu, Varshney, and Sog 2004). Incorporation of lactoperoxidase in *L. monocytogenes*-inoculated skim milk held at 25°C resulted in a delay of the pathogen's entry into logarithmic growth by approximately 47 h versus controls (Boussouel et al. 2000). Lactoperoxidase is also inhibitory toward *S. aureus, E. coli* O157:H7, *Salmonella*, and *Y. enterocolitica* in products such as beef and vegetable juices (Stopforth, Skandamis, et al. 2005).

5.2.3.1.3 *Lysozyme* Lysozyme (N-acetylmuramoyl hydrolase) is a 14.6-kDa lytic enzyme naturally present in avian eggs, mammalian milk, tears, and other secretions, affirmed as generally recognized as safe (GRAS) for direct addition in foods (FDA 1998). Hen egg white lysozyme (HEWL) is the primary antimicrobial in egg albumin, although its activity is enhanced by conalbumin, ovomucoid, and a pH of about 9.3. Lysozyme possesses stability to heating in acidic environments (30 min at 80°C), although its thermal stability is reduced as pH is raised. The enzyme is most active from 55 to 60°C, but it maintains approximately 50% activity at lower temperatures (10–25°C) (Johnson and Larson 2005). The enzyme hydrolyses the α-1,4 glycosidic bond between C1 of N-acetylmuramic acid and C4 of N-acetylglucosamine, degrading the cell wall in hypotonic systems. Lysozyme is most active against the Gram-positive bacteria, likely a result of the exposed murein in the bacterial cell wall. Species of *Bacillus*, *Micrococcus*, and *Clostridium* are especially sensitive to lysozyme's action (Johnson and Larson 2005). Gram-negative foodborne bacteria are not highly sensitive to lysozyme treatment (Johnson and Larson 2005). Susceptibility of Gram-negative organisms to lysozyme can be increased by coapplication of chelators (Boland et al. 2004; Boland, Davidson, and Weiss 2003).

Combination of 250–500 ppm lysozyme, 250–500 ppm nisin, and 5 mM EDTA suppressed aerobic mesophiles on ground ostrich patties over 7 days of storage at 4°C (Mastromatteo et al. 2010). Application of 0.5 U/cm^2 on turkey bologna followed by in-package pasteurization suppressed *L. monocytogenes* below detectable limits on samples stored under refrigerated vacuum (Mangalassary et al. 2008). Lysozyme is approved for prevention of late blowing of cheeses caused by excess anaerobic fermentation of gas by *Clostridium* spp. (FDA 1998).

5.2.3.1.4 Ovotransferrin Ovotransferrin (conalbumin) occurs in egg albumin; the enzyme chelates iron strongly with the presence of available bicarbonate, functioning in a manner similar to that observed for lactoferrin (Wilkins and Board 1989; Tranter 1994). *Micrococcus* and *Bacillus* spp. are especially sensitive to the activity of ovotransferrin, as are many other Gram-positive organisms, although the Gram-negative organisms are sensitive to iron chelation as it can significantly destabilize the LPS (Tranter 1994). Recent research has sought to enhance the antimicrobial efficiency of ovotransferrin by applying it in foods with other antimicrobials. Combination of ovotransferrin with EDTA on ham surfaces resulted in *E. coli* O157:H7 and *L. monocytogenes* being reduced over 13 days at 10°C, although not to a level that was significantly lower than controls (Ko et al. 2009; Ko, Mendonca, and Ahn 2008).

5.2.3.2 Microbial sources

5.2.3.2.1 Bacteriocins Bacteriocins are antimicrobial polypeptides synthesized and secreted by members of the LAB; they typically undergo some form of posttranslation modification and may be used to provide the producing organism a competitive advantage in a complex microbiological niche. Significant research has been completed on the bacteriocins and their antimicrobial activity in microbiological media and in differing foods, and a complete review is not warranted here. Currently, only the class IA bacteriocin nisin has been approved by the FDA for direct addition to differing food products, originally commercially developed by Applin and Barrett, Limited (United Kingdom) for addition to foods (FDA 1988). Nevertheless, numerous other bacteriocins have been identified and characterized for antimicrobial activity and mode of action (de Vuyst and Leroy 2007; Ennahar et al. 2000; Klaenhammer 1988; Venema, Venema, and Kok 1995).

Nisin is a class IA lantibiotic, named so for its inclusion of five lanthionine rings over its structure (Gross and Morell 1967; Gross and Morell 1971). Its stability to thermal processing is substantially increased due to this structure, with enhanced solubility and thermal process stability observed at reduced pH (Delves-Broughton 1990). Nisin is approved by the FDA for addition to cheeses at a level not exceeding 250 ppm nisin to inhibit the late blowing of various cheeses in accordance with good manufacturing practices (GMPs) (21CFR184.1538). Its addition to various dairy and meat products is also approved not only in the United States but also in many other nations, in some with no limits on dosage applied in a product-specific manner (Chen and Hoover 2003). Although highly active against the clostridia, most recent research involving nisin application to foods has investigated its antilisterial utility for the preservation of dairy and meat products (Delves-Broughton et al. 1996). Branen and Davidson (2004) reported synergistic inhibition of *L. monocytogenes* by the combined addition of nisin and EDTA in fluid medium, although antilisterial effects were not observed in whole-fat fluid milk, presumably due to partitioning of nisin at the fat/water interfaces (Jung, Bodyfelt, and Daeschel 1992). Similar reports of antilisterial activity, as well as inhibition of Gram-negative pathogenic bacteria, following the application of nisin and plant-derived or other animal-derived antimicrobials have also been documented (Yuste and Fung 2004; Tokarskyy and Marshall 2008; Singh, Falahee, and Adams 2001; Murdock et al. 2007). Multiple researchers have reported potent antilisterial efficacy of nisin applied alone or in combination with other antimicrobials on surfaces of various fresh and processed meat and poultry products (Barboza de Martinez, Ferrer, and Marquez Salas 2002; Fang and Tsai 2003; Mangalassary et al. 2008; Mastromatteo et al. 2010).

Like nisin, pediocin is a class I bacteriocin, although synthesized by species of *Pediococcus*. Its direct application in purified form is not allowed to foods, although its inclusion in various fermented foods is common through the secretion of pediocin by starter cultures applied to meat and produce products. Maks et al. (2010) described the combined application of sodium lactate (1.2–3.6%) and pediocin up to 7500 U resulted in significant decreases in the heat resistance of *L. monocytogenes* on bologna surfaces. Similar synergistic antimicrobial impacts were observed when pediocin was combined with citric acid, sorbate, or phytic acid for the inhibition of *Listeria* (Kim et al. 2001; Bari, Ukuku et al. 2005).

5.2.3.3 Plant sources

5.2.3.3.1 Spices and their essential oils   Spices are added to foods primarily as flavoring agents, although some spice compounds have been known to exhibit antimicrobial potential for thousands of years. Clove, cinnamon, oregano, thyme, and to a lesser extent sage and rosemary have repeatedly been investigated and their antimicrobial activity quantified more frequently than other spice-bearing plants.

Clove (*Syzgium aromaticum*) and cinnamon (*Cinnamomum zeylanicum*) are two commonly used spice-yielding plants that provide two phenolic compounds with antimicrobial activity that has been repeatedly investigated, those being eugenol [2-methoxy-4-(2-propenyl)-phenol] and cinnamic aldehyde (3-phenyl-2-propenal), respectively (Davidson and Naidu 2000). Cinnamon oil possesses approximately 75% cinnamic aldehyde and 8% eugenol. Likewise, clove volatile oil contains 95% eugenol. Cinnamic aldehyde and eugenol have demonstrated antimicrobial activity against multiple foodborne bacterial genera, including *Aeromonas, Bacillus, Campylobacter,* verotoxigenic *Escherichia, Lactobacillus, Listeria, Salmonella, Shigella, Staphylococcus,* and *Streptococcus* (Aureli, Costantini, and Zolea 1992; Friedman, Buick, and Elliott 2004; Friedman, Henika, and Mandrell 2002; Johny et al. 2010). Mytle et al. (2006) applied clove oil (1–2%) to surfaces of *Listeria*-inoculated frankfurters. Following incubation (5°C or 15°C), significant reductions of *Listeria* were observed on franks treated with essential oil. Eugenol inhibits pathogens on surfaces of cooked meats (beef, pork) and poultry (Hao, Brackett, and Doyle 1998a, 1998b). Application of 0.3% cinnamon extract in apple juice reduced *E. coli* O157:H7 by about 2.0 $\log_{10}$ CFU/mL; application of 0.1% sodium benzoate or potassium sorbate in addition to cinnamon extract resulted in the pathogen being reduced to nondetectable levels (Ceylan, Fung, and Sabah 2004).

Similar to clove and cinnamon, antimicrobial activities of oregano (*Origanum vulgare*) and thyme (*Thymus vulgaris*) extracts have been attributed to volatile oils containing carvacrol (2-methyl-5-[1-methyl]-phenol) and thymol (5-methyl-2-[1-methyl]-phenol), respectively. Bacterial and fungal microorganisms, including *A. hydrophila, Bacillus, C. jejuni, E. coli, Enterococcus,* the lactobacilli, *L. monocytogenes,* the pediococci, *Pseudomonas, Salmonella, Shigella, S. aureus, V. parahaemolyticus, Y. enterocolitica,* and *Aspergillus, Candida, Geotrichum, Penicillium, Pichia, Rhodotorula, Saccharomyces,* and *Schizosaccharomyces pombe* have been shown to be inhibited by the application of essential oil components or oils from oregano and thyme (Aligiannis et al. 2001; Bagamboula, Uyttendaele, and Debevere 2003; Chami, Chami, et al. 2005; Chami, Bennis, et al. 2005; Gutierrez, Barry-Ryan, and Bourke 2009; Gutierrez et al. 2008; López-Malo Vigil, Palou, and Alzamora 2005).

Mechanistic studies investigating the modes of action of spice essential oils have focused on the effect of compounds on the cytoplasmic membrane of targeted microorganisms. Ultee et al. (1998, 1999, 2002; Ultee, Kets, et al. 2000; Ultee, Slump, et al. 2000) reported several mechanisms of carvacrol against the toxigenic pathogen *B. cereus*, reporting the pathogen was inhibited via depletion of intracellular ATP, reduced membrane potential, and increased permeability, leading to cell death. Toxin synthesis was also reported to occur following the application of the phenolic antimicrobial (Ultee and Smid 2001). Studies have shown tea tree oil and β-pinene to increase leakage of ions (K⁺) in *E. coli* and yeasts (Burt 2004). Dissipation of PMF, ATP synthesis inhibition, and enzyme inhibition have also been described to occur (Gill and Holley 2006; Mann, Cox, and Markham 2000).

5.2.3.3.2 Onions and garlic Probably the most well-characterized antimicrobial systems obtained from plants are those from the juice and vapors of onion (*Allium cepa*) and garlic (*A. sativum*). Growth and toxin production of many microorganisms is inhibited by onion and garlic (López-Malo Vigil, Palou, and Alzamora 2005).

Cavallito and Bailey (1944) identified allicin (diallyl thiosulfinate; thio-2-propene-1-sulfanilic acid-5-allyl ester) and described it as the primary compound in garlic. Allicin has been reported to be inhibitory toward the naturally occurring microbiota in differing processed meats and sausages (Kim, Nahm, and Choi 2010; Sallam, Ishioroshi, and Samejima 2004). The mechanism of action of allicin may be tied to inhibition of sulfhydryl and disulfide-containing enzymes (Beuchat 1994). Researchers have reported that allicin inhibited many sulfhydryl enzymes, and that allicin inactivated proteins by oxidation of thiols to disulfides and inhibiting the intracellular reducing activity of glutathione and cysteine (Wills 1956).

5.2.3.3.3 Isothiocyanates Isothiocyanates (R–N=C=S) are derived from glucosinolates in plants belonging to the families *Brassicaceae* and *Cruciferae* (e.g., cabbage, cauliflower, broccoli, kale, horseradish, mustard, turnip, etc.); compounds are enzymatically synthesized by thioglucoside glucohydrolase when plant tissues are mechanically disrupted. Isothiocyanates inhibit both fungal and bacterial microbes (Isshiki et al. 1992; Mari et al. 1993; Schirmer and Langsrud 2010). Inhibitory activity against bacteria has been reported to be variable, but Gram-negative genera are generally more sensitive than are Gram-positive bacteria (Schirmer and Langsrud 2010; Obaidat and Frank 2009a, 2009b). Allyl isothiocyanate has been reported to be useful in reducing *E. coli* O157:H7 in fresh and further-processed beef (Chacon, Buffo, and Holley 2006; Muthukumarasamy, Han, and Holley 2003; Ward et al. 1998).

5.2.4   Encapsulation of food antimicrobials

Despite the variety of antimicrobial compounds available for application in a food and the development of novel compounds and novel chemistries for use in foods, opportunities remain for the innovation of new antimicrobial application technologies capable of preserving food quality and safety. One innovative method to enhance the efficacy of antimicrobials in foods relies on the micro- and nanoencapsulation of antimicrobial within another food-grade material. Antimicrobial encapsulation may assist in (1) stabilizing the antimicrobial against deleterious reactions with food components; (2) stabilizing volatile antimicrobials against rapid evaporation; (3) reducing the rate of the antimicrobial's release into the food, allowing lengthened exposure of microbes to antimicrobial pressure; and (4) protection of the antimicrobial during processing (Taylor et al. 2005).

5.2.4.1   Liposomes and micelles as food antimicrobial encapsulation technologies

Liposomes are constructed from phospholipids (amphipathic lipids), naturally occurring or synthetic analogues of natural phospholipids, mixed into an aqueous system. On adding in water or other aqueous buffer, lipids will spontaneously form as liposomes (Taylor et al. 2005). Liposomes have been utilized as models for biological membranes, as delivery systems for a multitude of pharmaceuticals, and for the encapsulation of food ingredients such as vitamins, enzymes, or colorants (Taylor et al. 2005).

Degnan and coworkers (Degnan, Buyong, and Luchansky 1993; Degnan and Luchansky 1992) reported antilisterial activity of liposomal pediocin acidilactici strain H (AcH) incorporated into slurries of beef tallow and meat slurries. Benech et al. (2002) reported that liposome encapsulation of nisin inhibited *Listeria innocua*; adding of liposomal nisin to a level of 300 IU nisin/g cheese resulted in reducing the organism by 3.0 $\log_{10}$ CFU/g in cheese over 6 months of ripening. Were et al. (2004) reported inhibition of *L. monocytogenes* via encapsulated nisin or lysozyme in liquid medium. Encapsulation of nisin in liposomes resulted in the inhibition of *L. monocytogenes* and *E. coli* O157:H7 in liquid medium and in fluid milk of differing fat contents (Taylor et al. 2008; Schmidt et al. 2009). Delivery of liposomal nisin to fluid milk resulted in bacteriostatic inhibition of *L. monocytogenes* (Da Silva Malheiros et al. 2010).

Micelles (and reverse micelles) are formed of amphiphilic detergents possessing both hydrophilic and hydrophobic regions; in the case of micelles, this results in a system in which hydrophobic surfactant tails are buried within and hydrophilic head groups protrude out in the water phase (Gaysinsky et al. 2005b; McClements et al. 2009). Application of micelle-encapsulated eugenol to *L. monocytogenes* and *E. coli* O157:H7 at pH 5.0–7.0 completely inhibited growth of both organisms (Gaysinsky et al. 2005a). Gaysinsky et al. (2007) reported that addition of eugenol-entrapping micelles to fluid milks inoculated with *L. monocytogenes* or *E. coli* O157:H7 produced inhibition of pathogens in a milk fat and pathogen-specific manner. Use of a sodium lactate and monolaurin-containing microemulsion (800 ppm) produced *in vitro* inhibition of *B. subtilis*, *S. aureus*, and *E. coli* O157:H7 (Zhang et al. 2008, 2009).

5.2.4.2   Edible films

In addition to the encapsulation of differing food antimicrobials in differing structures, the incorporation of antimicrobials into food-grade polymers has allowed for the development of various antimicrobial-bearing edible films. These technologies allow for preservation of antimicrobial activity prior to application to the food, increased opportunity for direct contact between antimicrobial and targeted microorganism, and long-term suppression of microbial growth during storage as a result of diffusion of antimicrobial from the film to the surface of the food product. Multiple polysaccharides and polypeptides have been explored for their utility to incorporate and deliver antimicrobials, although much research has been focused on chitosan, a polysaccharide obtained by deactylation of the naturally occurring polymer chitin. Chitosan is polycationic in nature and has been repeatedly reported to possess strong antimicrobial activity of its own, although observed antimicrobial efficacy has been shown to be increased when other antimicrobials are incorporated prior to casting of chitosan films (Devlieghere, Vermeulen, and Debevere 2004; Sudarshan, Hoover, and Knorr 1992).

Brown, Wang, and Oh (2008) reported that incorporation of 2.0 mg of lactoferrin into chitosan prior to casting resulted in statistically greater inhibition of both *E. coli* O157:H7 and *L. monocytogenes* on medium surfaces as compared to incorporation of nisin or lysozyme at equivalent concentrations, potentially due to more favorable charge interactions of antimicrobial with film. The combination of lactoferrin with lysozyme in films further increased the observed antimicrobial activity of chitosan films. Incorporation of garlic oil or nisin into chitosan films resulted in significant increases in inhibition of Gram-positive pathogens (*S. aureus*, *L. monocytogenes*, *B. cereus*), although inhibitory effects were not observed in Gram-negative pathogens, including *Salmonella* and *E. coli* (Pranato, Rakshit, and Salokhe 2005). These findings were in contradiction to reports by others that indicated that chitosan exhibited enhanced antimicrobial activity against Gram-negative organisms versus Gram-positive ones (Kong et al. 2010).

In addition to chitosan, alginates, whey-derived proteins, zein proteins, and other polymers have all been investigated for their utility in formulating antimicrobial-bearing edible films (Cagri, Ustunol, and Ryser 2004). Spice essential oils (eugenol, cinnamic aldehyde) were incorporated into alginate films that were subsequently applied for the inhibition of spoilage microorganisms and *Salmonella* Enteriditis (Raybaudi-Massilia, Mosqueda-Megar, and Martín-Belloso 2008). In addition to reductions observed in numbers of mesophilic and psychrotrophic bacteria, numbers of *S.* Enteriditis were significantly reduced by antimicrobial-bearing films over 21 days of refrigerated storage on melon surfaces. Antilisterial effects were observed on cold-smoked salmon fillets treated with alginate films incorporating organic acids. Use of 2.4% sodium lactate and 0.125% diacetate in combination in films resulted in an approximate 4.0 $\log_{10}$ difference in *L. monocytogenes* numbers on fillets after 30 days of 4°C storage versus controls (Neetoo, Ye, and Chen 2010).

5.3   Innovation and food antimicrobials

5.3.1   Intellectual property development

Development of intellectual property concerning food antimicrobials is undertaken mostly by food industry processors, ingredient companies, and allied food industrial suppliers. Sometimes, other non-food-related industries develop antimicrobial systems or technologies that have application in food and beverages as preservatives. Protecting the intellectual property concerning an antimicrobial or how it is produced and applied to a food or beverage ideally leads to the exclusive right to market and sell the antimicrobial as a food and beverage preservative or the exclusive right to license the technology. Two main approaches to protecting developed intellectual property are (1) maintaining the information as trade secrets or (2) publicly disclosing the inventions as patents.

A properly protected and maintained trade secret enables a company to use a technology in the marketplace without divulging certain details about it that would allow competitors to effectively steal the technology (U.S. Patent and Trademark Office 2010). This sort of stealth technology can be difficult to maintain and lends itself only to certain situations and company cultures. The main advantage is that the length of time for exclusive use of the technology is indefinite—as long as secrecy is maintained, the technology can be used exclusively by the trademark holder. This does not, however, provide businesses a legal right to exclusive use, and other competing companies may eventually figure out the technological approach either by reverse engineering or by accident. If information is already in the public domain, such as through journal publication, operating a technology as a trade secret may be the only option as a patent application would not likely result in a patent (see the following discussion).

Patents offer some distinct advantages over trade secrets. They can be used offensively or defensively to compete in a certain technological field. They can be considered assets and contribute to the value of a company.

They can publically establish an entity's leadership in a particular technological field. They also enable the possibility of licensing a technology to generate revenue from royalties. All patents, and even the patent application itself (whether the patent is actually awarded or not), create "prior art," which can preclude future patent awards in that area. Thus, by applying for a patent, whether it is accepted or not, an entity can prevent competitors from receiving patents in that area. This would be considered a defensive tactic.

The patent approach to protecting intellectual property gives the legal right to exclude others from "making, using, or selling" a patented "invention" (35 U.S.C. § 271). However, a patent does not grant an affirmative right to make, use, or sell the invention. There are three types of patents: utility, design, and plant. When people think of patents, they usually have utility patents in mind. Utility patents protect utilitarian structure, function, method, or composition of an invention. These patents have a term of 20 years from the effective filing date, and there are fees associated with maintaining the patent and most certainly legal fees with monitoring this process and addressing possible infringement. Utility patents use word claims to identify the invention, and these can be found as the numbered paragraphs at the end of a patent.

Design patents are for new, original, and ornamental design for an article of manufacture. They have a term of 14 years and do not have associated maintenance fees. However, design patents protect only the ornamental appearance of the article and not its structure or function. If appearance is dictated by function, appearance is not "ornamental." The infringement test is whether an accused design is "confusingly similar" to the patented design. Design patents have minimal relevance to food antimicrobials except with regard to specialized equipment designed for preparation of the antimicrobial as a food ingredient or delivery of the ingredient to foods during processing. One example would be the patented hot water and antimicrobial carcass spray cabinet of the Chad Corporation (Anderson and Gangel 1998).

Plant patents protect new and distinct varieties of asexually reproducing plants. They have a term of 20 years from effective filing date. Plant patents include photographs of plants to identify and protect the invention. Distinctiveness is determined by characteristics such as habitat, color, flavor, odor, and so on. One could envision how plant patents could relate to natural, plant-derived antimicrobials. For example, certain cultivars might produce higher-than-normal levels of a certain antimicrobial compound.

According to 35 U.S.C. § 101–103, there are certain requirements for acceptance of a patent. The subject matter must be useful, novel, and nonobvious. Subject matter is not considered novel if the invention was known or used by others in the United States or patented or described in a printed publication more than 1 year prior to the date of application for the patent. Obviousness means that the subject matter would have been obvious to a person having ordinary skill in the art to which the subject matter pertains.

Patents contain drawings, background information of the invention (such as description of the technical field and of the prior art), a summary of the invention, brief description, detailed description of the preferred embodiment of the invention, any claims, abstract, specifications, and drawings. While preferred embodiments are often disclosed in the patent, protection by U.S. patent law is given only for what is claimed. These claims can be independent (i.e., they stand alone) or dependent on another element or qualified by an existing element. The goal of the inventor is to make claims as broadly as the prior art allows, although one must be careful not to overstate the claims to the point that jeopardizes acceptance of the patent by the examiner. Patents must specify details that are enabling, such that one skilled in the art would be able to make use of the invention without undue experimentation.

International Patent Classification (IPC) can be applied for at the World Intellectual Property Organization (http://www.wipo.int) of the United Nations. The European Patent Office (http://www.epo.org) also assigns international patents, including those of Europe, Africa, and Asia. Country-specific patent rights are also available in many European countries.

5.3.1.1 Recently patented preservative systems

Several preservative systems have been patented in a wide range of possible food or beverage applications, and some of the patents awarded by the U.S. Patent and Trade Office (USPTO) are summarized here. Many of these patented technologies are in use as of the writing of this chapter. International patents are not summarized here, although the number of international filings has increased greatly in the past few years.

One major string of patents has come from the Procter & Gamble Company, based in Cincinnati, Ohio. Scientists at the company originally wrote a patent for a preservative system for noncarbonated, acidic (pH 2.5 to 4.5) fruit juice beverages consisting of about 900 to 3000 ppm polyphosphate in combination with the weak acid preservatives sorbic acid and benzoic acid (Calderas et al. 1994). At the time, the company manufactured a fruit juice-based product that was cold filled and preserved (Anonymous 2011). The company later protected the patented technology further by expanding its scope with additional related patents (Bunger and Ekanayake 1999a, 1999b; Calderas et al. 1994). A related preservative system was commonly used in cold-filled beverages in North America and typically contained benzoic and sorbic acids (each at between 150 and 250 ppm), sodium hexameta-phosphate (500–1000 ppm), and EDTA at near the maximum allowable level. While some beverage-producing companies purchased licenses to allow their use of this preservative system, other companies did not and were subsequently sued by Procter & Gamble, ultimately leading to settlement (Anonymous 2006).

At roughly the same time, other beverage companies were developing and patenting preservative systems for tea-based beverages. These preservative systems included the use of fumaric acid in tea to specifically prevent spoilage by LAB (Cirigiano and Tiberio 1986), the use of cinnamic acid or its acidic derivatives in conjunction with weak acid preservatives (Anslow and Stratford 1998), combinations of cinnamic acid with weak acid preservatives (Blyth et al. 2001), and combinations of cinnamic acid with dimethyl dicarbonate (DMDC) and at least one essential oil (Kirby, Savage, and Stratford 2001). A patent with narrower claims was awarded for combining about 250 ppm sorbic acid and 250 ppm benzoic acid preservatives with 60 to 120 mg of ascorbic acid (vitamin C) in a beverage made from equal portions of orange and tomato juice concentrate (Johnson 1995).

A considerable quantity of patents was awarded around the use of growth-suppressive preservatives in conjunction with the microbiocidal treatment of beverage by a processing aid. A patent was awarded for DMDC plus cinnamic acid and at least one essential oil to make an ambient stable beverage (Kirby, Savage, and Stratford 2001). A separate group received patents for use of dialkyl dicarbonate (DADC) plus natamycin (also known as pimaricin) and sorbate for production of beverages, sauces, condiments, and other foods (Bunger and Ekanayake 1999a, 1999b).

Proteins and peptides also comprise some of the patented preservative systems. Certain bacteriocins have been patented for the control of pathogens like *L. monocytogenes* (Henderson and Vandenbergh 1993; King and Ming 2000). Perhaps one of the most significant patents of an antimicrobial polypeptide was awarded to Aplin and Barrett, Limited (United Kingdom) for a novel process to prepare a high-potency nisin preparation in dry/powdered form (Hawley 1955). As described previously, nisin is approved for use the world over in a variety of products for the inhibition of multiple Gram-positive spoilage and pathogenic microorganisms. The application of serine, particularly L-serine, to food for the purpose of preservation was awarded a patent (Yamada and Saito 2003). Another patent exists claiming efficacy of proteinaceous antimicrobials that include amino acid residues (Mor and Radzishevsky 2011).

Various patents were assigned to Hormel Foods Corporation for the use of lactate salts with phosphates and diacetate to preserve and improve the quality of fresh pork (Ruzek 1996, 1998a, 1998b). The patents protect the composition as well as the process of applying the composition to the meat and the packaging conditions. These patents have restricted other pork producers in the United States from using lactate and diacetate to extend shelf life of fresh pork outside a licensing agreement with Hormel.

A patent has been awarded for adding dried microorganisms to food products as biopreservation agents (Domingues and Hanlin 2004). The microorganisms, including several genera of LAB (particularly *Lactobacillus* and *Pediococcus*), are selected to survive pasteurization of the food product and then inhibit subsequent growth of pathogens (particularly *C. botulinum*) by producing acid and lowering the pH of the food system. Although the patent provides ample details on studies in food products like cheese-based pasta filling and sauce, the concept seems obvious given that these types of cultures have naturally preserved fermented foods for centuries. The use of *Propionibacterium* spp. in conjunction with sorbate or benzoate was patented for inhibiting mold growth in baked goods for shelf life extension (Poulos, Critchley, and Diaz 1999). Recently, multiple patents have been awarded for the development of a LAB-derived feed additive reported to prevent the carriage of foodborne pathogenic microorganisms, particularly *E. coli* O157:H7 (Brashears 2003; Garner and Ware 2004; Ware 2003.

Examples of plant-derived patented inventions of note include the use of isothiocyanate as a solid food product preservative (Ekanayake et al. 2004); a mixture of oregano, thyme, cinnamon, rosemary, lavender, golden seal, and olive leaf extracts (D'Amelio and Mirhom 2004); and the production of a natural preservative from taro root (Muller et al. 2002). Produce washes that combine surfactants with GRAS compounds are patented as antimicrobial no-rinse treatments (Trinh et al. 1999).

Fatty acid and fatty acid-derived antimicrobials have also received a number of patents. A composition of glycerol fatty acid ester and C6 to C18 carbon fatty acids and the method for applying to foods was patented in the early 2000s (Kabara 1994). A specific cationic molecule derived from lauric acid and arginine was patented as an antimicrobial for foods (Beltran and Bonaventura 2001; Seguer Bonaventura, Rocabayera Bonvila, and Martinez Rubio 2002). A combination of about 1–5% octanoic acid with an acidulant was patented as treatment of post-lethality exposed ready-to-eat (RTE) meats to kill *L. monocytogenes* (Herdt et al. 2006).

5.4   Methods for antimicrobial research

5.4.1   Laboratory methods

Several laboratory methods can be used to assess performance of antimicrobials against selected microorganisms *in vitro* and in foods. Microbiological procedures employed to measure efficacy often include broth dilution assays and optical density measurements and agar-based methods such as disk diffusion, agar wells, and radial colony growth for testing inhibition of molds. While there exist many standardized methods for testing antimicrobial efficacy in pharmaceuticals, cosmetics, and textiles, no such standard methods exist for food preservatives. However, these clinical and cosmetic standardized procedures can provide a good framework for methods that can be adapted to food systems. The reader may consult standardized preservative testing for cosmetics (ASTM [formerly American Society for Testing and Materials] E640-06), time-to-kill procedures (ASTM E2315), and antimicrobial effectiveness testing (USP [U.S. Pharmacopeia] 51) (The United States Pharmacopeial Convention, Inc. 1970).

The preparation of inocula for antimicrobial efficacy testing can greatly influence the outcome of results of even standard methods (Al-Hiti and Gilbert 1980; Beuchat et al. 2001). For instance, acid-adapted strains would be expected to exhibit more resistance to organic acid treatment than would nonadapted strains. Nutrient depletion, temperature of growth, and growth phase are other examples of conditions that may affect the inocula and the resulting outcome of testing. Guidance and considerations for preparation and use of inocula for antimicrobial susceptibility testing can be gleaned from the work of Gilbert, Brown, and Costerton (1987). Contact time between antimicrobial and targeted microorganisms is often one of the most significant parameters in antimicrobial efficacy testing, especially for microbiocidal treatments. Therefore, effective neutralization of the antimicrobial prior to measuring surviving cells is a key, but often overlooked, step in the process of quantifying the antimicrobial activity of a chemical preservative/antimicrobial. Dey/Engley (D/E) neutralizing buffer (Dey and Engley 1994) is the most commonly used neutralizing agent for this purpose, although its utility is limited to specific types of antimicrobials or disinfectants, including quaternary ammonium compounds (QACs), substituted phenolics, halide preparations (iodine, chlorine preparations), mercurials, formaldehye, and glutaraldehyde (Zimbro et al. 2009). Typically, when weak organic acids are applied or pH control is of interest in antimicrobial testing, various pH-buffering agents are used as neutralizers (e.g., phosphates, carbonates). In addition, other neutralizing compounds may be required for the neutralization of a particular disinfectant (e.g., use of catalase to degrade hydrogen peroxide). Trials should be conducted to verify that the antimicrobial agent is being properly neutralized prior to determination of surviving microorganisms; these trials should be conducted prior to initiation of antimicrobial screening tests. There is a USP method specifically for neutralization validation (USP 1227). Despite the availability of D/E medium, there remains a need for effective neutralization systems for the degradation of plant-derived antimicrobials (e.g., phenolics, sulfur-containing, anthocyanins, etc.) in food preservation testing.

In vitro analysis can be used to determine several things about an antimicrobial: its spectrum of antimicrobial activity (the spectrum of sensitive microorganisms), interacting treatments, and the influence of temperature or time of exposure on observed microbial inhibition. The ability of the antimicrobial substance to induce bacteriostatic, fungistatic (growth inhibited without multiple log cycles being inactivated), or microbiocidal (killing of microbial cells) effects may also be assessed. Perhaps the most important piece of data that may be gleaned from *in vitro* antimicrobial efficacy testing is an estimation of the MIC. The MIC is commonly defined as the lowest concentration of an antimicrobial applied in a test medium or food product that prevents the growth of a specific microorganism under the conditions of experimentation (López-Malo Vigil, Palou, Parish, et al. 2005). However, demonstration of the efficacy of an antimicrobial in food or beverage system is the real test. Once a proof of concept has been established *in vitro*, testing should be repeated or at least validated in food systems of interest. Refer to previously published reviews and chapters for detailed descriptions of useful methods for the *in vitro* analysis of antimicrobial activity (López-Malo Vigil, Palou, Parish, et al. 2005; Davidson and Parish 1989).

5.4.2   Data comparison and analysis

Once proper analytical methods are used to measure antimicrobial efficacy *in vitro* and in food systems, the work of analyzing data and drawing comparisons between controls and treatments begins. Many of the assays used for *in vitro* analysis of antimicrobial efficacy, especially those used in MIC identification (e.g., agar diffusion, broth dilution), are "snapshot" tests that do not require intense statistical analysis. This is because these tests require the gathering of one piece of data, that being the quantification or estimation of growth or absence of growth of a microorganism. For example, in an agar diffusion assay, an inert disk is treated with an antimicrobial, and the disk placed on a petri plate filled with solidified medium previously inoculated with the target microorganism to produce a "lawn" of growth over the plate's surface. Following incubation at a temperature allowing growth of the microorganism, the researcher removes the petri dish and inspects it for evidence of antimicrobial diffusion, observed as a zone of clear medium that indicates suppression of the microorganism's growth. The zone will be surrounded by a larger zone of microbial growth; generally, this boundary is measured by either calipers or other device to provide the diameter or radius of inhibition. Such values may then be subjected to some moderate analysis if differing levels of antimicrobial were applied or the same level of one antimicrobial was applied to multiple microbial organisms. Nevertheless, the researcher must determine at what radius or diameter the zone of inhibition will be declared as indicating the microorganism is inhibited. Likewise, for broth dilution assays, the user must inspect tubes containing inoculated microorganism and antimicrobial for evidence of microbial growth, ultimately choosing the tube with the lowest amount of applied antimicrobial that

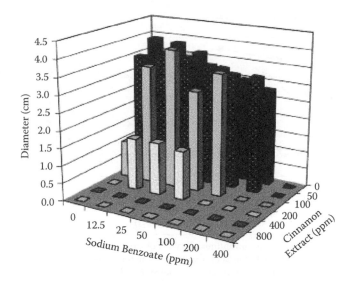

Figure 5.1 Effect of cinnamon extract and sodium benzoate individually or in combination at a_w 0.98 and pH 3.5 on *Aspergillus flavus* colony diameter after 7 days of incubation at 25°C. (From Lopez-Malo, Barreto-Valdivieso, Palou, and Martin, 2007. *Food Control* 18 (11):1358–1362. With permission from Elsevier.)

has no evidence of microbial growth as the MIC. This produces a type of binary response model, with each tube identified as inhibitory or non-inhibitory. This does not permit heavy statistical analysis of resulting data (López-Malo Vigil, Palou, Parish, et al. 2005).

Since antimicrobial efficacy data are often dependent on two or more factors, one of the simplest and best ways to gain understanding of an antimicrobial compound's utility *in vitro* or in a food is to depict results graphically in charts. An example of a bar chart is shown in Figure 5.1; radial growth of *Aspergillus* as affected by concentrations of combinations of sodium benzoate and cinnamon extract is shown.

Figure 5.1 depicts the impact of combining sodium benzoate with extract of cinnamon, indicating that application of less than 400 ppm cinnamon extract allowed growth of *Aspergillus* at all concentrations of organic acid except the highest applied concentration (400 ppm). Such graphical depictions can aid the researcher in determining optimal combinations of compatible preservatives for the inhibition of targeted microorganisms *in vitro* and can be adapted from testing in food products.

Line charts are also helpful, particularly for the quantification of antimicrobial impacts on microbial populations as determined by direct plating. Such testing is carried out routinely in antimicrobial development and validation procedures and produces data that might be used to describe the log cycle reductions that may be expected to result from application

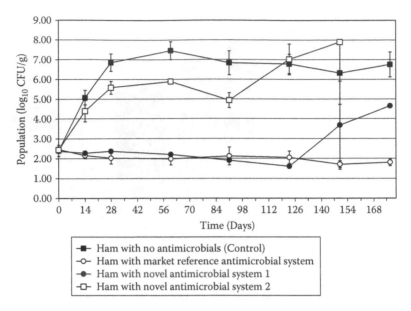

Figure 5.2 Behavior of *L. monocytogenes* on vacuum-packaged ham slices as affected by antimicrobial agents in formulas during storage at 4.4°C.

of the antimicrobial during processing. Figure 5.2 shows an example of a line chart for growth of an inoculum in food as affected by the presence of antimicrobials in the formula.

Note that positive-growth control samples (no antimicrobials applied in the formula) and market reference samples are helpful in assessing the relative performance of experimental antimicrobial systems. Such controls should always be included in antimicrobial testing for the construction of such plots and the determination of the antimicrobial's effectiveness in inhibiting/inactivating microorganisms in a food or beverage. Whether bar or line charts or other formats are used to display data, the chart should allow easy determination of the most efficacious treatments and the apparent impacts of control of relevant parameters such as time, temperature, and concentrations.

Comparison of treatment effects often requires some form of statistical analysis. Descriptive statistics can be used to generate simple comparisons between treatments. However, more advanced statistical procedures like the general linear model (GLM) or analysis of variance (ANOVA) are recommended to elucidate the most significant contributing or confounding factors, levels of significance caused by multiple treatments, and any potential interactive effects. Analysis of changes in growth or death rates over time resulting from antimicrobial application may also be necessary to compare treatments and make determination of useful antimicrobial levels. There are several models available for measuring lag phase and growth rates of microbial populations (see Chapter 8). For death rates, the use of D_{10} values that describe the concentration of applied antimicrobial that produces a 1.0 $\log_{10}$ decline in a microorganism's numbers may not be appropriate since most antimicrobial and biocidal inactivation curves do not follow first-order kinetics (i.e., the inactivation curves are nonlinear) (Hines, McKelvey, and Bodnaruk 2010).

5.4.2.1 Determining antagonistic, additive, or synergistic antimicrobial combinations

One researching in this area will quickly learn that there are few antimicrobial agents that have not already been researched. However, novel combinations of these already-known antimicrobials may be a way toward developing novel systems. Establishing the occurrence of antagonistic, additive, or synergistic effects with combinations of antimicrobials is important and synergistic effects are often imperative to establishing a novel invention. The method of establishing fractional inhibitory concentrations (FICs) (Figure 5.3) and the fractional inhibitory concentration indices (FIC$_I$'s) is useful for this purpose, as described by Davidson and Parish (1989):

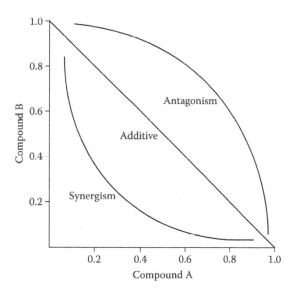

Figure 5.3 Three theoretical outcomes of fractional inhibitory concentration (FIC) testing and pictured as an FIC isobologram. (From Davidson and Parish, 1989. *Food Technology*, 43 (1):148–155. With permission from the Institute of Food Techologists.)

$$FIC_A = \frac{\text{MIC of A in presence of B}}{\text{MIC of A alone}}$$

$$FIC_B = \frac{\text{MIC of B in presence of A}}{\text{MIC of B alone}}$$

$$FIC_I = FIC_A + FIC_B$$

An example of an FIC isobologram demonstrating synergistic effects of potassium sorbate and vanillin combinations against mold is shown in Figure 5.4. The diagonal line connecting the maximal FICs for tested antimicrobials indicates an additive/indifferent interaction of antimicrobials, while FICs below the line indicate synergistic inhibition of a microorganism by combined antimicrobials. Researchers have begun to utilize more rigorous definitions of synergistic antimicrobial interactions. Brandt et al. (2010, 2011), while adhering to the definitions previously reported for interaction identification, reported the optimal inhibitory and bactericidal concentrations of antimicrobial pairings, further describing the optimal pairing of antilisterial agents *in vitro*. Zhou et al. (2007) and Pei et al. (2009) described the effect of combination (EC) term when testing combined antimicrobials against foodborne microorganisms. The EC provides a quantification of the degree to which two antimicrobials function to inhibit a foodborne microorganism synergistically, providing greater understanding of the antimicrobials' utility in food preservation.

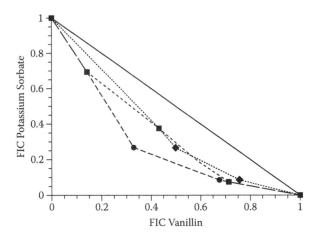

Figure 5.4 FIC isobolograms showing synergism for combinations of potassium sorbate and vanillin against *P. digitatum* (●), *P. glabrum* (■), and *P. italicum* (◆). (From Matamoros, Argaiz, and Pez-Malo, 1999. *Journal of Food Protection*, 62 (5):541–543. With permission. © The International Association for Food Protection, Des Moines, Iowa, U.S.A.)

In addition to binary combinations of antimicrobials, FICs and FIC_I's are useful for describing ternary (three) mixtures of antimicrobials or the application of two antimicrobials in a system with artificially modified pH or water activity; such data may be visualized using three-dimensional (3D) isobolograms as well (Lopez-Malo et al. 2006; Santiesteban-López, Palou, and López-Malo 2007).

5.4.3   Managing phases of research and development

Antimicrobial research leading to novel preservative systems for foods and beverages can be thought of as having three separate phases. Methodology will likely vary at each phase of the research and development process for food antimicrobials. Figure 5.5 provides a possible workflow and work output for different phases of research and development on antimicrobials for foods.

Phase 1, exploratory research, includes comprehensive literature review to gather information on the existing data on antimicrobials of

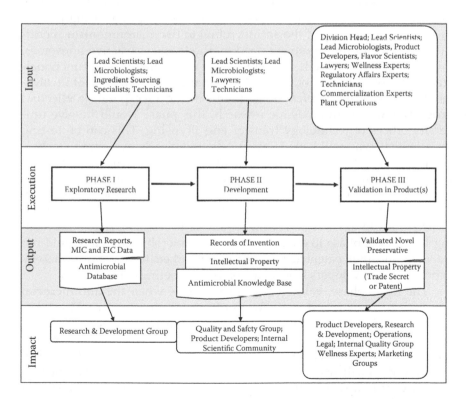

Figure 5.5 An example relational work diagram for research and development on novel antimicrobials for foods.

interest, susceptibility of targeted/relevant microorganisms, and product-specific information. This phase also includes preliminary laboratory testing such as MIC and FIC data development. Some of the *in vitro* assays previously discussed would be useful at this phase of the process. Data should be documented thoroughly both in laboratory notebook form (in case patent applications are pursued in the future) and, if the volume of data justifies, in database format for ease in searching for information. In phase 2, research transitions into development of useful inventions. It is hoped the data developed in phase 1 are sufficient to support various embodiments of the antimicrobial system (e.g., different levels, combinations, and delivery systems). If the antimicrobial system has not yet been evaluated in a food or beverage during phase 1, it is imperative that such evaluation is completed in phase 2. Efficacy of antimicrobials in microbiological media does not ensure efficacy in food and beverage systems. In fact, a promising new preservative system tested *in vitro* can often be nonefficacious when incorporated into a food or beverage or require significantly higher dosages to gain efficacy such that it becomes impractical or impossible to utilize the antimicrobial in the food or beverage of interest (e.g., aromatic essential oil components of spice-bearing plants). In this phase, delivery of the antimicrobial to the microorganism is critical, and some of the aforementioned methods (surface-active liposomes, micelles, or other methods of antimicrobial encapsulation and functionalization) become important. Phase 2 also marks the beginning of involvement of nontechnical professionals, in this instance, corporate intellectual property lawyers. In academic research, this phase would involve university offices of technology transfer and licensing. This can be testing for scientists who will have to accurately and clearly explain the technical advancements caused by the new preservative system in layperson's terms. It can also be a test for the discoveries themselves since lawyers would be keen to potential shortcomings of the work (such as nonnovelty or obviousness). Whether an invention is ultimately developed into a record of invention, a legal review would be necessary prior to proceeding to the next phase to ensure that the antimicrobial system would not infringe on other patented work or trademarked technologies, thereby exposing the developers to legal action by competing companies.

The third phase, validation in products, is when the antimicrobial system is truly morphed into a food or beverage preservative. This takes a team-based approach to antimicrobial product evaluation, requiring the expertise of food scientists, microbiologists, product developers, flavor scientists, and others expert in bringing a new technology to market. Senior leaders will also be involved since a strategy for the use of the preservative system will be needed to maximize return on the investment in research and development. Many new advancements in preservative technology die at this phase due to issues with scaling up and commercializing. From product compatibility issues and negative impacts on the product's sensorial characteristics as determined by descriptive or consumer panel testing, to processing obstacles created by the new preservatives, to ingredient sourcing problems, there are a number of problems that may be encountered that result in the death of new preservative systems. If, however, a preservative system is shown to be effective and usable, it can create many benefits, such as new product development opportunities and new distribution possibilities.

5.5 Future of antimicrobial research for foods and beverages

Opportunities abound for further innovation in the area of food and beverage antimicrobials. Such research will continue to be driven by regulations concerning foodborne pathogens. Negative economic impact of spoilage microorganisms can also justify research and development of novel antimicrobial systems. Novel antimicrobial systems can also help create new ways to process and distribute foods and beverages, thereby creating a market advantage.

Research and development in this area are also driven by market demands for fresh, minimally processed foods without synthetic chemical preservatives (Zink 1997). Negative consumer attitude toward food technology and a relatively positive attitude toward nature do influence consumers' perceived risks about foods (Klaus 2002). However, while exploration into natural antimicrobials for use in foods has been undertaken for years (Gould 1996), there have been few examples of successful implementations of natural preservative systems in the marketplace. Recently, examples in the United States include the use of fermentation products of LAB in RTE meats and other food products (e.g., nisin addition to cheese) and the use of natamycin in salad dressing and cheeses.

Methods for research will continue to improve as techniques such as integrated genomics (Figure 5.6) gain wider acceptance in applied research laboratories.

The use of computer technology to analyze data and even simulate cellular responses to multiple factors should help drive down the costs of research and development in this area as well as speed the time to develop preservative systems. Gould (1992) described an ecosystem-based approach toward food preservation. Such an approach toward food preservation that accounts for intrinsic and extrinsic properties of foods is likely to result in enhanced success in the development of new and innovative food antimicrobial technologies.

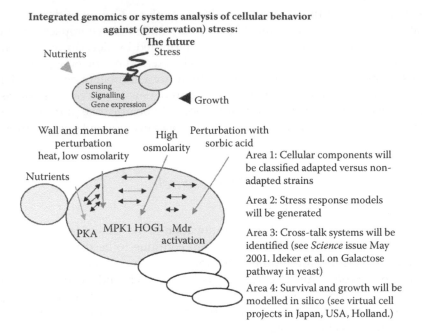

Figure 5.6 A schematic outline of cellular signal transduction and response mechanisms in yeast cells toward environmental factors. The combination of a given nutrient availability and a set of environmental (stress) conditions will determine whether a cell can restart growth. Depicted are the response against hyper and hypoosmotic stress, heat stress, and stress with the food preservative sorbic acid. A few future scenarios are indicated for the system analysis of cellular behavior against (food preservation) stresses. (From Brul et al., 2002. *International Journal of Food Microbiology*, 79 (1–2):54–64). With permission from Elsevier.)

References

Alakomi, H.-L., E. Skyttä, M. Saarela, T. Mattila-Sandholm, K. Latva-Kala, and I. M. Helander. 1999. Lactic acid permeabilizes gram-negative bacteria by disrupting the outer membrane. *Applied and Environmental Microbiology* 66:2001–2005.

Al-Hiti, M. M. A., and P. Gilbert. 1980. Changes in preservative sensitivity for the USP antimicrobial agents effectiveness test micro-organisms. *Journal of Applied Bacteriology* 49:119–126.

Aligiannis, N., E. Kalpoutzakis, S. Mitaku, and I. B. Chinou. 2001. Composition and antimicrobial activity of the essential oils of two *Origanum* species. *Journal of Agricultural and Food Chemistry* 49 (9):4168–4170.

Al-Nabulsi, A. A., and R. A. Holley. 2005. Effect of bovine lactoferrin against *Carnobacterium viridans*. *Food Microbiology* 22 (2–3):179–187.

Ananou, S., A. Baños, M. Maqueda, M. Martínez-Bueno, A. Gálvez, and E. Valdivia. 2010. Effect of combined physico-chemical treatments based on enterocin AS-48 on the control of *Listeria monocytogenes* and *Staphylococcus aureus* in a model cooked ham. *Food Control* 21:478–486.

Anderson, G. W., and M. G. Gangel. 1998. Method and apparatus for antimicrobial treatment of animal carcasses. USA Patent 5,980,375 filed 4/13/1998 and issued 11/9/1999.

Anonymous. 2006. P&G, Coca-Cola settle patent infringement lawsuits. *The Associated Press*, April 5, 2006.

Anonymous. 2011. SunnyD. Available from http://en.wikipedia.org/wiki/SunnyD; accessed September 19, 2011.

Anslow, P. A., and M. Stratford. 1998. Preservative and flavoring system. U.S. Patent 6,042,861 filed October 22, 1998, and issued March 28, 2000.

Arseneault, M., S. Bédard, M. Boulet-Audet, and M. Pézolet. 2010. Study of the interaction of lactoferricin B with phospholipid monolayers and bilayers. *Langmuir* 26 (5):3468–3478.

Aureli, P., A. Costantini, and S. Zolea. 1992. Antimicrobial activity of some plant essential oils against *Listeria monocytogenes*. *Journal of Food Protection* 55 (5):344–348.

Babu, R., M. L. Varshney, and D. S. Sog. 2004. Preservation of raw milk with lactoperoxidase system. *Journal of Food Science and Technology—Mysore* 41 (1):42–46.

Bagamboula, C. F., M. Uyttendaele, and J. Debevere. 2003. Antimicrobial effect of spices and herbs on *Shigella sonnei* and *Shigella flexneri*. *Journal of Food Protection* 66 (4):668–673

Bala, M. F. A., and D. L. Marshall. 1996a. Testing matrix, inoculum size, and incubation temperature affect monolaurin activity against *Listeria monocytogenes*. *Food Microbiology* 13 (6):467–473.

Bala, M. F. A., and D. L. Marshall. 1996b. Use of double-gradient plates to study combined effects of salt, pH, monolaurin, and temperature on *Listeria monocytogenes*. *Journal of Food Protection* 59 (6):601–607.

Banks, J. G., and R. G. Board. 1982. Sulfite-inhibition of Enterobacteriaceae including *Salmonella* in British fresh sausage and in culture systems. *Journal of Food Protection* 45 (14):1292–1297.

Barboza de Martinez, Y., K. Ferrer, and E. Marquez Salas. 2002. Combined effects of lactic acid and nisin solution in reducing levels of microbiological contamination in red meat carcasses. *Journal of Food Protection* 65 (11):1780–1783.

Bari, M. L., D. O. Ukuku, T. Kawasaki, Y. Inatsu, K. Isshiki, and S. Kawamoto. 2005. Combined efficacy of nisin and pediocin with sodium lactate, citric acid, phytic acid, and potassium sorbate and EDTA in reducing the *Listeria monocytogenes* population of inoculated fresh-cut produce. *Journal of Food Protection* 68 (7):1381–1387.

Barmpalia, I. M., I. Geornaras, K. E. Belk, J. A. Scanga, P. A. Kendall, G. C. Smith, and J. N. Sofos. 2004. Control of *Listeria monocytogenes* on frankfurters with antimicrobials in the formulation and by dipping in organic acid solutions. *Journal of Food Protection* 67 (11):2456–2464.

Beltran, J. B. U., and J. S. Bonaventura. 2001. Use of cationic preservative in food products. U.S. Patent 7,407,679 filed October 25, 2001, and issued August 5, 2008.

Benech, R.-O., E. E. Kheadr, C. Lacroix, and I. Fliss. 2002. Antibacterial activities of nisin Z encapsulated in liposomes or produced in situ by mixed culture during cheddar cheese ripening. *Applied and Environmental Microbiology* 68:5607–5619.

Beuchat, L. R. 1994. Antimicrobial properties of spices and their essential oils. In *Natural antimicrobial systems and food preservation*, ed. V. M. Dillon and R. G. Board. Wallingford, CT: CAB International, pp. 167–179.

Beuchat, L. R., J. M. Farber, E. H. Garrett, L. J. Harris, M. E. Parish, T. V. Suslow, and F. F. Busta. 2001. Standardization of a method to determine the efficacy of sanitizers in inactivating human pathogenic microorganisms on raw fruits and vegetables. *Journal of Food Protection* 64 (7):1079–1084.

Blyth, M., A. Y. Kanu, R. M. Kirby, and M. Stratford. 2001. Ambient stable beverage. U.S. Patent 6,761,919 filed May 14, 2001, and issued July 13, 2004.

Boland, J. S., P. M. Davidson, B. Bruce, and J. Weiss. 2004. Cations reduce antimicrobial efficacy of lysozyme-chelator combinations. *Journal of Food Protection* 67:285–294.

Boland, J. S., P. M. Davidson, and J. Weiss. 2003. Enhanced inhibition of *Escherichia coli* O157:H7 by lysozyme and chelators. *Journal of Food Protection* 66:1783–1789.

Boussouel, N., F. Mathieu, A.-M. Revol-Junelles, and J.-B. Millière. 2000. Effects of combinations of lactoperoxidase system and nisin on the behaviour of *Listeria monocytogenes* ATCC 15313 in skim milk. *International Journal of Food Microbiology* 61:169–175.

Brandt, A. L., A. Castillo, K. B. Harris, J. T. Keeton, M. D. Hardin, and T. M. Taylor. 2010. Inhibition of *Listeria monocytogenes* by food antimicrobials applied singly and in combination. *Journal of Food Science* 75 (9):M557–M563.

Brandt, A. L., A. Castillo, K. B. Harris, J. T. Keeton, M. D. Hardin, and T. M. Taylor. 2011. Synergistic inhibition of *Listeria monocytogenes* in vitro through the combination of octanoic acid and acidic calcium sulfate. *Journal of Food Protection* 74 (1):122–125.

Branen, J. K., and P. M. Davidson. 2004. Enhancement of nisin, lysozyme, and monolaurin antimicrobial activities by ethylenediaminetetraacetic acid and lactoferrin. *International Journal of Food Microbiology* 90:63–74.

Brashears, M. M. 2003. Lactic acid bacteria cultures that inhibit food-borne pathogens. U.S. Patent 7323166 filed June 18, 2003, and issued January 29, 2008.

Brown, C. A., B. Wang, and J.-H. Oh. 2008. Antimicrobial activity of lactoferrin against foodborne pathogenic bacteria incorporated into edible chitosan film. *Journal of Food Protection* 71 (2):319–324.

Brul, S., and P. Coote. 1999. Preservative agents in foods: mode of action and microbial resistance mechanisms. *International Journal of Food Microbiology* 50 (1–2):1–17.

Brul, S., P. Coote, S. Oomes, F. Mensonides, K. Hellingwerf, and F. Klis. 2002. Physiological actions of preservative agents: prospective of use of modern microbiological techniques in assessing microbial behaviour in food preservation. *International Journal of Food Microbiology* 79 (1–2):55–64.

Bullerman, L. B. 1983. Effects of potassium sorbate on growth and aflatoxin production by *Aspergillus parasiticus* and *Aspergillus flavus*. *Journal of Food Protection* 46:940–942.

Bullerman, L. B. 1984. Effects of potassium sorbate on growth and patulin production by *Penicillium patulum* and *Penicillium roqueforti*. *Journal of Food Protection* 47:312–315.

Bunger, J. R., and A. Ekanayake. 1999a. Antimicrobial combinations of a sorbate preservative, natamycin and a dialkyl dicarbonate useful in treating beverages and other food products and process of making. U.S. Patent 6,136,356 filed April 14, 1999, and issued October 24, 2000.

Bunger, J. R., and A. Ekanayake. 1999b. Antimicrobial composition for food and beverage products. U.S. Patent 6,376,005 filed April 14, 1999, and issued April 23, 2002.

Burt, S. 2004. Essential oils: their antibacterial properties and potential applications in foods—a review. *International Journal of Food Microbiology* 94:223–253.

Cagri, A., Z. Ustunol, and E. T. Ryser. 2004. Antimicrobial edible films and coatings. *Journal of Food Protection* 67:833–848.

Calderas, J. J., T. R. Graumlich, L. Jenkins, and R. P. Sabin. 1994. Preparation of noncarbonated beverage products with improved microbial stability. U.S. Patent 5,431,940 filed February 24, 1994, and issued July 11, 1995.

Calderas, J. J., T. R. Graumlich, L. Jenkins, and R. P. Sabin. 1997. Noncarbonated beverage products with improved microbial stability and processes for preparing. United States Patent 6,268,003 filed August 29, 1997 and issued July 31, 2001.

Cals, M.-M., P. Mailliart, G. Brignon, P. Anglade, and B. R. Dumas. 1991. Primary structure of bovine lactoperoxidase, a fourth member of a mammalian heme peroxidase family. *European Journal of Biochemistry* 198:733–739.

Canadian Department of Health. 2008. Regulations Amending the Food and Drug Regulations (1614-Food Additives).

Cavallito, C. J., and J. H. Bailey. 1944. Allicin, the antibacterial principal of *Allium sativum*. I. Isolation, physical properties and antibacterial action. *Journal of the American Chemists Society* 16:1950–1951.

Červenka, L., Z. Malíková, I. Zachová, and J. Vytřasová. 2004. The effect of acetic acid, citric acid, and trisodium citrate in combination with different levels of water activity on the growth of *Arcobacter butzleri* in culture. *Folia Microbiologica* 49 (1):8–12.

Ceylan, E., D. Y. C. Fung, and J. R. Sabah. 2004. Antimicrobial activity and synergistic effect of cinnamon with sodium benzoate or potassium sorbate in controlling *Escherichia coli* O157:H7 in apple juice. *Journal of Food Science* 69 (4):M102–M106.

Chacon, P. A., R. A. Buffo, and R. A. Holley. 2006. Inhibitory effects of microencapsulated allyl isothiocyanate (AIT) against *Escherichia coli* O157:H7 in refrigerated, nitrogen packed, finely chopped beef. *International Journal of Food Microbiology* 107 (3):231–237.

Chami, F., N. Chami, S. Bennis, T. Bouchikhi, and A. Remmal. 2005. Oregano and clove essential oils induce surface alteration of *Saccharomyces cerevisiae*. *Phytotherapy Research* 19 (5):405–408.

Chami, N., S. Bennis, F. Chami, A. Aboussekhra, and A. Remmal. 2005. Study of anticandidal activity of carvacrol and eugenol in vitro and in vivo. *Oral Microbiology and Immunology* 20 (2):106–111.

Chang, S.-S., M. Redondo-Solano, and H. Thippareddi. 2010. Inactivation of *Escherichia coli* O157:H7 and *Salmonella* spp. on alfalfa seeds by caprylic acid and monocaprylin. *International Journal of Food Microbiology* 144:141–146.

Chen, H., and D. G. Hoover. 2003. Bacteriocins and their food applications. *Comprehensive Reviews in Food Science and Food Safety* 2:82–100.

Cherrington, C. A., M. Hinton, and I. Chopra. 1990. Effect of short-chain organic acids on macromolecular synthesis in *Escherichia coli*. *Journal of Applied Bacteriology* 68:69–74.

Cherrington, C. A., M. Hinton, G. S. Mead, and I. Chopra. 1991. Organic acid: chemistry, antibacterial activity and practical applications. *Advanced Microbial Physiology* 32:87–108.

Chipley, J. R. 2005. Sodium benzoate and benzoic acid. In *Antimicrobials in foods*, ed. P. M. Davidson, J. N. Sofos, and A. L. Branen. New York: CRC Press, pp. 11–48.

Cirigiano, M. C., and J. E. Tiberio. 1986. Acid preservation systems for food products. U.S. Patent 4,756,919 filed July 14, 1986, and issued July 12, 1988.

Comes, J. E., and R. B. Beelman. 2002. Addition of fumaric acid and sodium benzoate as an alternative method to achieve a 5-log reduction of *Escherichia coli* O157:H7 populations in apple cider. *Journal of Food Protection* 65 (3):476–483.

Cutter, C. N., and G. R. Siragusa. 1994. Efficacy of organic acids against *Escherichia coli* O157:H7 attached to beef carcass tissue using a pilot scale model carcass washer. *Journal of Food Protection* 57 (2):97–103.

Dai, Y., M. D. Normand, J. Weiss, and M. Peleg. 2010. Modeling the efficacy of triplet antimicrobial combinations: yeast suppression by lauric arginate, cinnamic acid, and sodium benzoate or potassium sorbate as a case study. *Journal of Food Protection* 73 (3):515–523.

D'Amelio, F. S., Sr., and Y. W. Mirhom. 2004. Process and composition for inhibiting growth of microorganisms. U.S. Patent 7,214,392 filed February 24, 2004, and issued May 8, 2007.

Da Silva Malheiros, P., D. J. Daroit, N. Pesce da Silveira, and A. Brandelli. 2010. Effect of nanovesicle-encapsulated nisin on growth of *Listeria monocytogenes* in milk. *Food Microbiology* 27:175–178.

Davidson, P., T. Taylor, and L. Santiago. 2005. Pathogen resistance and adaptation to natural antimicrobials. In *Understanding pathogen behaviour: virulence, stress response and resistance*, ed. M. Griffiths. Cambridge, UK: Woodhead, pp. 460–483.

Davidson, P. M., and A. L. Branen. 2005. Food antimicrobials—an introduction. In *Antimicrobials in food*, ed. P. M. Davidson, J. N. Sofos, and A. L. Branen. New York: CRC Press, pp. 1–10.

Davidson, P. M., and M. A. Harrison. 2002. Resistance and adaptation to food antimicrobials, sanitizers, and other process controls. *Food Technology* 56 (11):69–78.

Davidson, P. M., and A. S. Naidu. 2000. Phyto-phenols. In *Natural food antimicrobial systems*, ed. A. S. Naidu. Boca Raton, FL: CRC Press, pp. 266–294.

Davidson, P. M., and M. E. Parish. 1989. Method for testing the efficacy of food antimicrobials. *Food Technology* 43 (1):148–155.

Davidson, P. M., J. N. Sofos, and A. L. Branen, eds. 2005. *Antimicrobials in Food.* 3rd ed. Boca Raton, FL: CRC Press.

Davidson, P. M., and T. M. Taylor. 2007. Chemical preservatives and natural antimicrobial compounds. In *Food microbiology: fundamentals and frontiers*, ed. M. P. Doyle and L. R. Beuchat. Washington, DC: ASM Press, pp. 713–745.

Degnan, A. J., and J. B. Luchansky. 1992. Influence of beef tallow and muscle on the antilisterial activity of pediocin AcH and liposome-encapsulated pediocin AcH. *Journal of Food Protection* 55:552–554.

Degnan, A. J., N. Buyong, and J. B. Luchansky. 1993. Antilisterial activity of pediocin AcH in model food systems in the presence of an emulsifier or encapsulated within liposomes. *International Journal of Food Microbiology* 18 (2):127–138.

Del Olmo, A., J. Calzada, and M. Nuñez. 2010. Antimicrobial efficacy of lactoferrin, its amidated and pepsin-digested derivatives, and their combinations, on *Escherichia coli* O157:H7 and *Serratia liquefaciens. Letters in Applied Microbiology* 52 (1):9–14.

del Río, E., B. González de Caso, M. Prieto, C. Alonso-Calleja, and R. Capita. 2008. Effect of poultry decontaminants concentration on growth kinetics for pathogenic and spoilage bacteria. *Food Microbiology* 25:888–894.

del Río, E., R. Muriente, M. Prieto, C. Alonso-Calleja, and R. Capita. 2007. Effectiveness of trisodium phosphate, acidified sodium chlorite, citric acid, and peroxyacids against pathogenic bacteria on poultry during refrigerated storage. *Journal of Food Protection* 70 (9):2063–2071.

Delves-Broughton, J. 1990. Nisin and its uses as a food preservative. *Food Technology* 44:110–117.

Delves-Broughton, J., P. Blackburn, R. J. Evans, and J. Hugenholtz. 1996. Applications of the bacteriocin, nisin. *Antonie Van Leeuwenhoek International Journal of General and Molecular Microbiology* 69:193–202.

Devlieghere, F., A. Vermeulen, and J. M. Debevere. 2004. Chitosan: antimicrobial activity, interactions with food components and applicability as a coating on fruit and vegetables. *Food Microbiology* 21:703–714.

de Vuyst, L., and F. Leroy. 2007. Bacteriocins from lactic acid bacteria: production, purification, and food applications. *Journal of Molecular Microbiology and Biotechnology* 13:194–199.

Dey, B. P., and J. F. B. Engley. 1994. Neutralization of antimicrobial chemicals by recovery media. *Journal of Microbiological Methods* 19:51–58.

Domingues, D. J., and J. H. Hanlin. 2004. Food products with biocontrol preservation and method. U.S. Patent 7,579,030 filed February 3, 2004, and issued August 25, 2009.

Doores, S. 2005. Organic acids. In *Antimicrobials in foods*, ed. P. M. Davidson, J. N. Sofos, and A. L. Branen. New York: CRC Press.

Drosinos, E. H., P. N. Skandamis, and M. Mataragas. 2009. Antimicrobials treatment. In *Safety of meat and processed meat*, ed. F. Toldrá. New York: Springer, pp. 255–298.

Duncan, C. L., and E. M. Foster. 1968. Effect of sodium nitrite, sodium chloride, and sodium nitrate on germination and outgrowth of anaerobic spores. *Applied Microbiology* 16 (2):406–411.

Ekanayake, A., J. J. Kester, J. J. Li, G. N. Zehenbauer, and P. R. Bunke. 2004. Isothiocyanate preservatives and methods of their use. U.S. Patent 7,658,961 filed August 9, 2004, and issued February 9, 2010.

Eklund, T. 1983. The antimicrobial effect of dissociated and undissociated sorbic acid at different pH levels. *Journal of Applied Bacteriology* 54:383–389.

Eklund, T. 1985a. Inhibition of microbial growth at different pH levels by benzoic and propionic acids and esters of p-hydroxybenzoic acid. *International Journal of Food Microbiology* 2:159–167.

Eklund, T. 1985b. The effect of sorbic acid and esters of p-hydroxybenzoic acid on the proton motive force in *Escherichia coli* membrane vesicles. *Journal of General Microbiology* 131:73–76.

Ennahar, S., T. Sashihara, K. Sonomoto, and A. Ishizaki. 2000. Class IIa bacteriocins: biosynthesis, structure and activity. *FEMS Microbiology Reviews* 24 (1):85–106.

Enrique, M., P. Manzanares, M. Yuste, M. Martínez, S. Vallés, and J. F. Marcos. 2009. Selectivity and antimicrobial action of bovine lactoferrin derived peptides against wine lactic acid bacteria. *Food Microbiology* 26:340–346.

Fang, T. J., and H.-C. Tsai. 2003. Growth patterns of *Escherichia coli* O157:H7 in ground beef treated with nisin, chelators, organic acids and their combinations immobilized in calcium alginate gels. *Food Microbiology* 20:243–253.

FDA. 1988. Nisin preparation: affirmation of GRAS status as a direct human food ingredient. *Federal Register* 53:11247–11251.

FDA. 1998. Direct food substances affirmed as generally recognized as safe: egg white lysozyme. *Federal Register* 63 (4):12421–12426.

Fernández-Segovia, I., I. Escriche, A. Fuentes, and J. A. Serra. 2007. Microbial and sensory changes during refrigerated storage of desalted cod (*Gadus morhua*) preserved by combined methods. *International Journal of Food Microbiology* 116:64–72.

Freese, E. 1978. Mechanism of growth inhibition by lipophilic acids. In *The pharmacological effect of lipids*, ed. J. J. Kabara. Champaign, IL: American Oil Chemists Society, pp. 321–327.

Freese, E., C. W. Sheu, and E. Galliers. 1973. Function of lipophilic acids as antimicrobial food additives. *Nature* 241:321–327.

Friedman, M., R. Buick, and C. T. Elliott. 2004. Antibacterial activities of naturally occurring compounds against antibiotic-resistant *Bacillus cereus* vegetative cells and spores, *Escherichia coli*, and *Staphylococcus aureus*. *Journal of Food Protection* 67 (8):1774–1778.

Friedman, M., P. R. Henika, and R. E. Mandrell. 2002. Bactericidal activities of plant essential oils and some of their isolated constituents against *Campylobacter jejuni, Escherichia coli, Listeria monocytogenes*, and *Salmonella enterica*. *Journal of Food Protection* 65 (10):1545–1560.

Fujita, K.-I., and I. Kubo. 2005. Naturally occurring antifungal agents against *Zygosaccharomyces bailii* and their synergism. *Journal of Agricultural and Food Chemistry* 53:5187–5191.

Garner, B. E., and D. R. Ware. 2004. Compositions and methods for inhibiting pathogenic growth. U.S. Patent 7291327 filed December 21, 2004, and issued November 6, 2007.

Gaysinsky, S., P. M. Davidson, B. D. Bruce, and J. Weiss. 2005a. Growth inhibition of *Escherichia coli* O157:H7 and *Listeria monocytogenes* by carvacrol and eugenol encapsulated in surfactant micelles. *Journal of Food Protection* 68 (12):2559–2566.

Gaysinsky, S., P. M. Davidson, B. D. Bruce, and J. Weiss. 2005b. Stability and antimicrobial efficiency of eugenol encapsulated in surfactant micelles as affected by temperature and pH. *Journal of Food Protection* 68 (7):1359–1366.

Gaysinsky, S., T. M. Taylor, P. M. Davidson, B. D. Bruce, and J. Weiss. 2007. Antimicrobial efficacy of eugenol microemulsions in milk against *Listeria monocytogenes* and *Escherichia coli* O157:H7. *Journal of Food Protection* 70 (11):2631–2637.

Geornaras, I., K. E. Belk, J. A. Scanga, P. A. Kendall, G. C. Smith, and J. N. Sofos. 2005. Postprocessing antimicrobial treatments to control *Listeria monocytogenes* in commercial vacuum-packaged bologna and ham stored at 10°C. *Journal of Food Protection* 68 (5):991–998.

Gerez, C. L., M. I. Torino, M. D. Obregozo, and G. Font de Valdez. 2010. A ready-to-use antifungal starter culture improves the shelf life of packaged bread. *Journal of Food Protection* 73 (4):758–762.

Gibson, A. M., and T. A. Roberts. 1986a. The effect of pH, water activity, sodium nitrite and storage temperature on the growth of enteropathogenic *Escherichia coli* and salmonellae in laboratory medium. *International Journal of Food Microbiology* 3:183–194.

Gibson, A. M., and T. A. Roberts. 1986b. The effect of pH, sodium chloride, sodium nitrite and storage temperature on the growth of *Clostridium perfringens* and faecal streptococci in laboratory media. *International Journal of Food Microbiology* 3:195–210.

Gilbert, P., M. R. W. Brown, and J. W. Costerton. 1987. Inocula for antimicrobial sensitivity testing: a critical review. *Journal of Antimicrobial Chemotherapy* 20:147–154.

Gill, A. O., and R. A. Holley. 2000. Inhibition of bacterial growth on ham and bologna by lysozyme, nisin and EDTA. *Food Research International* 33 (2):83–90.

Gill, A. O., and R. A. Holley. 2003. Interactive inhibition of meat spoilage and pathogenic bacteria by lysozyme, nisin and EDTA in the presence of nitrite and sodium chloride at 24°C. *International Journal of Food Microbiology* 80:251–259.

Gill, A. O., and R. A. Holley. 2006. Inhibition of membrane bound ATPases of *Escherichia coli* and *Listeria monocytogenes* by plant oil aromatics. *International Journal of Food Microbiology* 111 (2):170–174.

Glass, K., D. Preston, and J. Veesenmeyer. 2007. Inhibition of *Listeria monocytogenes* in turkey and pork-beef bologna by combinations of sorbate, benzoate, and propionate. *Journal of Food Protection* 70 (1):214–217.

Glass, K. A., L. M. McDonnell, R. C. Rassel, and K. L. Zierke. 2007. Controlling *Listeria monocytogenes* on sliced ham and turkey products using benzoate, propionate, and sorbate. *Journal of Food Protection* 70 (10):2306–2312.

González-Chávez, S. A., S. Arévalo-Gallegos, and Q. Rascón-Cruz. 2009. Lactoferrin: structure, function and applications. *International Journal of Antimicrobial Agents* 33:301.e1–301.e8.

Gould, G. W. 1964. Effect of food preservatives on the growth of bacteria from spores. In *Microbial inhibitors in food*, ed. N. Molin. Stockholm: Almqvist and Miksell, pp. 17–24.

Gould, G. W. 1989. *Mechanisms of action of food preservation procedures*. London: Elsevier Applied Sciences.

Gould, G. W. 1992. Ecosystem approaches to food preservation. *Journal of Applied Bacteriology Symposium Supplement* 73:58–68.

Gould, G. W. 1996. Industry perspectives on the use of natural antimicrobials and inhibitors for food applications. *Journal of Food Protection* supplement:82–86.

Gould, G. W. 2000. The use of other chemical preservatives: sulfite and nitrite. In *The microbiological safety and quality of food*, ed. B. M. Lund, T. C. Baird-Parker and G. W. Gould. Gaithersburg, MD: Aspen, pp. 200–213.

Gould, G. W., and N. J. Russell. 1991. Sulphite. In *Food preservatives*, ed. N. J. Russell and G. W. Gould. Glasgow, Scotland: Blackie and Son, pp. 72–88.

Gross, E., and J. L. Morell. 1967. The presence of dehydroalanine in the antibiotic nisin and its relationship to activity. *Journal of the American Chemical Society* 89 (11):2791–2792.

Gross, E., and J. L. Morell. 1971. The structure of nisin. *Journal of the American Chemical Society* 93:4634–4635.

Guan, W., and X. Fan. 2010. Combination of sodium chlorite and calcium propionate reduces enzymatic browning and microbial population of fresh-cut "Granny Smith" apples. *Journal of Food Science* 75 (2):M72–M77.

Gutierrez, J., C. Barry-Ryan, and B. Bourke. 2009. Antimicrobial activity of plant essential oils using food model media: efficacy, synergistic potential and interactions with food components. *Food Microbiology* 26:142–150.

Gutierrez, J., G. Rodriguez, C. Barry-Ryan, and P. Bourke. 2008. Efficacy of plant essential oils against foodborne pathogens and spoilage bacteria associated with ready-to-eat vegetables: antimicrobial and sensory screening. *Journal of Food Protection* 71 (9):1846–1854.

Hao, Y. Y., R. E. Brackett, and M. P. Doyle. 1998a. Efficacy of plant extracts in inhibiting *Aeromonas hydrophila* and *Listeria monocytogenes* in refrigerated, cooked poultry. *Food Microbiology* 15:367–378.

Hao, Y. Y., R. E. Brackett, and M. P. Doyle. 1998b. Inhibition of *Listeria monocytogenes* and *Aeromonas hydrophila* by plant extracts in refrigerated cooked beef. *Journal of Food Protection* 61 (3):307–312.

Harris, K., M. F. Miller, G. H. Loneragan, and M. M. Brashears. 2006. Validation of the use of organic acids and acidified sodium chlorite to reduce *Escherichia coli* O157 and *Salmonella* Typhimurium in beef trim and ground beef in a simulated processing environment. *Journal of Food Protection* 69 (8):1802–1807.

Hawley, H. B. 1955. A dry nisin preparation. U.K. Patent GB738655. Filed May 4, 1953 and issued March 12, 1957.

Henderson, J. T., and P. A. Vandenbergh. 1993. Synthetically derived peptide. U.S. Patent 5,861,376 filed March 15, 1993, and issued January 19, 1999.

Herdt, J. G., J. H. Chopskie, S. Burnett, L., T. C. Podtburg, and T. A. Gutzmann. 2006. Antimicrobial compositions for use on food products. U.S. Patent 7,915,207 filed July 21, 2006, and issued March 29, 2011.

Hines, J. D., P. J. McKelvey, and P. W. Bodnaruk. 2010. Inappropriate use of D-values for determining biocidal activity of various antimicrobials. *Journal of Food Science* 76 (1):M8–M11.

Hughes, A. H., and J. C. McDermott. 1989. The effect of phosphate, sodium chloride, sodium nitrite, storage temperature and pH on the growth of enteropathogenic *Escherichia coli* in a laboratory medium. *International Journal of Food Microbiology* 9 (3):215–223.

Isshiki, K., K. Tokuora, R. Mori, and S. Chiba. 1992. Preliminary examination of allyl isothiocyanate vapor for food preservation. *Bioscience, Biotechnology, and Biochemistry* 56:1476–1477.

Johnson, E. A., and A. E. Larson. 2005. Lysozyme. In *Antimicrobials in foods*, ed. P. M. Davidson, J. N. Sofos, and A. L. Branen. New York: CRC Press, pp. 361–388.

Johnson, E. L. 1995. Fruit and vegetable juice beverage and process of making. U.S. Patent 6,099,889 filed February 2, 1995, and issued August 8, 2000.

Johny, A. K., M. J. Darre, A. M. Donoghue, D. J. Donoghue, and K. Venkitanarayana. 2010. Antibacterial effect of *trans*-cinnamaldehyde, eugenol, carvacrol, and thymol on *Salmonella* Enteriditis and *Campylobacter jejuni* in chicken cecal contents in vitro. *Journal of Applied Poultry Research* 19:237–244.

Jones, E. M., A. Smart, G. Bloomberg, L. Burgess, and M. R. Millar. 1994. Lactoferricin, a new antimicrobial peptide. *Journal of Applied Bacteriology* 77:208–214.

Juneja, V. K., and B. S. Eblen. 1999. Predictive thermal inactivation model for *Listeria monocytogenes* with temperature, pH, NaCl, and sodium pyrophosphate as controlling factors. *Journal of Food Protection* 62 (9):986–993.

Juneja, V. K., H. M. Marks, and T. Mohr. 2003. Predictive thermal inactivation model for effects of temperature, sodium lactate, NaCl, and sodium pyrophosphate on *Salmonella* serotypes in ground beef. *Applied and Environmental Microbiology* 69 (9):5138–5156.

Juneja, V. K., B. S. Marmer, J. G. Phillips, and S. A. Palumbo. 1996. Interactive effects of temperature, initial pH, sodium chloride, and sodium pyrophosphate on the growth kinetics of *Clostridium perfringens*. *Journal of Food Protection* 59 (9):963–968.

Jung, D.-S., F. W. Bodyfelt, and M. A. Daeschel. 1992. Influence of fat and emulsifiers on the efficacy of nisin in inhibiting *Listeria monocytogenes* in fluid milk. *Journal of Dairy Science* 75:387–393.

Jung, Y. J., K. J. Min, and K. S. Yoon. 2009. Responses of acid-stressed *Salmonella* Typhimurium in broth and chicken patties to subsequent antimicrobial stress with ε-polylysine and combined potassium lactate and sodium diacetate. *Food Microbiology* 26:467–474.

Kabara, J. J. 1994. Antimicrobial preservative compositions and methods. U.S. Patent 6,638,978 filed January 26, 1994, and issued October 28, 2003.

Karapinar, M., and S. A. Gonul. 1992. Removal of *Yersinia enterocolitica* from fresh parsley by washing with acetic acid or vinegar. *International Journal of Food Microbiology* 16:261–264.

Kilinc, B., S. Cakli, T. Dincer, and S. Tolasa. 2009. Microbiological, chemical, sensory, color, and textural changes of rainbow trout fillets treated with sodium acetate, sodium lactate, sodium citrate, and stored at 4°C. *Journal of Aquatic Food Product Technology* 18:3–17.

Kim, C. R., J. O. Hearnsberger, A. P. Vickery, C. H. White, and D. L. Marshall. 1995. Extending shelf life of refrigerated catfish fillets using sodium acetate and monopotassium phosphate. *Journal of Food Protection* 58 (6):644–647.

Kim, S. Y., Y. M. Lee, S. Y. Lee, Y. S. Lee, J. H. Kim, C. Ahn, B. C. Kang, and G. E. Ji. 2001. Synergistic effect of citric acid and pediocin K1, a bacteriocin produced by *Pediococcus* sp K1, on inhibition of *Listeria monocytogenes*. *Journal of Microbiology and Biotechnology* 11 (5):831–837.

Kim, Y., M. Kim, and K. B. Song. 2009a. Combined treatment of fumaric acid with aqueous chlorine dioxide or UV-C irradiation to inactivate *Escherichia coli* O157:H7, *Salmonella enterica*, serovar Typhimurium, and *Listeria monocytogenes* inoculated on alfalfa and clover sprouts. *LWT Food Science and Technology* 42:1654–1658.

Kim, Y. J., M. H. Kim, and K. B. Song. 2009b. Efficacy of aqueous chlorine dioxide and fumaric acid for inactivating pre-existing microorganisms and *Escherichia coli* O157:H7, *Salmonella typhimurium*, and *Listeria monocytogenes* on broccoli sprouts. *Food Control* 20 (11):1002–1005.

Kim, Y. J., B. A. Nahm, and I. H. Choi. 2010. An evaluation of the antioxidant and antimicrobial effectiveness of different forms of garlic and BHA in emulsion-type sausages during refrigerated storage. *Journal of Muscle Foods* 21 (4):813–825.

King, W., and X. Ming. 2000. Antibacterial composition for control of gram positive bacteria in food applications. U.S. Patent 6,620,446 filed July 14, 2000, and issued September 17, 2002.

Kirby, R. M., D. Savage, and M. Stratford. 2001. Ambient stable beverage and process of making. U.S. Patent 6,562,387 filed May 14, 2001, and issued May 13, 2003.

Klaenhammer, T. R. 1988. Bacteriocins of lactic-acid bacteria. *Biochimie* 70:337–349.

Klaus G, G. 2002. Current issues in the understanding of consumer food choice. *Trends in Food Science and Technology* 13 (8):275–285.

Ko, K. Y., A. F. Mendonca, and D. U. Ahn. 2008. Effect of ethylenediaminetetraacetate and lysozyme on the antimicrobial activity of ovotransferrin against *Listeria monocytogenes*. *Poultry Science* 87:1649–1658.

Ko, K. Y., A. F. Mendonca, H. Ismail, and D. U. Ahn. 2009. Ethylenediaminetetraacetate and lysozyme improves antimicrobial activities against ovotransferrin against *Escherichia coli* O157:H7. *Poultry Science* 88:406–414.

Kong, M., X. G. Chen, K. Xing, and H. J. Park. 2010. Antimicrobial properties of chitosan and mode of action: a state of the art review. *International Journal of Food Microbiology* 144:51–63.

Kussendrager, K. D., and A. C. M. Van Hooijdonk. 2000. Lactoperoxidase: physicochemical properties, occurrence, mechanism of action and applications. *British Journal of Nutrition* 84 (Suppl. 1):S19–S25.

Laury, A. M., M. V. Alvarado, G. Nace, C. Z. Alvarado, J. C. Brooks, A. Echeverry, and M. M. Brashears. 2009. Validation of a lactic acid- and citric acid-based antimicrobial product for the reduction of *Escherichia coli* O157:H7 and *Salmonella* on beef tips and whole chicken carcasses. *Journal of Food Protection* 72 (10):2208–2211.

Leistner, L. 2000. Basic aspects of food preservation by hurdle technology. *International Journal of Food Microbiology* 55 (1–3):181–186.

Leistner, L., and L. G. M. Gorris. 1995. Food preservation by hurdle technology. *Trends in Food Science and Technology* 6 (2):41–46.

Lihono, M. A., A. F. Mendonca, J. S. Dickson, and P. M. Dixon. 2003. A predictive model to determine the effects of temperature, sodium pyrophosphate, and sodium chloride on thermal inactivation of starved *Listeria monocytogenes* in pork slurry. *Journal of Food Protection* 66 (7):1216–1221.

Lopez-Malo, A., J. Barreto-Valdivieso, E. Palou, and F. S. Martín. 2007. *Aspergillus flavus* growth response to cinnamon extract and sodium benzoate mixtures. *Food Control* 18 (11):1358–1362.

Lopez-Malo, A., E. Palou, R. Leon-Cruz, and S. M. Alzamora. 2006. Mixtures of natural and synthetic antifungal agents. In *Advances in food mycology*, ed. A. D. Hocking, J. I. Pitt, Robert A. Samson, and U. Thrane. New York: Springer, pp. 261–286.

López-Malo Vigil, A., E. Palou, and S. M. Alzamora. 2005. Naturally occurring compounds—plant sources. In *Antimicrobials in foods*, ed. P. M. Davidson, J. N. Sofos, and A. L. Branen. New York: Taylor and Francis, pp. 429–452.

López-Malo Vigil, A., E. Palou, M. E. Parish, and P. M. Davidson. 2005. Methods for activity assay and evaluation of results. In *Antimicrobials in food*, ed. P. M. Davidson, J. N. Sofos, and A. L. Branen. New York: Taylor and Francis, pp. 659–680.

Maks, N., L. Zhu, V. K. Juneja, and S. Ravishankar. 2010. Sodium lactate, sodium diacetate and pediocin: effects and interactions on the thermal inactivation of *Listeria monocytogenes* on bologna. *Food Microbiology* 27:64–69.

Mangalassary, S., I. Han, J. Rieck, J. Acton, and P. Dawson. 2008. Effect of combining nisin and/or lysozyme with in-package pasteurization for control of *Listeria monocytogenes* in ready-to-eat turkey bologna during refrigerated storage. *Food Microbiology* 25:866–870.

Mann, C. M., S. D. Cox, and J. L. Markham. 2000. The outer membrane of *Pseudomonas aeruginosa* NCTC 6749 contributes to its tolerance to the essential oil of *Melaleuca alternifolia* (tea tree oil). *Letters in Applied Microbiology* 30:294–297.

Mansour, M., and J.-B. Milliere. 2001. An inhibitory synergistic effect of a nisin-monolaurin combination on *Bacillus* sp. vegetative cells in milk. *Food Microbiology* 18:87–94.

Mari, M., R. Iori, O. Leoni, and A. Marchi. 1993. In vitro activity of glucosinolate derived isothiocyanates against postharvest fruit pathogens. *Annals of Applied Biology* 123:155–164.

Mastromatteo, M., A. Lucera, M. Sinigaglia, and M. R. Corbo. 2010. Use of lysozyme, nisin, and EDTA combined treatments for maintaining quality of packed ostrich patties. *Journal of Food Science* 75 (3):M178–M186.

Matamoros, L., B., A. Argaiz, and A. Pez-Malo. 1999. Individual and combined effects of vanillin and potassium sorbate on *Penicillium digitatum, Penicillium glabrum,* and *Penicillium italicum* growth. *Journal of Food Protection* 62 (5):541–543.

McClements, D. J., E. A. Decker, Y. Park, and J. Weiss. 2009. Structural design principles for delivery of bioactive components in nutraceuticals and functional foods. *Critical Reviews in Food Science and Nutrition* 49 (6):577–606.

Miller, A. J., J. E. Call, and R. C. Whiting. 1993. Comparison of organic acid salts for *Clostridium botulinum* control in an uncured turkey product. *Journal of Food Protection* 56 (11):958–962.

Mor, A., and I. Radzishevsky. 2011. Antimicrobial agents. U.S. Patent 7,915,223 filed August 8, 2006 and issued March 29, 2011.

Mountney, G. J., and J. O'Malley. 1965. Acids as poultry meat preservatives. *Poultry Science* 44:582–586.

Muller, W. S., A. L. Allen, A. Sikes, and A. Senecal. 2002. Method for making a food preservative and for preserving food. U.S. Patent 6,753,024 filed March 20, 2002, and issued June 22, 2004.

Murdock, C. A., J. Cleveland, K. R. Matthews, and M. L. Chikindas. 2007. The synergistic effect of nisin and lactoferrin on the inhibition of *Listeria monocytogenes* and *Escherichia coli* O157:H7. *Letters in Applied Microbiology* 44 (3):255–261.

Muthukumarasamy, P., J. H. Han, and R. A. Holley. 2003. Bactericidal effects of *Lactobacillus reuteri* and allyl isothiocyanate on *Escherichia coli* O157:H7 in refrigerated ground beef. *Journal of Food Protection* 66 (11):2038–2044.

Mytle, N., G. L. Anderson, M. P. Doyle, and M. A. Smith. 2006. Antimicrobial activity of clove (*Syzgium aromaticum*) oil in inhibiting *Listeria monocytogenes* on chicken frankfurters. *Food Control* 17 (2):102–107.

Nakai, S. A., and K. J. Siebert. 2004. Organic acid inhibition models for *Listeria innocua, Listeria ivanovii, Pseudomonas aeruginosa,* and *Oenococcus oeni. Food Microbiology* 21:67–72.

Neetoo, H., M. Ye, and H. Chen. 2010. Bioactive alginate coatings to control *Listeria monocytogenes* on cold-smoked salmon slices and fillets. *International Journal of Food Microbiology* 136:326–331.

Nobmann, P., A. Smith, J. Dunne, G. Henehan, and P. Bourke. 2009. The antimicrobial efficacy and structure activity relationship of novel carbohydrate fatty acid derivatives against *Listeria* spp. and food spoilage microorganisms. *International Journal of Food Microbiology* 128:440–445.

Nuñez de Gonzalez, M. T., J. T. Keeton, G. R. Acuff, L. J. Ringer, and L. M. Lucia. 2004. Effectiveness of acidic calcium sulfate with propionic and lactic acid and lactates as postprocessing dipping solutions to control *Listeria monocytogenes* on frankfurters with or without potassium lactate and stored vacuum packaged at 4.5°C. *Journal of Food Protection* 67 (5):915–921.

Obaidat, M. M., and J. F. Frank. 2009a. Inactivation of *Escherichia coli* O157:H7 on the intact and damaged portions of lettuce and spinach leaves by using allyl isothiocyanate, carvacrol, and cinnamaldehyde in vapor phase. *Journal of Food Protection* 72 (10):2046–2055.

Obaidat, M. M., and J. F. Frank. 2009b. Inactivation of *Salmonella* and *Escherichia coli* O157:H7 on sliced and whole tomatoes by allyl isothiocyanate, carvacrol, and cinnamaldehyde in vapor phase. *Journal of Food Protection* 72 (2):315–324.

O'Bryan, C. A., P. G. Crandall, and S. C. Ricke. 2008. Organic poultry pathogen control from farm to fork. *Foodborne Pathogens and Disease* 5 (6):709–720.

Orsi, N. 2004. The antimicrobial activity of lactoferrin: current status and perspectives. *BioMetals* 17:189–196.

Ough, C. S., and L. Were. 2005. Sulfur dioxide and sulfites. In *Antimicrobials in foods*, ed. P. M. Davidson, J. N. Sofos, and A. L. Branen. Boca Raton, FL: CRC Press, pp. 143–168.

Over, K. F., N. Hettiarachchy, M. G. Johnson, and B. Davis. 2009. Effect of organic acids and plant extracts on *Escherichia coli* O157:H7, *Listeria monocytogenes*, and *Salmonella* Typhimurium in broth culture model and chicken meat systems. *Journal of Food Science* 74 (9):M515–M521.

Pandey, A., F. Bringel, and J.-M. Meyer. 1994. Iron requirement and search for siderophores in lactic acid bacteria. *Applied Microbiology and Biotechnology* 40:735–739.

Papadopoulos, L. S., R. K. Miller, G. R. Acuff, C. Vanderzant, and H. R. Cross. 1991. Effect of sodium lactate on microbial and chemical composition of cooked beef during storage. *Journal of Food Science* 56:341–347.

Pei, R.-S., F. Zhou, B.-P. Ji, and J. Xu. 2009. Evaluation of combined antibacterial effects of eugenol, cinnamaldehyde, thymol, and carvacrol against *E. coli* with an improved method. *Journal of Food Science* 74 (7):M379–M383.

Porto-Fett, A. C. S., S. G. Campano, J. L. Smith, A. Oser, B. Shoyer, J. E. Call, and J. B. Luchansky. 2010. Control of *Listeria monocytogenes* on commercially-produced frankfurters prepared with and without potassium lactate and sodium diacetate and surface treated with lauric arginate using the Sprayed Lethality in Container (SLIC®) delivery method. *Meat Science* 85:312–318.

Poulos, P. G., J. Critchley, and R. E. Diaz, Jr. 1999. Long-term mold inhibition in intermediate moisture food products stored at room temperature. U.S. Patent 6,132,786 filed March 17, 1999, and issued October 17, 2000.

Pranato, Y., S. K. Rakshit, and V. M. Salokhe. 2005. Enhancing antimicrobial activity of chitosan films by incorporating garlic oil, potassium sorbate and nisin. *Food Science and Technology* 38:859–865.

Raybaudi-Massilia, R. M., J. Mosqueda-Megar, and O. Martín-Belloso. 2008. Edible alginate-based coating as carrier of antimicrobials to improve shelf-life and safety of fresh-cut melon. *International Journal of Food Microbiology* 121 (3):313–327.

Ricke, S. C. 2003. Perspectives on the use of organic acids and short chain fatty acids as antimicrobials. *Poultry Science* 82:632–639.

Roberts, T. A., and M. Ingram. 1966. The effect of sodium chloride, potassium nitrate and sodium nitrite on the recovery of heated bacterial spores. *Journal of Food Technology* 1:147–163.

Roller, S., ed. 2003. *Natural antimicrobials for the minimal processing of foods.* Cambridge, UK: Woodhead.

Russell, J. B. 1991. Resistance of *Streptococcus bovis* to acetic acid at low pH: relationship between intracellular pH and anion accumulation. *Applied and Environmental Microbiology* 57 (1):255–259.

Russell, J. B. 1992. Another explanation for the toxicity of fermentation acids at low pH: anion accumulation versus uncoupling. *Journal of Applied Bacteriology* 73:363–370.

Ruzek, D. C. 1996. Chemical treatment and packaging process to improve the appearance and shelf life of fresh pork. U.S. Patent 5,780,085 filed October 4, 1996, and issued July 14, 1998.

Ruzek, D. C. 1998a. Chemical treatment and packaging process to improve the appearance and shelf life of fresh meat. U.S. Patent 5,989,610 filed July 13, 1998, and issued November 23, 1999.

Ruzek, D. C. 1998b. Chemical treatment and packaging system to improve the appearance and shelf life of fresh pork. U.S. Patent 5,976,593 filed April 17, 1998, and issued November 16, 1999.

Sallam, K. I., M. Ishioroshi, and K. Samejima. 2004. Antioxidant and antimicrobial effects of garlic in chicken sausage. *Lebensmittel-Wissenschaft Und-Technologie-Food Science and Technology* 37:849–855.

Santiesteban-López, A., E. Palou, and A. López-Malo. 2007. Susceptibility of food-borne bacteria to binary combinations of antimicrobials at selected a_w and pH. *Journal of Applied Microbiology* 102:486–497.

Schirmer, B. C., and S. Langsrud. 2010. Evaluation of natural antimicrobials on typical meat spoilage bacteria in vitro and in vacuum-packed pork meat. *Journal of Food Science* 75 (2):M98–M102.

Schmidt, S. E., G. Holub, J. M. Sturino, and T. M. Taylor. 2009. Suppression of *Listeria monocytogenes* Scott A in fluid milk by free and liposome-entrapped nisin. *Probiotics and Antimicrobial Proteins* 1 (2):152–158.

Seguer Bonaventura, J., X. Rocabayera Bonvila, and M. A. Martinez Rubio. 2002. Preservatives and protective systems. U.S. Patent 7,662,417 filed May 8, 2002, and issued February 16, 2010.

Shelef, L. A. 1994. Antimicrobial effects of lactates: a review. *Journal of Food Protection* 57 (5):445–450.

Sheu, C. W., and E. Freese. 1972. Effects of fatty acids on growth and envelope proteins of *Bacillus subtilis. Journal of Bacteriology* 111:516–524.

Sheu, C. W., W. N. Konings, and E. Freese. 1972. Effects of acetate and other short-chain fatty acids on sugars and amino acid uptake of *Bacillus subtilis. Journal of Bacteriology* 111:525–530.

Sheu, C. W., D. Salomon, J. L. Simmons, T. Sreevalsan, and E. Freese. 1975. Inhibitory effects of lipophilic acids and related compounds on bacteria and mammalian cells. *Antimicrobial Agents and Chemotherapy* 7:349–363.

Singh, A., N. R. Korasapati, V. K. Juneja, and H. Thippareddi. 2010. Effect of phosphate and meat (pork) types on the germination and outgrowth of *Clostridium perfringens* spores during abusive chilling. *Journal of Food Protection* 73 (5):879–887.

Singh, B., M. B. Falahee, and M. R. Adams. 2001. Synergistic inhibition of *Listeria monocytogenes* by nisin and garlic extract. *Food Microbiology* 18 (2):133–139.

Siragusa, G. R., and J. S. Dickson. 1993. Inhibition of *Listeria monocytogenes, Salmonella typhimurium* and *Escherichia coli* O157:H7 on beef muscle tissue by lactic or acetic acid contained in calcium alginate gels. *Journal of Food Safety* 13 (2):147–158.

Stopforth, J. D., P. N. Skandamis, P. M. Davidson, and J. N. Sofos. 2005. Naturally occuring compounds—animal sources. In *Antimicrobials in foods*, ed. P. M. Davidson, J. N. Sofos, and A. L. Branen. Boca Raton, FL: Taylor and Francis, pp. 453–505.

Stopforth, J. D., J. N. Sofos, and F. F. Busta. 2005. Sorbic acid and sorbates. In *Antimicrobials in foods*, ed. P. M. Davidson, J. N. Sofos, and A. L. Branen. Boca Raton, FL: Taylor and Francis, pp. 49–90.

Stopforth, J. D., D. Visser, R. Zumbrink, L. Van Dijk, and E. W. Bontebal. 2010. Control of *Listeria monocytogenes* on cooked cured ham by formulation with a lactate-diacetate blend and surface treatment with lauric arginate. *Journal of Food Protection* 73 (3):552–555.

Stratford, M., and P. A. Anslow. 1998. Evidence that sorbic acid does not inhibit yeast as a classic "weak acid preservative." *Letters in Applied Microbiology* 27:203–206.

Suarez, V. B., L. Frison, M. Z. De Basilico, M. Rivera, and J. A. Reinheimer. 2005. Inhibitory activity of phosphates on molds isolated from foods and food processing plants. *Journal of Food Protection* 68 (11):2475–2479.

Sudarshan, N. R., D. G. Hoover, and D. Knorr. 1992. Antibacterial action of chitosan. *Food Biotechnology* 6 (3):257–272.

Taormina, P. J., and W. J. Dorsa. 2009a. Inactivation of *Listeria monocytogenes* on hams shortly after vacuum packaging by spray application of lauric arginate. *Journal of Food Protection* 72 (12):2517–2523.

———. 2009b. Short-term bactericidal efficacy of lauric arginate against *Listeria monocytogenes* present on the surface of frankfurters. *Journal of Food Protection* 72 (6):1216–1224.

Tassou, C. C., K. Lambropoulou, and G.-J. E. Nychas. 2004. Effect of prestorage treatments and storage conditions on the survival of *Salmonella* Eneriditis PT4 and *Listeria monocytogenes* on fresh marine and freshwater aquaculture fish. *Journal of Food Protection* 67 (1):193–198.

Taylor, T. M., B. D. Bruce, J. Weiss, and P. M. Davidson. 2008. *Listeria monocytogenes* and *Escherichia coli* O157:H7 inhibition in vitro by liposome-encapsulated nisin and ethylene diaminetetraacetic acid. *Journal of Food Safety* 28 (2):183–197.

Taylor, T. M., P. M. Davidson, B. D. Bruce, and J. Weiss. 2005. Liposomal nanocapsules in food science and agriculture. *Critical Reviews in Food Science and Nutrition* 45 (7–8):587–605.

Teraguchi, S., K. Shin, T. Ogata, M. Kingaku, A. Kaino, H. Miyauchi, Y. Fukuwatari, and S. Shimamura. 1995a. Orally administered bovine lactoferrin inhibits bacterial translocation in mice fed bovine milk. *Applied and Environmental Microbiology* 61:4131–4134.

Teraguchi, S., K. Shin, K. Ozawa, S. Nakamura, Y. Fukuwatari, S. Tsuyuki, H. Namihira, and S. Shimamura. 1995b. Bacteriostatic effect of orally administered bovine lactoferrin on proliferation of *Clostridium* species in the gut of mice fed bovine milk. *Applied and Environmental Microbiology* 61:501–506.

Theron, M. M., and J. F. R. Lues. 2007. Organic acids and meat preservation: a review. *Food Reviews International* 23:141–158.

The United States Pharmacopeial Convention, Inc. 1970. *The United States Pharmacopeia, 18th Revision*. Rockville, MD.

Tokarskyy, O., and D. L. Marshall. 2008. Mechanism of synergistic inhibition of *Listeria monocytogenes* growth by lactic acid, monolaurin, and nisin. *Applied and Environmental Microbiology* 74 (23):7126–7129.

Toldrá, F., ed. 2009. *Safety of meat and processed meat*. New York: Springer.

Tompkin, R. B. 2005. Nitrite. In *Antimicrobials in foods*, ed. P. M. Davidson, J. N. Sofos, and A. L. Branen. Boca Raton, FL: Taylor and Francis, pp. 169–236.

Tong, C.-H., and F. A. Draughon. 1985. Inhibition by antimicrobial food additives of ochratoxin A production by *Aspergillus sulphureus* and *Penicillium viridicatum*. *Journal of Food Protection* 49 (6):1407–1411.

Tranter, H. S. 1994. Lysozyme, ovotransferrin and avidin. In *Natural antimicrobial systems and food preservation*, ed. V. M. Dillon and R. G. Board. Wallingford, CT: CAB International, pp. 65–97.

Trinh, T., B. J. Roselle, A. H. Chung, P. A. Geis, T. E. Ward, and D. K. Rollins. 1999. Microorganism reduction methods and compositions for food cleaning. U.S. Patent 6,455,086 filed December 16, 1999, and issued September 24, 2002.

Tsai, S., and C. Chou. 1996. Injury, inhibition and inactivation of *Escherichia coli* O157:H7 by potassium sorbate and sodium nitrite as affected by pH and temperature. *Journal of the Science of Food and Agriculture* 71:10 12.

Ukuku, D. O., and W. F. Fett. 2004. Effect of nisin in combination with EDTA, sodium lactate, and potassium sorbate for reducing *Salmonella* on whole and fresh-cut cantaloupe. *Journal of Food Protection* 67:2143–2150.

Ultee, A., M. H. J. Bennik, and R. Moezelaar. 2002. The phenolic hydroxyl group of carvacrol is essential for action against the food-borne pathogen *Bacillus cereus*. *Applied and Environmental Microbiology* 68:1561–1568.

Ultee, A., L. G. M. Gorris, and E. J. Smid. 1998. Bactericidal activity of carvacrol towards the food-borne pathogen *Bacillus cereus*. *Journal of Applied Microbiology* 85:211–218.

Ultee, A., E. P. W. Kets, M. Alberda, F. A. Hoekstra, and E. J. Smid. 2000. Adaptation of the food borne pathogen *Bacillus cereus* to carvacrol. *Archives in Microbiology*. 174:233–238.

Ultee, A., E. P. W. Kets, and E. J. Smid. 1999. Mechanisms of action of carvacrol on the food borne pathogen *Bacillus cereus*. *Applied and Environmental Microbiology* 65:4606–4610.

Ultee, A., R. A. Slump, G. Steging, and E. J. Smid. 2000. Antimicrobial activity of carvacrol toward *Bacillus cereus* on rice. *Journal of Food Protection* 63:620–624.

Ultee, A., and E. J. Smid. 2001. Influence of carvacrol on growth and toxin production by *Bacillus cereus*. *International Journal of Food Microbiology* 64:373–378.

U.S. Department of Agriculture Food Safety and Inspection Service (USDA-FSIS). 2011. Safe and suitable ingredients used in the production of meat and poultry products. Directive 7120.1, Rev. 7. Available from www.fsis.usda.gov/OPPDE/rdad/FSISDirectives/7120.1Rev2.pdf; accessed June 27, 2011.

U.S. Patent and Trademark Office. 2010. *Basic facts about trademarks*. Alexandria, VA: U.S. Patent and Trademark Office.

Vasavada, M., C. E. Carpenter, and D. P. Cornforth. 2003. Sodium levulinate and sodium lactate effects on microbial growth and stability of fresh pork and turkey sausages. *Journal of Muscle Foods* 14:119–129.

Venema, K., G. Venema, and J. Kok. 1995. Lactococcal bacteriocins: mode of action and immunity. *Trends in Microbiology* 3 (8):299–304.

Venkitanarayanan, K. S., T. Zhao, and M. P. Doyle. 1999. Antibacterial effect of lactoferricin B on *Escherichia coli* O157:H7 in ground beef. *Journal of Food Protection* 62:747–750.

Walker, M., and C. A. Phillips. 2008. The effect of preservatives on *Alicyclobacillus acidoterrestris* and *Propionibacterium cyclohexanicum* in fruit juice. *Food Control* 19:974–981.

Ward, S. M., P. J. Delaquis, R. A. Holley, and G. Mazza. 1998. Inhibition of spoilage and pathogenic bacteria on agar and pre-cooked roast beef by volatile horseradish distillates. *Food Research International* 31:19–26.

Ware, D. R. 2003. Compositions and methods for reducing the pathogen content of meat and meat products. U.S. Patent 7,291,326 filed December 31, 2003, and issued November 6, 2007.

Weaver, R., and L. A. Shelef. 1993. Antilisterial activity of sodium, potassium or calcium lactate in pork liver sausage. *Journal of Food Safety* 13:133–146.

Were, L. M., B. Bruce, P. M. Davidson, and J. Weiss. 2004. Encapsulation of nisin and lysozyme in liposomes enhances efficiency against *Listeria monocytogenes*. *Journal of Food Protection* 67:922–927.

Wilkins, K. M., and R. G. Board. 1989. Natural antimicrobial systems. In *Mechanisms of action of food preservation procedures*, ed. G. W. Gould. New York: Elsevier Applied Science, pp. 285–362.

Wills, E. D. 1956. Enzyme inhibition by allicin, the active principal of garlic. *Biochemical Journal* 63:514–520.

Yamada, K., and K. Saito. 2003. Process of preserving food and food preservative. U.S. Patent 6,602,532 filed March 2, 2001 and issued August 5, 2003.

Yousef, A. E., M. A. El-Shenawy, and E. H. Marth. 1989. Inactivation and injury of *Listeria monocytogenes* in a minimal medium as affected by benzoic acid and incubation temperature. *Journal of Food Science* 54 (3):650–652.

Yuste, J., and D. Y. C. Fung. 2004. Inactivation of *Salmonella* Typhimurium and *Escherichia coli* O157:H7 in apple juice by a combination of nisin and cinnamon. *Journal of Food Protection* 67:371–377.

Zaika, L. L., and J. S. Fanelli. 2003. Growth kinetics and cell morphology of *Listeria monocytogenes* Scott A as affected by temperature, NaCl, and EDTA. *Journal of Food Protection* 66 (7):1208–1215.

Zhang, H., Y. Shen, Y. Bao, Y. He, F. Feng, and X. Zheng. 2008. Characterization and synergistic antimicrobial activities of food-grade dilution-stable microemulsions against *Bacillus subtilis*. *Food Research International* 41:495–499.

Zhang, H., Y. Shen, P. Weng, G. Zhao, F. Feng, and X. Zheng. 2009. Antimicrobial activity of a food-grade fully dilutable microemulsion against *Escherichia coli* and *Staphylococcus aureus*. *International Journal of Food Microbiology* 135:211–215.

Zhao, T., P. Zhao, and M. P. Doyle. 2009. Inactivation of *Salmonella* and *Escherichia coli* O157:H7 on lettuce and poultry skin by combinations of levulinic acid and sodium dodecyl sulfate. *Journal of Food Protection* 72 (5):928–936.

Zhao, T., P. Zhao, and M. P. Doyle. 2010. Inactivation of *Escherichia coli* O157:H7 and *Salmonella* Typhimurium DT104 on alfalfa seeds by levulinic acid and sodium dodecyl sulfate. *Journal of Food Protection* 73 (11):2010–2017.

Zhou, F., B. Ji, H. Zhang, H. Jiang, Z. Yang, J. Li, J. Li, Y. Ren, and W. Yan. 2007. Synergistic effect of thymol and carvacrol combined with chelators and organic acids against *Salmonella* Typhimurium. *Journal of Food Protection* 70 (7):1704–1709.

Zimbro, M. J., D. A. Power, S. M. Miller, G. E. Wilson, and J. A. Johnson, eds. 2009. *Difco and BBL manual: manual of microbiological culture media.* 2nd ed. Sparks, MD: Becton, Dickinson and Company.

Zink, D. L. 1997. The impact of consumer demands and trends on food processing. *Emerging Infectious Diseases* 3 (4):467–469.

chapter six

Competitive research and development on food-processing sanitizers and biocides

Junzhong Li and Scott L. Burnett

Contents

6.1 Introduction

Research exploring the best application practices and limitations of sanitizers for food-processing equipment began in earnest during the 1940s. The demand for milk and dairy products with extended shelf life boomed, partially as a result of World War II, which drove the advancement of understanding in sanitation programs and sanitizer technologies. Development of new chemistries and improvements to existing products was a focus of research laboratories in the chemical supply industry as food sanitarians conducted research to understand better how these products are best applied. Research expanded knowledge regarding the impact of sanitizers on product color and taste, the contribution of sanitizers to equipment corrosion, the enhancement of methods used to evaluate efficacy of sanitizers, the personnel and environmental safety of sanitizer chemicals, and best application practices in a plant's sanitation program.

The field of research investigating sanitation practices and application gave shape to the prophetic attributes of an ideal sanitizer. It has been toward these aspirational properties that researchers in the field of chemical product development have devoted their resources. These attributes include (Marriot and Gravani 2006)

- Broad antimicrobial spectrum (bacteria, viruses, yeasts, molds) and rapid kill
- Nontoxic/low toxicity (safe for consumers) and readily biodegradable in the environment
- No adverse effects on food
- Environmental tolerance (efficacy not impacted by water conditions, soil load, temperature, detergents)
- Good material compatibility (noncorrosive to the materials applied)
- Stable under concentrate and use-solution application conditions
- No odor or acceptable odor
- Ease of measurement and monitoring
- Inexpensive

The ideal sanitizer is yet to be discovered. Today's commercial chemical sanitizers can be classified into two categories: oxidative and nonoxidative chemistries. Products within both classifications have advantages and disadvantages as outlined in Tables 6.1 and 6.2. Nonetheless, each of these commercially available and utilized sanitizers has undergone extensive research and development to overcome hurdles to bring improved products to market. Innovation was wrought by addressing the disadvantages and limitations identified in the performance of the known sanitizers. This chapter illustrates the research and development involved in the major classes of chemical sanitizer products employed by the food-processing industry today as a part of an overall sanitation program.

6.2   Nonoxidative chemistry

6.2.1   Quaternary ammonium compounds

Quaternary ammonium compounds (quats) are synthetic surface active agents, which are poor detergents but excellent germicides. Their chemical structure can be represented as shown in Figure 6.1, where R_1, R_2, R_3, and R_4 are alkyls, substituted alkyls, aromatics, or substituted aromatics, respectively; X^- are anions, such as chloride, bicarbonate, carbonate, and so on. Not all quats are good disinfectants. Biocidal activity is strongly related to the balance of the hydrophilic and lipophilic moieties of the compound (i.e., the alkyl or aromatic structures attached to the nitrogen). As surface-active agents, quats are generally high foaming and have only

Table 6.1 Attribute Comparison of Common Nonoxidative-Based Sanitizers

Property	Quat	Acid anionic	Fatty acid
Toxicity	Low	Low	Non
Efficacy	Very active to Gram-positive bacteria and yeast; low active to Gram-negative bacteria, molds, and virus	Very active to Gram-negative bacteria; low active to Gram-positive bacteria; not active to virus and molds	Effective against Gram-positive and -negative bacteria; less effective against yeasts and molds
Environmental tolerance	Low sensitivity to pH and organic soil load; some are impacted by water hardness	Less active above pH 3; no sensitivity to organic material	Less active above pH 4; not sensitive to organic material
Material compatibility	Noncorrosive	Noncorrosive	Corrosive to plastics and some rubbers
Odor	Non	Non	Low
Stability	Stable	Stable	Stable
Cost	High	Moderate	Moderate

slight wetting and detergency capability at typical sanitizer use-solution concentration (150–250 ppm).

The mechanism of bactericidal action of quats is generally related to adsorption by the negatively charged bacteria cell to the positively charged quat molecule. Following adsorption, quat penetrates the cell membrane, disrupting its function. As a result, K^+ and other cytoplasmic constituents are released, resulting in the death of the cell.

The advantages of quats as a sanitizer include (1) they are stable under a broad range of storage and distribution conditions, providing a comparably long shelf life; (2) they have a low odor profile and contribute minimally to flavor carryover when utilized at sanitizer use-solution concentrations; (3) they are noncorrosive to common hard surface materials and nonirritating to skin; (4) they exhibit bacteriostatic film properties on surfaces after treatment; (5) they are effective over a wide pH range, although most effective in slightly alkaline solutions.

The limitations of quats include that (1) they require a comparably high concentration of active material for bactericidal or bacteriostatic action; (2) they are more efficacious against Gram-positive bacteria than Gram-negative bacteria; (3) their efficacy can be severely impacted by water hardness; (4) they are not compatible with anionic surfactants, which are common ingredients in detergents; and (5) they degrade slowly to known nonactive or nontoxic species in the environment.

Table 6.2 Attribute Comparison of Common Oxidative-Based Sanitizers

Sanitizers	Chlorine	Chlorine dioxide	Iodophors	Peroxycarboxylic acid
Toxicity	High	High	Moderate	None
Efficacy	Effective against bacteria, viruses, and yeasts; less effective against molds	More effective and rapid kill than chlorine against all micro-organisms	Effective against bacteria, viruses, and yeasts; less effective against molds	Effective against bacteria and viruses; less effective against molds
Environmental tolerance	Sensitive to organic soils; no activity at pH greater than 9; practical pH condition is 6.0 to 7.5	Not sensitive to organic soils; less sensitive to pH; high pH levels enhance the activity	Less sensitive to organic soils than chlorine; active under acidic pH (less than 5)	Not sensitive to organic soils and pH conditions; more active under acidic pH
Material compatibility	Corrosive to metals and elastomers	Corrosive to metals and elastomers	Corrosive to some metals; stain plastics	Noncorrosive to metals and normal surface applications
Odor	Strong	Strong	Low	Strong
Stability	Poor	Poor	Stable below 110°F	Stable under equilibrium conditions
Cost	Low	Moderate	High	Moderate

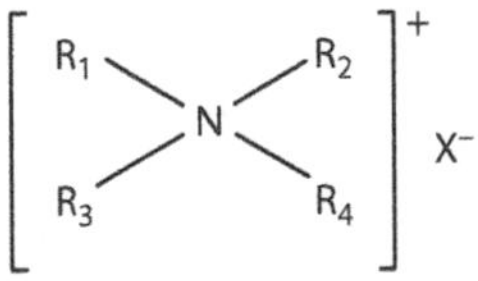

Figure 6.1 General chemical structure of quaternary ammonium compounds.

While some of the inherent limitations to the early discovered quat molecules have proven difficult to overcome, research and development activities from academia and the chemical supply industry have improved their performance since the most notable development of quats as antimicrobials by Domagk (1935). Since then, several generations of structurally variable quat-based antimicrobials of commercial importance have been developed (Merianos, 2001).

The first-generation quat is alkyldimethylbenzyl ammonium chloride, and the general structure is shown in Figure 6.2.

It was established that the long alky chain is the key to achieve antimicrobial activity; when the alkyl chain is 14, the molecule has the best biocidal efficacy (Hansch and Fujuta, 1964). These quats could be easily synthesized from readily available raw materials and were demonstrated to have biocidal activity but were identified with minimal tolerance to hard water conditions.

Modification to the aromatic ring of the first-generation quats led to the development of the second-generation quats (Figure 6.3). Substitution of the hydrogen atoms of the aromatic ring with methyl-, ethyl-, or chlorine resulted in improved biocidal efficacy. A commercial example of this group of quats is BTC 471 from the Stepan Company.

The third-generation quats, the so-called dual quats, are a mixture of equal portions of alkyldimethylbenzyl ammonium chloride and

$$R = C_{12}-C_{18}$$

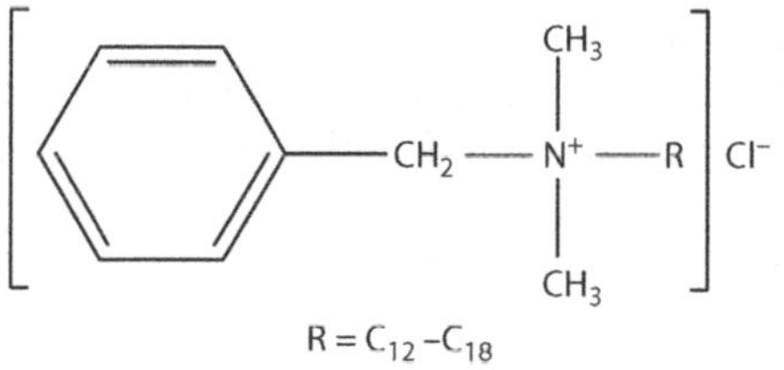

Figure 6.2 General structure of first generation quaternary ammonium compounds.

$$R = C_{12}-C_{18}$$
$$X = CH_3, C_2H_5, Cl$$

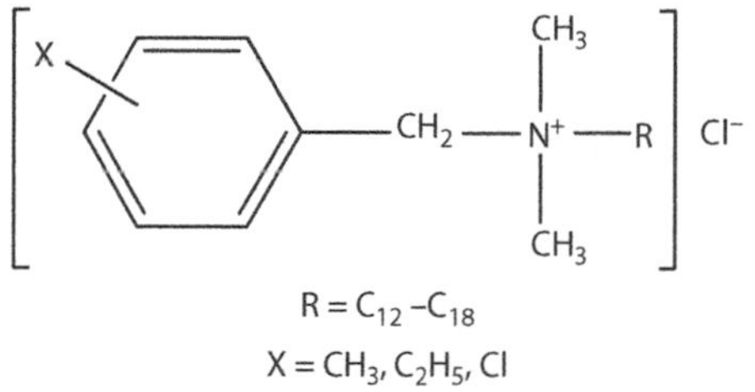

Figure 6.3 General structure of second generation quaternary ammonium compounds.

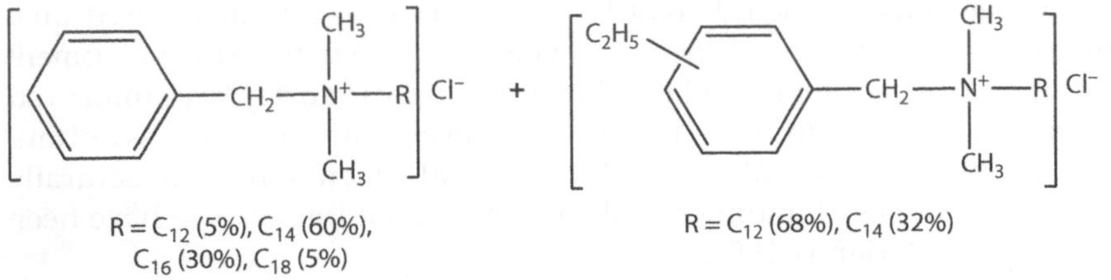
CH3
CH2
N+
R
Cl−
CH3
+
C2H5
CH2
N+
R
Cl−
CH3
CH3
R = C12 (5%), C14 (60%),
C16 (30%), C18 (5%)
R = C12 (68%), C14 (32%)

Figure 6.4 Structure of the third generation quaternary ammonium compounds (BTC 2125M by Stepan Co.).

alkyldimethylethylbenzyl ammonium chloride. By combining these two quats at equal ratios, significant synergistic biocidal performance was achieved; in addition, the toxicity of the product was dramatically decreased. The typical example of a third-generation quat is BTC 2125M (Figure 6.4), developed in 1955 by the Stepan Company; it has achieved by far the greatest commercial success of quat-based sanitizer products.

In the 1960s, the development of a chemical process for long-chain alcohol amination made it commercially feasible to manufacture dialkylmethyl amines, which could be subsequently quaternized with methyl chloride to generate the dialkydimethyl ammonium chloride (Ditoro, 1969). This new class of quats, the so-called twin-chain quats, represents the fourth-generation quat products. They are structurally quite different from previously developed and commercialized quats (Figure 6.5).

The dialkyldimethyl ammonium quats have significant overall performance improvements compared with the aromatic quats of previous commercial products. Twin-chain quats displayed not only outstanding biocidal properties but also unusual hard water and anionic surfactant tolerance. In addition, these quats demonstrated lower foaming characteristics, which make them more suitable for use where the mechanic action is involved in application (Petrocci et al., 1974).

Applying the concept of synergistic combination of the third-generation quats by mixing dialkyldimethyl ammonium chloride with

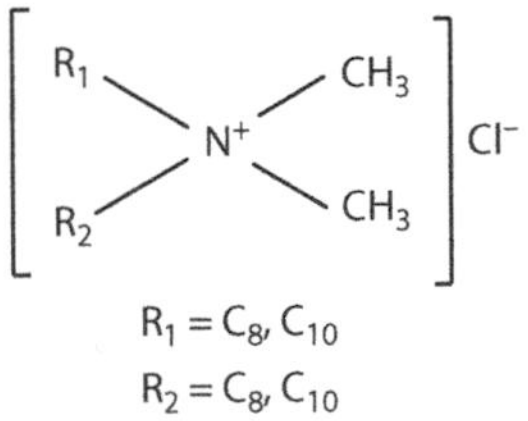
R1
CH3
N+
Cl−
R2
CH3
R1 = C8, C10
R2 = C8, C10

Figure 6.5 General structure of fourth generation quaternary ammonium compounds.

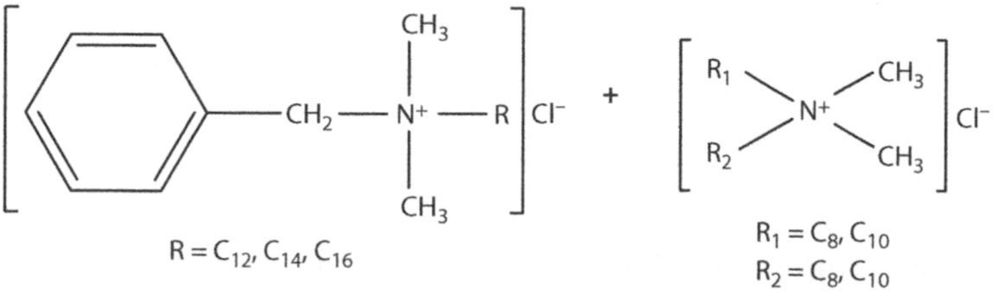

Figure 6.6 General structure of fifth generation quaternary ammonium compounds.

alkyldimethylbenzyl ammonium chloride resulted in the fifth-generation quats (Figure 6.6) (Schaeufele, 1984). The blend generated synergistic bio-cidal efficacy, remained active under the most hostile conditions, and was less toxic and more cost effective.

Subsequent research and development activities on quats were mainly focused on decreasing their toxicity profile rather than improving their biocidal efficacy owing to environmental safety concerns, impact on wastewater treatment, and related regulatory restrictions beginning in the 1980s. This research and development effort led to the invention of polymeric quats (Stark, 1983). Polymeric quats are less environmentally toxic than small-molecule quats but are also less powerful than the third-generation quats in terms of biocidal activity. The polymeric quats have been used as active ingredients in formulations for environmental disinfectants, but not as sanitizers in the food industry.

The counteranion of quats is almost always chloride, owing to its abundant availability and the low cost of the raw material in manufacturing quats. For most applications, the level of chloride in the use-solution, including sanitizer use-solutions employed in the food industry, will not contribute to equipment corrosion. However, for some specific applications, such as wood treatment, the presence of chloride in quats causes corrosion issues (i.e., degradation of nails in wood). This promulgated the development of a quat with carbonate or bicarbonate as the counteranion, the so-called Carboquat® by the Lonza Group, Limited. The availability of a commercial quat with an alternative to chloride as the counteranion makes it possible to blend quats with peroxygen compounds, such as hydrogen peroxide and peracetic acid. It has been demonstrated that the mixing of quats with both hydrogen peroxide and peracetic acid results in significant synergistic biocidal efficacy, especially against yeast and mold (Hilgren et al., 2000).

More recently, research related to germini quats is often seen in the scientific literature (Shukla and Tyagi 2006). Although research and development efforts are primarily aimed at improving detergency attributes over

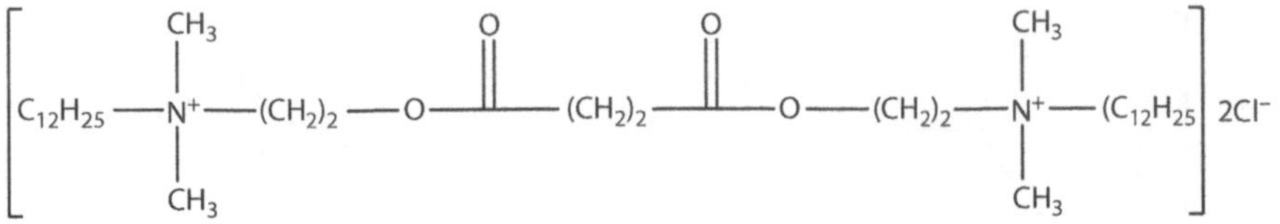

$$\left[C_{12}H_{25}\!-\!\!\overset{\overset{\displaystyle CH_3}{|}}{\underset{\underset{\displaystyle CH_3}{|}}{N^+}}\!\!-\!(CH_2)_2\!-\!O\!\!\overset{\overset{\displaystyle O}{\|}}{}\!\!(CH_2)_2\!\!\overset{\overset{\displaystyle O}{\|}}{}\!\!-\!O\!-\!(CH_2)_2\!-\!\!\overset{\overset{\displaystyle CH_3}{|}}{\underset{\underset{\displaystyle CH_3}{|}}{N^+}}\!\!-\!(C_{12}H_{25} \right] 2Cl^-$$

Figure 6.7 Structure of Germini quaternary ammonium compound.

traditional quats, germini quats are also reported to have biocidal activity. These quats have distinct structures with dual hydrophilic quaternized ammonium heads and a hydrophobic chain in the middle (Figure 6.7). Their unique structure results in exceptional application properties, such as very low critical micelle concentration (CMC), outstanding wetting capability, and improved biocidal activity compared with the traditional quats.

6.2.2   Acid anionic surfactants

Certain anionic surfactants demonstrate antimicrobial activity under acidic conditions and have been employed as equipment sanitizers in food processing. Advantages of this class of sanitizer include (1) low odor; (2) stability throughout distribution and storage; (3) broad antimicrobial spectrum; (4) tolerance to a broad range of application temperatures; and (5) good detergency and wetting properties. Acid anionic surfactants are microbially active only in an acidic milieu, typically less than pH 3. It is believed that under acidic pH, the net charge of the bacterial cell envelope changes from negative to positive; thus, the anionic surfactants adsorb to the cell membrane in a similar manner as cationic surfactants, such as quats, do at neutral pH.

The biocidal activity of anionic surfactants under certain conditions has long been realized (Cowles, 1938; Birkeland and Steinhaus, 1939) and is strongly dependent on the nature of the hydrophobic and hydrophilic groups composing the structure of the surfactant molecule. The hydrophilicity-lipophilicity balance of the compound plays the key role in determining its biocidal activity. The anionic surfactants found in most applications as sanitizers are linear alkylbenzene sulfonate (LAS) and sodium dodecyl sulfate (SDS). These are typical surfactants, and they have high foam characteristics. For applications in food-processing equipment sanitizing that involve mechanical operation, such as clean in place (CIP), the high degree of foam limits their use. Therefore, foam suppressants, such as fatty alcohol and polyvalent salt (e.g., aluminum sulfate), were subsequently added to the acid anionic sanitizing formulation to combat the foam (Carandang and Dychdala, 1974). The effort that has achieved the most successful CIP application of acid anionic surfactant is the development of a low-foam anionic surfactant, that is, sulfonated oleic acid. Sulfonated oleic acid has slightly lower antimicrobial activity compared with LAS but has a much lower propensity to foam, which makes it more suitable for CIP applications (Sedliar et al., 1968).

Acid anionic surfactants are generally less effective against Gram-negative than Gram-positive bacteria, and the activity is strongly pH dependent. Research efforts aimed at enhancing the antimicrobial efficacy and decreasing the pH dependence of anionic surfactants led to the discovery that certain cations, such as La^{3+}, could enhance the biocidal efficacy of anionic surfactants (i.e., LAS), and the efficacy was observed even under neutral pH conditions (Berger and Zimmerer, 1987). It was demonstrated that the presence of such cations assists the adsorption of the anionic surfactant molecule to the cell envelope, thereby enhancing the biocidal activity of the surfactants.

6.2.3   Fatty carboxylic acids

The early discovery of antimicrobial activity of fatty acids coincided with that of their corresponding polyol derivatives, such as fatty acid glycerides (Adams, 1931; Harris, 1941a, 1941b). As a result, there was confusion regarding whether the esters or their parent fatty acids are active. Later, Kabara (1975) tried to clarify the antimicrobial profiles of fatty acids and their corresponding esters by performing microbial efficacy studies of various fatty acids and their esters, concluding that the activity of fatty acids is unpredictable, thus discouraging attempts at using these materials as antimicrobial agents. As a result, Kabara believed the esters are more useful antimicrobial agents and claimed the esters in his patent.

The work of Wang led to the applications of fatty acids as CIP sanitizers for food-processing equipment (Wang, 1981). The important clarification from Wang was that the fatty acid in the protonated neutral form is responsible for biocidal activity. This discovery also lent clarification to earlier controversial and confounding results, as previous investigations were not conducted under the same pH conditions, making comparisons difficult. Other discoveries were key contributors to the commercial success of fatty acid-based sanitizer formulations. For example, it was found that nonionic surfactants tend to diminish the biocidal efficacy of fatty acids, while anionic surfactants could enhance their efficacy. In addition, it was found that by mixing short-chain fatty acids (C_6 to C_9) with longer-chain fatty acids (C_{10} to C_{14}) the resulting product holds particularly enhanced antimicrobial activity.

As sanitizers, fatty acids share similar advantages as acid anionic surfactants. In addition, fatty acids are practically nontoxic as they naturally occur in various food sources. For applications in CIP sanitizing, fatty acids are nonfoaming under typical application conditions and have a wider pH tolerance range (pH less than 5) than acid anionic surfactants.

6.3   Oxidative chemistry

6.3.1   Chlorine and related compounds

Chlorine is among the earliest used sanitizers in food-processing sanitation. In 1945, Vaughn and Stadtman (1945) recognized that the use of chlorinated sprays at selected points in food-processing lines resulted in lower bacterial counts in the finished product and reduced buildup of bacteria slimes. By the 1950s and 1960s, in-plant chlorination was already standard practice (Troller, 1983).

In general, chlorine has excellent bactericidal activity against a broad range of microorganisms and is generated economically. There are three major species in the chlorine solution: chlorine (Cl_2), hypochlorous acid (HOCl), and hypochlorite ion (OCl^-). The three forms of chlorine exist together in equilibrium (Figure 6.8). The relative proportions of each are determined by pH and temperature. Figure 6.9, generated with the equilibrium constants of K_1 and K_2 at 25°C (Asano, 2007), shows the effect of pH value on the form of HOCl in water.

Among the three species, HOCl is responsible for chlorine's antimicrobial properties; neither Cl_2 nor OCl^- is particularly active against microorganisms (Mercer and Somers, 1957). Thus, the pH of the chlorine solution should be kept between 4 and 5, where virtually all the chlorine is in the form of HOCl, to give the maximum antimicrobial performance. In

$$Cl_2 + H_2O \rightleftharpoons HOCl + H^+ + Cl^- \quad K_1 = 4.5 \times 10^{-4} \text{ (25°C)}$$

$$HOCl \rightleftharpoons H^+ + OCl^- \quad\quad\quad\quad K_2 = 3.0 \times 10^{-8} \text{ (25°C)}$$

Figure 6.8 Equilibrium reactions in chlorine solutions.

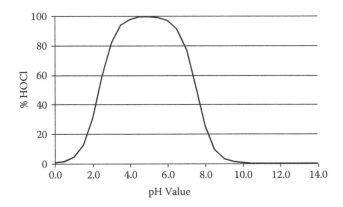

Figure 6.9 The effect of pH on the hypochlorus acid concentration within a chlorine solution.

practice, however, the pH of chlorine sanitizer use-solution is adjusted to the range of pH 6 to pH 7.5, where corrosion and irritancy are minimal but the percentage of HOCl is still significant.

Chlorine-based products are commercially available in two forms: organic- and inorganic-based compounds. The inorganic chlorine compounds, such as sodium and calcium hypochlorite, release hypochlorite ion into water, the ultimate concentration of which depends on the pH of the solution. Some hypochlorite ion will be protonated and exist as the antimicrobially active hypochlorous acid. The most commonly used and least-expensive inorganic chlorine compound is sodium hypochlorite, which is commercially available as an aqueous alkaline solution to maintain stability, while other inorganic chlorine compounds such as calcium and magnesium hypochlorite are available as solid powder. The organic chlorine compounds were later developed to release HOCl directly into water by hydrolysis. The biggest advantage of organic chlorine compounds is that HOCl can be directly generated in a controlled manner under neutral pH (7), thereby eliminating the related corrosion and irritation characteristics of alkaline solutions. Figure 6.10 shows the structure of trichloroisocyanuric acid, one of the commonly used organic chlorine compounds.

The dangerous nature of chlorine products in regard to transportation and handling promoted research and development efforts toward its on-site generation from early-stage NaOCl generators, to the relatively recent development of so-called electrolyzed (EO) water (Huang et al., 2008). The NaOCl generators produce NaOCl from NaCl solutions by electrolysis, utilizing a direct current applied to electrodes immersed in salt solution. Initially developed in Japan (Shimizu and Hurusawa, 1992), the fundamental difference of EO water generators versus NaOCl generators is the presence of a membrane between the anode and cathode in the electrolytic cell in the latter, which allows the separation of the electrolytic products in the anode and the cathode regions, as expressed in Figure 6.11 (Huang et al., 2008).

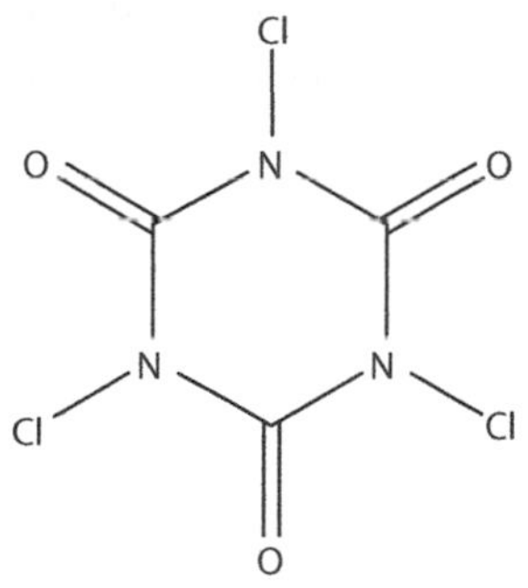

Figure 6.10 Structure of trichloroisocyanuric acid.

Reaction in anode: $2H_2O \longrightarrow 4H^+ + O_2 + 4e^-$

$2NaCl \longrightarrow Cl_2 + 2Na^+ + 2e^-$

$Cl_2 + H_2O \longrightarrow HCl + HClO$

Reaction in cathode: $2H_2O + 2e^- \longrightarrow 2OH^- + H_2$

$2NaCl + 2OH^- \longrightarrow 2NaOH + 2Cl^-$

Figure 6.11 Products in the anode and the cathode regions of an electrolytic chlorine generator.

The product from the anode, which contains HOCl and free chlorine Cl_2 with low pH and high oxidation-reduction potential, is the so-called EO water. The pH of the EO water could be set and controlled by the generator to produce a maximum percentage of HOCl for disinfection.

The biggest disadvantage of the various chlorine-based sanitizers, regardless of whether they are inorganic or organic based, including EO water, is incompatibility with organic substances. Chlorines rapidly lose antimicrobial activity in the presence of organic matter, which results in formation of potentially carcinogenic compounds. This drawback promoted research and development activities with the aim of commercializing alternative chlorine-based antimicrobials with the biocidal advantages of chlorine, but with improved organic substance tolerance. These efforts resulted in the applications of chlorine dioxide (ClO_2) and acidified sodium chlorite (ASC) in food-processing sanitation.

Chlorine dioxide was discovered in the early 1800s, and its first use as an antimicrobial agent was in water treatment (Simpson, 2005). Chlorine dioxide is a gas under ambient conditions with limited water solubility and is highly explosive if compressed as a liquid. As a result, ClO_2 is generated and used at the point of application as a dilute aqueous solution for essentially all applications. Depending on the scale of the application, ClO_2 can be generated by the reduction of chlorate ion (ClO_3^-) or the oxidation of chlorite ion (ClO_2^-). For applications related to disinfection, ClO_2 is generally produced on site by the acid activation of chlorite (Figure 6.12).

The major advantages of ClO_2 over chlorine are that it retains its antimicrobial activity in the presence of organic matter and does not form chlorinated substances. Moreover, its biocidal efficacy is less pH dependent, its action is more rapid than chlorine, and it is significantly more virucidal than chlorine. The improved efficacy of ClO_2 relative to chlorine

$$4ClO_2^- + 2H^+ \longrightarrow 2ClO_2 + Cl^- + ClO_3^- + H_2O$$

Figure 6.12 Acid activation of chlorite to form chlorine dioxide.

is ascribed to better cell penetration as ClO_2 is less polar, less reactive, and uncharged. ClO_2 is cleared by the U.S. Food and Drug Administration (FDA) as a disinfectant for use in poultry chilling water systems and has been investigated as a microbial control agent in various food-processing applications (Mir, 1984; Drechsler et al., 1990). However, for most food applications, the use of chlorine continues to be preferred because it is less expensive than chlorine dioxide. Chlorine dioxide is mainly used when organic material is a concern.

In addition to higher cost over chlorine, another important disadvantage of ClO_2 is off-gassing at all use-solution concentrations. Worker exposure limits established by the U.S. Occupational Safety and Health Administration (OSHA) and ACGH (American Conference of Governmental Hygienists) are 0.1 mg/L ClO_2 for 8-h TWA (time-weighted average) and 0.3 mg/L ClO_2 for short-term exposure limit (STEL) (National Institute for Occupational Safety and Health, 1994). These low concentrations indicate there is a fair concern regarding potential worker exposure to ClO_2 during treatment. The off-gassing issue of ClO_2 promoted the development of ASC to be used as the alternative disinfectant (Davidson and Kross, 1990).

ASC chemistry is principally that of chlorous acid ($HClO_2$), the metastable oxychlorine species, which forms on acidification of chlorite. Once formed, chlorous acid slowly decomposes to form chlorate ion, chlorine dioxide, and chloride ion (Figure 6.13).

Under certain conditions, ASC can generate small quantities of chlorine dioxide. However, by the careful selection of reaction parameters, such as chlorite concentration, nature and concentration of acid employed, and the total acidity (pH 2.5 to 3.2) targeted, chlorine dioxide is typically minimized in true ASC solutions (Davidson and Kross, 1990). ASC is a highly effective, broad-spectrum antimicrobial that has been approved by the USFDA as a "secondary direct food additive permitted in food for human consumption" specifically as an antimicrobial intervention treatment for poultry carcasses, poultry carcass parts, red meat carcasses, red meat parts and organs, seafood, and raw agricultural commodities.

The mechanism of microbial action of ASC is not clearly understood. Like other oxidative chemistries such as chlorine, it was thought that the attack on microbial proteins by chlorination of the amino groups and oxidation of the thio (–SH) groups is responsible for the biocidal efficacy.

$$H^+ + ClO_2^- \rightleftharpoons HClO_2$$

$$4HClO_2 \longrightarrow 2ClO_2 + HCl + HClO_3 + H_2O$$

Figure 6.13 Equilibrium and decomposition reactions associated with acidified sodium chlorite.

The uncharged chlorous acid is able to penetrate bacterial cell membranes and disrupt protein synthesis by virtue of its reaction with thio-containing amino acids and nucleotides. In addition, the undissociated acid is thought to facilitate proton leakage into cells and thereby increase energy output of the cells to maintain their normal internal pH, adversely affecting amino acid transport.

Iodine, like chlorine, is bactericidal against a wide spectrum of bacteria, viruses, and protozoa, including bacterial and fungal spores (Gottardi, 2001). Elemental iodine, a nonmetallic purple solid, is only slightly soluble in water and generates various species in aqueous solution: I^-, I_2, I_3^-, I_5^-, I_6^{2-}, HOI, OI$^-$, HI$_2$O$^-$, I$_2$O^{2-}, H$_2$OI$^+$, and IO$_3^-$ (Clough and Starke, 1985). Among all of these species, only free iodine (I_2) has been demonstrated to have bactericidal activity. The drawback of the low solubility of I_2 was overcome by either adding iodide (I^-) in aqueous solution or using organic solvents (such as ethanol) to make disinfection formulas. However, these formulas are only used in clinical applications as ready-to-use (RTU) disinfectants since their first introduction in the early 1800s (Halliday, 1821), owing to their instability, unpleasant odor, and tendency to stain hard surfaces. These limitations promoted the development of alternative iodine-based preparations designed to improve the properties without a significant loss of germicidal efficiency. The iodophors were the first such compounds achieving significant improvements and broad commercial use in various fields, including food plant sanitation.

Iodophors are polymeric organic molecules (i.e., carriers) that are capable of complexing iodine species, resulting in iodine preparations with significantly improved solubility and stability (Yamada and Kozima, 1960; Schmulbach and Drago, 1960). In addition, the equilibrium concentrations of the iodine species compared with those of pure aqueous solutions with the same total iodine and iodide concentrations are also reduced. In most iodophor preparations, the carrier is usually a nonionic surfactant, in which the iodine is present as micellar aggregates. When an iodophor is diluted with water, dispersion of the micelles occurs, and most of the iodine is slowly liberated. The presence of a surface-active agent as a carrier also improves the wetting capacity.

Iodine is chemically less reactive than chlorines and thus less affected by the presence of organic substances. Hence, in sanitizing in the presence of organic soils, iodine is much more efficient than chlorine. The activity of iodine is greater at acid than at alkaline pH and is most active at approximately pH 3.0. As a result, most iodophor preparations are formulated to be acidic under application conditions. This property has made iodophors especially useful in the dairy industry, whereas the acidity of the use-solution prevents the buildup of milk stone.

Iodophors are more costly than chlorine-based sanitizers, although they normally are used at much lower concentrations (12.3–25 ppm).

Other disadvantages include instability at high temperature (release of free iodine occurs at temperatures > 43.3°C), staining of some plastics (e.g., polyvinyl chloride [PVC] and epoxy), and discoloration of starchy food. Despite these disadvantages, iodophors remain widely used as biocides, including as sanitizers for food-processing plants, due to their ability to penetrate organic soil, tolerance to hard water, and minimal irritancy.

Because of its innocuous end products (i.e., oxygen and water), hydrogen peroxide has been favorably viewed for use in food plant sanitation. However, the low biocidal activity of H_2O_2 limits its applications to sterilization of containers for aseptically preserved foods at very high concentration (35%) and evaluated temperature. The substitution of one hydrogen atom in H_2O_2 with an acyl or aroyl group generates peroxycarboxylic acids (Figure 6.14). Peroxycarboxylic acids have the attributes of H_2O_2 but with significantly improved germicidal activity. Among all biocides known, peroxycarboxylic acids enjoy most of the properties of an ideal sanitizer or disinfectant for food plant sanitation. In addition to being rapidly active at low concentration against a wide spectrum of microorganisms, peroxycarboxylic acids are virtually nontoxic because they decompose to carboxylic acid, oxygen, and water in the environment.

The mechanism of action of peroxycarboxylic acids is related to their strong oxidation power and low activation energy. The research by Clapp et al. (1994) using electron paramagnetic resonance (EPR) spin-trapping technology revealed the direct correlation of bacterial kill and hydroxyl radicals produced inside the bacteria cell. Peroxycarboxylic acids first penetrate the cell membrane, producing hydroxyl radicals in the presence of ion species, resulting in microbial lethality through the subsequent oxidation of cellular components. Peroxycarboxylic acids have much lower activation energy than H_2O_2, facilitating the production of hydroxyl radicals inside the cell. This may explain the significantly improved germicidal efficacy demonstrated by peroxycarboxylic acids relative to H_2O_2.

Peroxyacetic acid (POAA) is the most important and widely used peroxycarboxylic acid. Antimicrobial activity of POAA was first identified in 1902 (Freer and Novy, 1902) and later disclosed by Hutchings and Xezones (1949) to be superior to 23 compounds tested. The commercial use of POAA was slow because of stability hurdles initially identified.

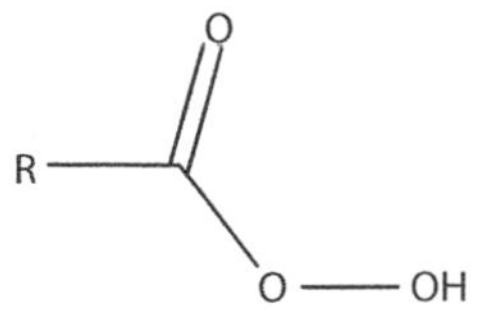

Figure 6.14 Structure of peroxycarboxylic acid.

$$CH_3COOH + H_2O_2 \rightleftharpoons CH_3COOOH + H_2O$$

Figure 6.15 Preparation of peroxyacetic acid.

$$CH_3COOOH \longrightarrow CH_3COOH + O_2 \qquad (1)$$

$$CH_3COOOH + H_2O \longrightarrow CH_3COOH + H_2O_2 \qquad (2)$$

$$CH_3COOOH \longrightarrow CH_3COOH + O_2 + CO_2 \qquad (3)$$

Figure 6.16 Decomposition of peroxyacetic acid.

POAA is generally prepared by treating H_2O_2 with acetic acid in the presence of an acid catalyst (Figure 6.15). This reaction is an equilibrium reaction, with the equilibrium constant of about 2.7 at 25°C. Pure POAA is a high-energy explosive compound, which is a colorless liquid with a pungent odor, and is freely water soluble. As a result, POAA is always commercially sold as an equilibrium solution. POAA is intrinsically unstable, and decomposes through three major routes (Figure 6.16): (1) spontaneous decomposition to yield the corresponding carboxylic acid and oxygen; (2) hydrolysis under strongly acidic or basic conditions to give the carboxylic acid and H_2O_2; and (3) catalytic decomposition by transition metals and their salts to yield mainly carbon dioxide, oxygen, and carboxylic acid. As a result, stable POAA compositions need to be balanced on the level of individual ingredients, have a suitable pH level, and contain a stabilizer to sequester transition metals (e.g., iron). Stable POAA compositions developed by Boewing et al. (1976a, 1976b) represent the early successful examples that eliminated most of the handling issues identified previously for commercial applications.

POAA has excellent antimicrobial properties against bacteria, viruses, and fungi. It is germicidal at low temperatures and remains effective in the presence of organic material. As a weak acid ($pK_a = 8.5$), POAA is more active under acidic conditions but is germicidal with higher concentration in the alkaline range. The powerful antimicrobial action of POAA along with the absence of toxic residuals has led to a wide range of applications. It has been accepted worldwide in the food-processing and beverage industries, where its water solubility and no-foam properties make it ideal to be used in CIP systems. In addition, its low corrosion profile also makes it suitable to be used as a terminal disinfectant or sterilant for stainless steel and glass tanks, piping, tank trucks, and railroad tankers.

Successful elimination of bacterial spores with antimicrobial agents can be difficult and present many operation hurdles (e.g., corrosion, worker safety, etc.). The unique biochemical composition of spore layers affords resistance to the antimicrobial effects of chemical agents. POAA has a broad spectrum of antimicrobial properties, but its activity against spores is less desirable. Research efforts have been ongoing to improve the antimicrobial activity of POAA toward bacterial spores, those produced from the *Bacillus cereus* group, which has been demonstrated to have higher tolerance to POAA inactivation than other bacterial spores.

It is a common understanding that POAA and H_2O_2 work cooperatively in antimicrobial use-solutions to reduce microbial populations. Research by Hilgren et al. (2000) aimed at increasing efficacy of POAA against bacterial spores uncovered surprising results. It was demonstrated that H_2O_2 facilitates the resistance of bacterial spores, particularly of the *Bacillus cereus* group, toward POAA. As a result, by increasing the ratio of POAA to H_2O_2, the efficacy of POAA use-solution against *Bacillus cereus* spores was significantly improved. The results are summarized in Table 6.3.

Traditional synergistic approaches for enhancing the efficacy of antimicrobials have been applied to POAA chemistry, especially toward the improvement of its sporicidal activity. Hei et al. (2002) discovered that the combination of certain solvents with POAA generated significant synergistic activity against *Bacillus cereus* spores (Table 6.4). Hilgren et al.

Table 6.3 Sporicidal Efficacy of Various POAA/H_2O_2 Ratios against *Bacillus cereus* Spores (40°C, pH 3.7)

Test solutions		POAA:H_2O_2	Contact	Log
POAA	H_2O_2	(wt ratio)	time (h)	reduction
150 ppm	31 ppm	4.8:1	0.5	>6.43
150 ppm	275 ppm	0.5:1	0.5	0.65
150 ppm	529 ppm	0.3:1	0.5	0.26
150 ppm[a]	31 ppm	4.8:1	0.5	0.61

[a] Spores treated with 300 ppm H_2O_2 for 1 min, followed by exposure to the following composition.

Table 6.4 Biocidal Effects of POAA and Solvents against *Bacillus cereus* Spores

Test solution	Exposure time (s)	Temperature (°C)	Log CFU reduction
1000 ppm POAA	10	60	0.2
4000 ppm POAA	10	60	0.8
3% Benzyl alcohol	10	60	0.1
3% Diester blend[a]	10	60	0.3
3% Benzyl alcohol + 1000 ppm POAA	10	60	2.4
2.5% Diester blend + 1500 ppm POAA	10	60	>6.3

[a] DBE-31™ (Dupont Nylon).

Table 6.5 Antifungal Effects of POAA and Quaternary Ammonium
Compounds (Exposure: 30 s at 25°C)

Test solution	Microorganism	Log CFU reduction
128 ppm POAA	*C. albicans*	0.18
20 ppm ADBAC[a]	*C. albicans*	1.24
128 ppm POAA + 20 ppm ADBAC	*C. albicans*	>5.00
128 ppm POAA	*S. cerevisiae*	0.00
20 ppm ADBAC	*S. cerevisiae*	0.05
128 ppm POAA + 20 ppm ADBAC	*S. cerevisiae*	>5.00
128 ppm POAA	*G. candidum*	0.00
20 ppm ADBAC	*G. candidum*	0.00
128 ppm POAA + 20 ppm ADBAC	*G. candidum*	>5.00

[a] Alkyl (C_{12}-C_{16}) dimethyl benzyl ammonium chloride.

(2000) found that adding a small amount of quaternary ammonium compounds to POAA could enhance its antifungal activity (Table 6.5).

Research aimed to further enhance the germicidal efficacy of peroxy-carboxylic acids without introducing other chemical agents resulted in the highly potent mixed peroxycarboxylic acid systems (Oakes et al., 1992). Antimicrobial compositions with high activity allow for low use-solution concentrations, thereby minimizing use cost, surface corrosion, odor, carryover of biocide into foods, and potential toxic effects to the user.

The invention of the mixed peroxycarboxylic acid compositions was based on two independent discoveries (Oakes et al., 1992): (1) the peroxy-fatty acids (C_6–C_{18}) are very potent biocides under acidic pH (see Table 6.6),

Table 6.6 Comparison of Antimicrobial Activity
of Peroxyfatty Acids (20°C)

Peroxyfatty acid	pH	Minimum concentration (ppm) required for a 5-log reduction	
		S. aureus	*E. coli*
Peroxyhexanoic acid (C_6)	3.5	15	15
	5.0	20	15
Diperoxyadipic acid (C_6)	3.5	>50	40
	5.0	>60	35
Peroxyoctanoic acid (C_8)	3.5	5	5
	5.0	10	15
Peroxydecanoic acid (C_{10})	3.5	3	10
	5.0	1	30
Diperoxysebacic acid (C_{10})	3.5	15	15
	5.0	10	50

Table 6.7 Efficacy of Peroxyfatty Acids and POAA against *E. coli* in Suspension[a]

POAA (ppm)	Peroxyoctanoic acid (ppm)	Peroxydecanoic acid (ppm)	Log reduction[a]
25	0	0	0
30	0	0	0.7
0	5	0	0.1
0	0	6	0
25	5	0	3.8
30	0	6	2.6

[a] Exposure conditions: 30 seconds, pH 5.0, 20°C.

and (2) the combination of peroxycarboxylic acids (C_1–C_5) and peroxyfatty acids (C_6–C_{18}) generates significant synergistic effects, providing a much more potent biocide than can be obtained by using these components separately (Table 6.7).

The mixed peroxycarboxylic acids share such advantages with POAA as hard water tolerance, minimal deleterious effect by organic material relative to other oxidative biocides, and effectiveness over a broad pH and temperature range. In addition, the mixed peroxycarboxylic acid compositions are generally more effective against various yeasts and molds than POAA and can be applied at a lower use-solution concentration than POAA to achieve sanitization.

The conventional peroxycarboxylic acid compositions typically include short-chain peroxycarboxylic acids, such as POAA, or mixtures of short-chain peroxycarboxylic acids and peroxyfatty acids. Short-chain peroxycarboxylic acids such as POAA have a strong pungent odor in concentrate, which may be offensive or cause irritation to the users and limit its applications in some cases. It has been an ongoing effort to seek alternative peroxycarboxylic acid compositions that have the same advantages of the conventional peroxycarboxylic acids but with reduced or eliminated odor.

Peroxyfatty acids (C_5–C_{18}) have much lower vapor pressure compared with POAA and thus significantly less odor. As discussed, peroxyfatty acids are very potent biocides under acidic conditions. Peroxycarboxylic acid compositions with peroxyfatty acids and a small amount, or absence of, POAA have dramatically reduced odor but retain similar biocidal efficacy of conventional peroxycarboxylic acid compositions (Man et al., 2004). Compositions need to be carefully balanced in terms of the levels of fatty acids, coupler, and acidity to obtain a stable composition while delivering the required results once diluted for sanitizing applications. The higher acidity of these compositions is beneficial in combining sanitizing

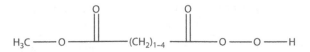

Figure 6.17 Structure of ester peroxycarboxylic acid.

and an acid rinse, which reduces mineral film buildup in the processing of certain foods (e.g., dairy). Also, the surfactants present in the formula for coupling peroxyfatty acids reduce surface tension and improve the wetting of treated surfaces.

An alternative approach to low-odor peroxycarboxylic acids resulted in ester peroxycarboxylic acid compositions (Carr et al., 1997). An ester peroxycarboxylic acid is mixed peroxycarboxylic acids comprised of monomethyl ester peroxyacids of glutaric, succinic, and adipic acids (Figure 6.17). The ester group of the peroxycarboxylic acids reduces the typical pungent odor of peroxycarboxylic acids while imparting a fruity odor to the compounds. These ester peroxycarboxylic acids are water soluble but less potent than POAA. The pleasing odor of ester peroxycarboxylic acids represents an advantage when used in closed areas or in critical care environments such as hospital rooms. These peroxycarboxylic acids are commercially available in Europe for use as environmental disinfectants.

The peroxycarboxylic acids of other common odorless carboxylic acids, especially food-grade carboxylic acids such as lactic and citric acid, were also investigated as nonodorous peroxycarboxylic acid. However, despite continuing reports (Mckenzie et al., 2001; Ferdousi et al., 2006) on efforts in making such peroxycarboxylic acids, it was reported (Ray, 1924) that α-hydroxycarboxylic acids, such as lactic and citric acid, were not stable toward the oxidation of H_2O_2 under the investigated conditions.

6.4   Summary

Research and development efforts aimed toward commercialization of products with the attributes of an ideal sanitizer continue. As innovation and regulations in the food-processing industry drive the need for additional sanitation requirements, academia and the chemical supply industry will be poised to respond. Over the past few decades, sanitizer and biocide research and development efforts have amounted to incremental improvements to efficacy, delivery, and means of application. The future of sanitizer and biocide research and development will likely be marked by technologies that address and mitigate specific needs such as new foodborne pathogens or new routes of contamination and foodborne illness.

References

Adams, R. 1931. Polyhydric alcohol ester of aliphatic carboxylic acid. U.S. Patent 1917681 filed January 21, 1931, and issued July 11, 1933.

Asano, T. 2007. *Water reuse: issues, technologies, and applications.* New York: McGraw-Hill Professional, Chapter 11, 624.

Berg, R. W., and Zimmerer, R. E. 1987. Effect of rare earth cations on bactericidal activity of anionic surfactants. *Journal of Industrial Microbiology* 1:377–381.

Birkeland, J. M., and Steinhaus, E. A. 1939. *Proceedings of the Society for Experimental Biology and Medicine* 40:86.

Boewing, W. G., Mrozek, H., Haan, H. J. S., Tinnefeld, B., and Voegele, P. 1976a. Peroxy-containing microbicides stable in storage. U.S. Patent 4051059B filed August 3, 1976, and issued September 27, 1977.

Boewing, W. G., Mrozek, H., Haan, H. J. S., Tinnefeld, B., and Voegele, P. 1976b. Stable peroxy-containing microbicides. U.S. Patent 4051058 filed August 3, 1976, and issued September 27, 1977.

Carandang, C. M., and Dychdala, G. R. 1974. Low foaming acid anionic surfactant sanitizer compositions. U.S. Patent 3969258 filed October 10, 1974, and issued July 13, 1976.

Carr, G., James, A. P., Morton, K. J., Sankey, J. P., and Lawton, V. 1997. Percarboxylic acid solutions. U.S. Patent 6274542 B1 filed December 16, 1997, and issued August 14, 2001.

Clapp, P. A., Davies, M. J., French, M S., and Gilbert, B. C. 1994. The bactericidal action of peroxides; an E.P.R. spin-trapping study. *Free Radical Research* 21:147–167.

Clough, P. N., and Starke, H. C. 1985. *European Applied Research Reports: Nuclear Science and Technology* 6:631.

Cowles, P. B. 1938. Alkyl sulfates: Their selective bacteriostatic action. *Yale Journal of Biology and Medicine* 11:33–38.

Davidson, E. A., and Kross R. D. 1990. Disinfection method and composition thereof. U.S. Patent 5185161 filed November 27, 1990, and issued February 9, 1993.

Ditoro, R. D. 1969. New generation biologically active quaternaries. *Soap Chemical Specialties* March: 47–52.

Domagk, G. 1935. A new class of disinfecting agents. *Deutsche Medizinische Wochenschrift*, 61:829–832.

Drechsler, P. A., Wildman, E. E., and Pankey, J. W. 1990. Evaluation of a chlorous experimental and natural acid-chlorine dioxide teat dip under experimental and natural exposure conditions. *Journal of Dairy Science* 73:2121–2128.

Ferdousi, B. N., Islam, M. M., Awad, M. I., Okajima, T., Kitamura, F., and Ohsaka, T. 2006. Preparation and potentiometric measurement of peroxycitric acid (E). *Electrochemistry* 8:606.

Freer, P. C., and Novy, F. G. 1902. *American Chemical Journal* 27:161.

Gottardi, W. 2001. Iodine and iodine compounds. In *Disinfection, sterilization, and preservation,* 5th ed., ed. S. S. Block. Philadelphia: Lippincott Williams & Wilkins, Chapter 8.

Halliday, A. 1821. *London Medical Reports* 16:199.

Hansch, C., and Fujuta, T. 1964. p-σ-π analysis, a method for the correlation of biological activity and chemical structure. *Journal of the American Chemical Society* 86:1616–1626.

Harris, B. R. 1941a. Amino carboxylic acid esters. U.S. Patent 2321595 filed February 5, 1941, and issued June 15, 1943.

Harris, B. R. 1941b. Amino carboxylic acid esters of higher molecular weight carboxylic monoesters of glycols. U.S. Patent 2321594 filed February 5, 1941, and issued June 15, 1943.

Hei, R. D., P., Herdt, B., and Grab, L. 2002. Two solvent antimicrobial compositions and methods employing them. U.S. Patent 6927237B2 filed February 28, 2002, and issued August 9, 2005.

Hilgren, J. D., Richter, F. L., Reinhardt, D. J., Salcerda, J. A., and Rahm, C. L. 2000. Peroxycarboxylic acid compositions and methods of use against microbial spores. U.S. Patent 6627657 B1 filed March 22, 2000, and issued September 30, 2003.

Huang, Y. R., Hung, Y. C., Hsu, S. Y., Huang, Y. W., and Hwang, D. F. 2008. Application of electrolyzed water in the food industry—review. *Food Control* 19:329–345.

Hutchings, I. J., and Xezones, H. 1949. Comparative evaluation of the bacterial efficiency of perocetic acid, quaternanes, and chlorine containing compounds. *The 49th Annual Meetings of the Society of American Bacteriologists* 50.

Kabara, J. J. 1975. Fatty acids and derivatives as antimicrobial agents. U.S. Patent 4002775 filed May 21, 1975, and issued January 11, 1977.

Man, V. F. P., Maguson, J. P., and Lentsch, S. E. 2004. Medium chain peroxycarboxylic acid compositions. U.S. Patent 7771737 B2 filed January 9, 2004, and issued August 10, 2010.

Marriot, N. G., and Gravani, R. B. 2006. Santizer. In *Principals of food sanitation*, 5th ed. New York: Springer, Chapter 10.

Mckenzie, K. S., Giletto, A., Hitchens, G. D., Hargis, B. M., and Herron, K. L. 2001. Control of microbial populations in the gastrointestinal tract of animals. U.S. Patent 6518307 B2 filed October 17, 2001, and issued February 11, 2003.

Mercer, W. A., and Somers, I. I. 1957. *Advances in Food Research* 7:129.

Merianos, J. J. 2001. Surface-active agents. In *Disinfection, sterilization, and preservation*, 5th ed., ed. S. S. Block. Philadelphia: Lippincott Williams & Wilkins, Chapter 14.

Mir, Z. 1984. Chlorine dioxide as a santizer for carbon filters. *Beverage World* 187.

National Institute for Occupational Safety and Health (NIOSH). 1994. *National Institute for Occupational Safety and Health Pocket Guide to Chemical Hazards.* Atlanta, GA: NIOSH.

Oakes, T. R., Stanley, P. M., and Keller, J. D. 1992. Peroxyacid antimicrobial composition. U.S. Patent 5314687 filed August 20, 1992, and issued May 24, 1994.

Petrocci, A. N., Green, H. A., and Merianos J. J. 1974. *CSMA Proceedings of the 60th Mid-year Meeting*, 87.

Ray, G. B. 1924. The oxidation of sodium lactate by hydrogen peroxide. *The Journal of General Physiology* 6 (5):509–529.

Schaeufele, P. J. 1984. Advances in quaternary ammonium biocides. *Journal of the American Oil Chemists' Society* 61:387.

Schmulbach, C. D., and Drago, R. S. 1960. Molecular addition compounds of iodine. III. An infrared investigation of the interaction between dimethylacetamide and iodine. *Journal of the American Chemical Society* 82:4484–4487.

Sedliar, R. M., Garvin, D. F., and Aepli, O. T. 1968. Low foam anionic acid sanitizer compositions. U.S. Patent 3650964 filed May 13, 1968, and issued March 21, 1972.

Shimizu, Y., and Hurusawa, T. 1992. Antiviral, antibacterial, and antifungal actions of electrolyzed oxidizing water through electrolysis. *Dental Journal* 37:1055–1062.

Shukla, D., and Tyagi, V. K. 2006. Cationic gemini surfactants: a review. *Journal of Oleo Science* 55 (8):381–390.

Simpson, G. D. 2005. *Practical chlorine dioxide.* Vol. 1. Colleyville, TX: Simpson.

Stark, R. L. 1983. Aqueous antimicrobial ophthalmic solutions. U.S. Patent 4525346 filed September 29, 1983, and issued June 25, 1985.

Troller, J. A. 1983. *Sanitation in food processing.* New York: Academic Press.

Vaughn, R. H., and Stadtman, T. C. 1977. A method of or control of sanitation in food processing plants. *American Journal of Public Health* 35(12).

Wang, Y. 1981. Short chain fatty acid sanitizing composition and methods. U.S. Patent 4404040 filed July 1, 1981, and issued September 13, 1983.

Yamada, H., and Kozima, K. 1960. The molecular complexes between iodine and various oxygen-containing organic compounds. *Journal of the American Chemical Society* 82:1543.

chapter seven

Research during microbial food safety emergencies and contaminant investigations

Jeffrey L. Kornacki

Contents

7.1 What constitutes a food safety emergency?

So, what is a food emergency? The primary definition of "emergency" in Merriam-Webster's *Online Dictionary* (http://www.merriam-webster.com/dictionary/emergency; accessed April 9, 2010) is "an unforeseen combination of circumstances or the resulting state that calls for immediate action." Few would doubt that acute foodborne illness constitutes an emergency. What about food spoilage or the presence of a pathogen in zone 3 (e.g., production environment away from the product stream)? Experience indicates that an emergency can also be in the "eye of the beholder" to some degree. But, who is the beholder? The beholder can be a regulatory agency and, of course, the victims and general public (including the press) in the case of foodborne illness. However, in some cases the emergency may be a pathogen or indicator organisms found by a concerned food processor in its processing environment but not on food contact or near food contact surfaces. Nevertheless, the manufacturer may well feel that "immediate action" is necessitated, especially when issues like a perception of a food safety risk, the potential for litigation, or lost market exists, even with a nonpathogenic spoilage or indicator organism.

Regardless of how and who initially defines the emergency, the food production facility may find itself in emergency mode at some time. This emergency usually results in the urgent need to answer several questions, including the following: (1) Did the organism in the food come from my product manufacturing environment (e.g., my plant) or my growing field? (2) How do I test for the organism? (3) Could it have originated from an ingredient? (4) Did my critical control points fail or are they valid?

7.2 Did the organism come from my plant?

The question of whether an organism comes from a specific plant involves several considerations, including the accuracy and reliability of the assay for the organism and subtype and the strength of the epidemiological association (as in the case of foodborne illness) and the quality systems of the testing laboratory. Reliability aspects of microbiological assays are covered in detail in the following discussion and in a subsequent chapters on microbiological methods of analysis (Chapters 9, 10, and 11).

7.2.1 Assay reliability

It is not advisable to accept the reliability of any finished product assay at face value, especially when it involves a qualitative assay (e.g., presence/absence of a low-infectious-dose pathogen like *Salmonella*, enterohemorrhagic *Escherichia coli*, etc.). Initial questions to ask of the laboratory doing the testing include the following: What is the method used? Is it officially approved, validated, or accepted by an authoritative body (e.g., AOAC approved, Food and Drug Administration [FDA] *Bacteriological Analytical Manual* method [AOAC, 2012], *Compendium of Methods for the Microbiological Examination of Foods* [Downes and Ito, 2001], International Organization for Standardization [ISO], etc.)? Has it been validated in my product matrix? If the answers to the last two questions are "no," then ask if it has been demonstrated to be effective in any peer-reviewed publications. Be sure to ask if the assay is a screening method (and thus a "presumptive" assay) or if the result is "confirmed." If the result is confirmed, be sure to ask the basis of the confirmation. Not all confirmation approaches are reliable. For example, despite extensive history of biochemical identification of organisms and many advances in molecular characterization of bacteria, *Salmonella* spp. remain serologically defined. Be sure that whatever technique you rely on for final confirmation of *Salmonella*, it is supported by serological confirmation of both somatic ("O") and flagellar ("H") antigens.

Useful questions for which to seek answers with reference to qualitative (enrichment-based) microbiological assays include those related to its inclusivity, exclusivity, sensitivity, and specificity. AOAC (2005) guidelines for method validations are available (http://www.eoma.aoac.org/app_d.pdf). As mentioned, research and development aspects of assay development are covered in detail in Chapters 9, 10, and 11.

7.2.2   Assay parameters

7.2.2.1   Inclusivity

Inclusivity attempts to answer the question, "How inclusive is the assay of my target organism group?" Hence, inclusivity relates to how broad the assay is with regard to the ability of the assay to recover all the organisms in the target group. Usually, an "inclusivity panel" of organisms has been screened by the assay to gain this type of information. However, such a panel is likely to be somewhat truncated with regard to genera that have many species or strains (e.g., *Salmonella* spp. have over 2400 serovars; Brenner et al., 2000). Hence, representative groups should have been selected. If the target organism's group has not been screened, this should be part of the assay validation.

7.2.2.2   Exclusivity

Exclusivity is related to how many organisms outside the target group are excluded. For example, an assay for *Salmonella* spp. should not recover *Staphylococcus* spp. An exclusivity panel of representative organisms outside the target group should have been selected for screening. Many screening assays for *Salmonella* spp. will also recover related species, such as *Citrobacter* spp. or *Proteus* spp., among others. Such organisms should be included in an exclusivity panel if the *Salmonella* test is being evaluated. This is also referred to as "selectivity," the accuracy with which the assay selects for the target organisms by inhibiting (i.e., not selecting) other similar nontarget organisms.

7.2.2.3 Sensitivity

The sensitivity of an assay can be defined as

[Number of true assay positives/(True assay positives + False assay negatives)] × 100%

This is typically determined with inoculation studies in food matrices with the target organism and how one recovery method (e.g., for *Salmonella*, *Listeria monocytogenes*, hemorrhagic *Escherichia coli*, etc.) compares to a reference method, sometimes referred to as a "gold standard." Typically, with qualitative pathogen assays it is desirable to have an assay that is capable of detecting 1 cell in 25 g. This type of determination will likely involve inoculation of food matrices with low (e.g., 1–5 CFU/25 g), medium (e.g., 10–50 CFU/25 g), and high (100 or more CFU/25 g) levels of the target population, although these targets can vary depending on expected values previously derived from other validated matrices.

Sensitivity in this context refers to the ability of the method (including any preenrichments and subsequent enrichments) to recover the target organism from the matrix. However, it does not address the sampling process or the sample size relation to the substrate lot.

7.2.2.4 Specificity

You want to answer the question, "How specific is the assay in selected food matrices?" Specificity relates to the ability of the assay to recover the target organism rather than closely related "false positives."

The specificity of an assay can be defined as

[Number of true assay negatives/(False assay positives + True assay negatives)] × 100%

This question relates to the level of false negatives and false positives (as described in the following) and can also be determined via a method validation study. There is usually a trade-off between specificity and sensitivity during the creation of an assay. Specificity can be enhanced at the expense of sensitivity and vice versa. Regulatory agencies typically desire an assay with the lowest false-negative rate possible (greatest sensitivity); hence, the assay would be less likely to miss true positives but might be more likely to elicit false positives than a less-sensitive method. The food industry typically seeks an assay with the lowest false-positive rate possible (highest specificity) since that influences the quantity of product that is involved in an extended hold for confirmation testing.

7.2.2.4.1   False-positive rate.   The false-positive rate can be described as

[Number of false assay positives/(Number of false assay positives + True assay negatives)] × 100%

or

$$100 - \text{Specificity} = \text{False-positive rate}$$

Desirable assays typically have false-positive rate percentiles in the upper 90s.

7.2.2.4.2   False-negative rate

[Number of false assay positives/(True assay positives + False assay negatives)] × 100

or

$$100 - \text{Sensitivity} = \% \text{ False negatives}$$

Desirable assays have false-negative rate percentiles also in the upper 90s.

Positive results of most methods are considered presumptive until and unless confirmed by acceptable nonlike methods. For example, certain polymerase (PCR) or immunocapture assays can detect presumptive *Salmonella* or presumptive *E. coli* O157:H7, and the cultures must then be subjected to biochemical confirmatory methods such as those outlined in the FDA BAM (*Bacteriological Analytical Manual* [USFDA, 2012]) and USDA (U.S. Department of Agriculture) microbiology guidebook (USDA, 2012).

7.2.3   Assay validation

Regardless of how well an assay performs in the published literature, responsible users of the assay will ideally validate the efficacy of the methods used in their particular food matrices. However, this should be done in noncrisis times (i.e., before a crisis occurs). Nevertheless, food contamination events are often unanticipated, and many surprises have occurred over the years (e.g., pathogens in dry foods). Thus, it is inevitable that some companies will need to perform (or have their contract laboratory perform) validation of a particular method for their unique product matrix during a crisis. In general, an ISO 17025-certified laboratory will have some guidelines for performing a cursory validation of established methods in a unique food matrix. Section 5.4 of the ISO 17025 standard addresses methods. Various sections address particular areas related to methods in the following manner (ISO/Conformity Assessment Policy Development Committee [CASCO], 2007):

5.4.1 Addresses methods in general.

5.4.2 Addresses sources and selection of methods and modifications of methods (and the need for their validation).

5.4.3 Relates to laboratory-developed methods.

5.4.4 This deals with nonstandard methods and the general components and addresses validation in a general way.

5.4.5 This section relates more directly and in more detail with validation of methods (e.g., when validation is needed for any reason, e.g., extension of an assay from one validated matrix to another).

5.4.6 Measurement uncertainty, which is part of validation, is addressed in this section.

5.4.7 Control of data is discussed here.

The AOAC has developed an "emergency response validation" (ERV) system for AOAC approvals of methods as applied to new, previously not validated, food matrices (AOAC, 2009). This was done during a recent peanut butter recall that affected large segments of the food industry in the United States. However, such an emergency validation is unlikely to occur apart from such a significant multi-industry event in my view. However, a company or an industry could consider adoption of a similar approach for its own unique food matrix preemptively or during its own crisis period. The ISO 17025 guidelines for method validations are fairly general given the large diversity of analytes and matrices (B. Stawick, Stawick Laboratory Management, personal communication, December 22, 2010). Consequently, in my opinion appropriate peer review of such approaches should be undertaken on a case-by-case basis.

7.3   Pathogen contamination of product

You have researched the test result and now believe the product is contaminated. This raises all sorts of additional questions, including

1. Did the contaminant come from my environment?
2. Did the contaminant come from my ingredients? If so, was it direct or indirect contamination from the ingredients?
3. Did the contaminant come from both my ingredient and my environment?

A comprehensive review of how to approach contamination investigations can be found elsewhere (Kornacki, 2010a).

7.3.1   Environmental considerations

The principal source of contamination of processed foods is their manufacturing environment (Kornacki, 2010a). It is unrealistic to expect sterility in the food-processing environment as there are potentially multiple sources of original microbial contamination to the facility (e.g., human beings' skin, nasal passages, gastrointestinal tract; negative pressure in filling/packaging rooms) (or the plant as a whole, drawing contaminants from external areas, e.g., bioaerosols from untrapped or poorly designed or maintained drains, foot traffic or wheeled vehicle traffic from the outside environment or through wet areas in the plant, poorly designed air-handling units, raw ingredients, etc.). However, vast populations of microorganisms are required to contaminate foods in a direct measurable way. For example, one needs 1 billion cells added to 100,000 pounds of product to achieve a level of about 20 cells/g in well-mixed product, assuming a stable contaminant population. Such levels are unlikely to occur from airborne routes. Nevertheless, such populations can exist in areas called "microbial growth niches." Such areas may be as small as or smaller than the volume of a droplet but could take up a good deal more space, such as the underside of a contaminated fibrous conveyor. Such niches occur when a constellation of events comes together and include (1) the presence of moisture (e.g., water activity above 0.63 for some osmophilic mold species); (2) nutrition (as may be supplied by dust particles, product residues, etc.); (3) time (in abundance in areas inaccessible for cleaning); (4) the target organism(s); and (5) an appropriate temperature for microbial growth (e.g., typically in the range of 0 to 55°C, although some organisms can grow outside these ranges).

7.3.2   Ingredients as a direct or indirect source of product contamination

Ingredients can be sources of cross contamination of the plant environment. Ingredients can be sources of food in processes that do not have a CCPm (a hazard analysis and critical control point [HACCP] critical control point for control of microbial contamination) or have one that only reduces, but not necessarily destroys, all populations of the pathogen (as may occur with some vegetable washes/flumes or organic acid treatments during the manufacture of cold pack cheese, etc.). Contaminated ingredients could also be sources of contamination in foods wherein the critical limit (CL) for the CCP is not achieved.

However, indirect contamination of foods from an ingredient can also occur in processes with effective CCPm's. Cross contamination of the plant environment from contaminated ingredients may occur and result in transient or persistent contamination of a facility. Pathogenic strains have been known to persist in food plant environments for years (Grocery Manufacturers of America [GMA], 2009; International Commission on the Microbiological Specifications for Foods [ICMSF], 2002; Miettinen,

Bjorkroth, and Korkeala, 1999; Lunden, Autio, and Korkeala, 2002; Martinez-Urtaza and Liebana, 2005), and I have frequently observed this. These strains may develop into microbial growth niches whether transient or persistent. These growth niches may develop into biofilms. Given the resistance of biofilms to environmental stresses (Johnson et al., 2002), such niches should be aggressively sought (i.e., in zones 3 and 4)* by plant personnel through microbiological swab testing and remediation performed and their effectiveness documented. Should a plant determine that its environmental contamination can be traced back to an ingredient (e.g., through molecular subtyping approaches), it does not absolve a plant from responsibility to diligently seek and destroy niches where the microbe is found. Often, plants will attempt to eliminate a microbial growth niche considering a subsequent (or several subsequent) negatives to provide verification that corrective actions have been effective. However, tracking results from sites over a long period of time (e.g., 6 months to a year) could result in different conclusions. A site that is positive several times a year may indicate the need for other corrective actions or investigation. Such actions can be informed by investigational swabbing as one "vectors out" from a positive site. Sampling areas surrounding positives should be done with a view toward establishing potential routes of contamination. This microbial "investigational mapping" has an ultimate goal of eliminating the identifiable sources. Molecular subtyping of isolates derived from vector-based sampling can further enhance hypothesis generation related to directionality of contamination and lead to more focused corrective actions. A more detailed description of the basic types of molecular subtyping approaches and their value can be found in the work of Moorman et al. (2010). An approach to tracking positive sites that could also be used with subtypes was described by Eifert and Arritt (2002).

7.4   Testing and scoping

Responsible food-processing companies endeavor to clean their processing environment effectively at appropriate frequencies. Documentation of the effectiveness of cleaning and sanitation is useful to establish a "break point" that can be used to focus an investigation on selected "lots" of product. A manufacturing *lot* is "a quantity of food or food units produced and handled under uniform conditions" (ICMSF, 2002). In my experience, this is interpreted by regulators as any product produced on a common line between appropriately validated cleaning and sanitization regimes. This definition may differ substantially from a food producer's lot definition.

---

* Zones of sampling in food-processing environments were defined by the International Commission on the Microbiological Specifications for Foods (ICMSF, 2002).

Food safety emergencies are nearly always based on a regulatory lot definition. Epidemiologic research may implicate product consumed prior to an initially implicated lot, despite apparently successful verification activities (e.g., acceptable zone 1 swabs and finished product tests on those prior lots). Such contamination may occur as a consequence of product contamination from persistent strains in occult zone 2 environments that slough sporadically into product streams and are not detected in product due to nonhomogeneous and (frequently) low-level contamination. In these cases, it is prudent to retest multiple product lots (e.g., obtained from retained or returned lots, as in the case of a recall) samples in a highly rigorous statistically based manner. These results can be compared to those from epidemiological studies and a more complete picture of the onset of contamination in the plant determined.

Also, as a cautionary note, lot separation by processing line can be muddled and difficult to demonstrate if there exists shared equipment, personnel, or postlethality exposed product flows between multiple lines. True line separation by proximity and ideally physical barriers (such as dedicated rooms for processing lines) is helpful toward maintaining and proving a break point or lot separation.

7.4.1   Lot testing of ingredients

In any food safety emergency, routine product and ingredient testing approaches are typically replaced by more rigorous approaches. The FDA BAM specifies three sampling categories for the testing of foods for *Salmonella*. These are listed as categories I, II, and III. The FDA BAM (2003) lists these as follows:

> **Food Category I.**—Foods that would *not* normally be subjected to a process lethal to *Salmonella* between the time of sampling and consumption and are intended for consumption by the aged, the infirm, and infants.
>
> **Food Category II.**—Foods that would *not* normally be subjected to a process lethal to *Salmonella* between the time of sampling and consumption.
>
> **Food Category III.**—Foods that would normally be subjected to a process lethal to *Salmonella* between the time of sampling and consumption.

Category I, II, and III foods are tested at $N = 60, 30$, and 15 samples per lot, respectively, with each sample 25 g. The FDA BAM states:

This sampling plan applies to the collection of finished products under surveillance and/or for determination of compliance for regulatory consideration. It also applies to the collection of factory samples of raw materials in identifiable lots of processed units and/or finished products where regulatory action is possible.

The FDA category I level of testing at $N = 60$ (or 1500 g) provides 95% confidence that the randomly sampled lot contains less than 1 cell/500 g. Stated that way, it seems to provide a measure of safety, until one considers quantities greater than 500 g or incidence rates. Table 7.1 indicates that at $N = 60$ there is a 30% chance that none of the 60×25 g samples will test positive if the lot is contaminated at a 2% level. These statistics could apply to some contemporary products. For example, Cutter and Henning (2001) found that between 1% and 6% of ready-to-eat meats were contaminated with *Listeria monocytogenes*. In crisis situations, it is useful to exceed the $N = 60$ testing level.

Table 7.2 has proven useful in the context of investigational activities provided that a compositing scheme has been developed. For example, both wet and dry *Salmonella* compositing schemes were validated by Silliker and Gabis (1973, 1974, respectively), which apparently provided a basis for acceptance of the scheme for dry compositing as prescribed by the FDA (2003) in which 15×25 g dry sample composites are composited up to individual 375 g samples. Hence, dry sample composite (i.e., 15×25 g samples per composite) testing of $N = 60 \times 25$ g samples (1500 g) results in

Table 7.1 Finished Product Sampling: A Statistical Perspective

Two-class plans ($c = 0$): Probabilities of acceptance						
Composition of lot		Number of sample units tested				
% Acceptable	% Defective	5	10	20	60	100
98	2	.90	.82	.67	.30	.13
95	5	.77	.60	.36	.05	.01
90	10	.59	.35	.12	<	<
80	20	.17	.11	.01		
70	30	.03	.03	<		
50	50	.01	<			
40	60	<				
30	70					

Source: Adapted from ICMSF. 2002. *Microorganisms in foods: microbiological testing in food safety management*, Vol. 7. New York: Springer, 2002, 125 (Table 7-1).

< = probabiity of acceptance is less than 0.005.

Table 7.2 Finished Product Sampling: A Statistical Perspective

Test number needed to detect one or more positives per lot			
	Number of analytical units to be tested (u)		
% Positive	90% Confidence	95% Confidence	99% Confidence
100	3	4	4
10	23	30	46
1	230	299	461
0.1	2,303	2,996	4,605
0.01	23,026	29,963	46,052

Source: Adapted from Downes, F. P., & K. Ito (eds.). 2001. *Compendium of methods for the microbiological examination of foods.* 4th ed. Washington, DC: American Public Health Association.

testing only four samples, each of which is 375 g. The case of compositing for *Listeria monocytogenes* has been shown to be more complex, being both method and food matrix dependent (Curiale, 2000; McMahon et al., 2005). Consequently research to validate compositing schemes related to (1) the analytical method (e.g., PCR, cultural, enzyme immunoassay, etc.); (2) the food matrix (sample preparation for methods for analysis of guar gum likely to be very different from analysis of fluid milk; (3) the target organism; and (4) the number and size of the sample composite may be required.

7.4.2   *Investigational swabbing*

Most postprocess contamination events occur as a result of zone 2 contamination in my experience. This is because zone 1 sites are relatively easy to clean and sanitize, whereas zones 3 and 4 sites are further away from the product contact area. However, zone 2 areas frequently contain sandwiched areas that entrap wet residues that can be dislodged into the product contact area (zone 1) via dripping, sloughing, aspiration, vibrations, and the like during production. A thorough investigation should focus on operational, postoperational, and preoperational observations and environmental samples, where possible. The investigator should bear in mind that the probability of product contamination is related to at least six factors (Gabis and Faust, 1988), as follows:

1. The number of growth niches in the environment
2. Microbial population in niches
3. The degree of niche disruption
4. The spatial relationship of niches to the product stream
5. The proximity of niches to the product stream
6. The exposure of the product stream to the environment

Table 7.3 Zone Concept

Zone 1	Product contact surfaces: conveyors, tables, racks, holding vats and tanks, utensils, pumps, valves, slicers, dicers, freezers, filling/packaging machines
Zone 2	Nonproduct contact surfaces in close proximity to product: exterior of equipment, refrigeration units
Zone 3	Telephones, forklifts, walls, drains, floors
Zone 4	Locker rooms, cafeteria, hallways

Source: Adapted from ICMSF. 2002. *Microorganisms in foods: microbiological testing in food safety management,* Vol. 7. New York: Springer, 212 (Figure 11-1).

7.4.2.1   The number of growth niches in the environment

The number of growth niches in the environment can be assessed by a thorough investigation of zones 3 and 4 and, if practical, zone 2. Often, zone 1 and 2 sites are sampled and tested by companies for sanitation indicators (e.g., aerobic plate count [APC], enterobacteriaceae [EB] count, esculin hydrolyzing Gram-positive bacteria, etc.) (Table 7.3). More detailed information for doing in-plant microbiological investigations and risk assessments have been described elsewhere (Kornacki, 2009a, 2009b, 2010a).

The plant should track all sites where positives have been found.

7.4.2.2   Microbial population in niches

The microbial population in niches can be accessed via quantitative measures such as APC and EB counts. Kornacki (2010b) reported an assay for a *"Salmonella*-like" organism referred to as the HTEB assay (an abbreviation for *h*ydrogen sulfide producing *"t*hermophilic" *e*ntero*b*acteriaceae). Isolates from this assay can also be subjected to molecular subtyping (e.g., repetitive extragenic palindromic sequences (REP) PCR, etc.; as shown by Kornacki, 2010b). It is rare that low-dose infectious pathogen counts are deliberately assayed on environmental food contact surfaces by food-manufacturing companies to avoid potential regulatory complications unless a recall is already in process and the line has not been subsequently restarted.

7.4.2.3   The degree of niche disruption

The degree of niche disruption is best assessed during operations when the degree of niche disruption is likely to be greatest (given traffic, operating equipment, vibrations, operating air-handling units, etc.). The number of niches found in the plant is often greatest during and after operations but before cleaning and sanitation. However, the opposite may occur in wet-cleaned dry product processing facilities due to recovery of injured cells from moisture added during wet cleaning and the ability of some wet-cleaning processes to spread microbes throughout a plant (e.g., through torn hoses, traffic patterns, uncontrolled moisture, etc.). This is the only reason why there is a great need for development and validation of appropriate dry-cleaning approaches.

One means of understanding the potential for microbial growth niche disruption is via simple observation of processes. Gaining a grasp of processes that vibrate, rotate, scrape, are subject to impacts (e.g., air hammers on spray dryers) or aspirate, and so on through visual observation is important. Systems to understand airflow range from the simple to complex, involving particle counters and pressure measurements. However, many companies do not do this. Solid or liquid aerosols will move from high to low pressure. Consequently, understanding the airflow in the plant environment is strongly recommended. For example, a heat-resistant mold problem in hot-filled beverages may be remedied by closing loading dock doors of the processing plant when not in use, thereby restricting influx of high spore load in outside air.

A change in the baseline level of microorganisms in the air should provide a measurement of microbial growth niche disruption. A number of plants perform routine air monitoring of general microbial population (e.g., APC and yeast and mold counts). Evancho et al. (2001) reviewed some of the available systems. However, it is impractical to monitor for specific pathogens for two reasons: (1) levels are likely to be extremely low due to dilution and injury, such that the likelihood of recovery from air is extremely small, even if the organisms are present; and (2) the presence of recovered pathogens in the air could have far-reaching implications for product disposition. Hence, a good deal of thought should be given to the value of such testing. It is my perspective that directly measurable food contamination from air is unlikely in most instances where the microbe cannot grow in the product. Tompkin (2002) reported that investigational works done over a 14-year period failed to implicate air as a chronic source of *Listeria* contamination to product contact surfaces. Furthermore, bioaerosols have only rarely been a source of direct measureable product contamination in my experience and only when the bioaerosol has been released in a confined space and in close proximity to the product.

7.4.2.4 Spatial relationship and proximity of niches to the product
The ability of microbes to migrate through the plant environment is remarkable and dynamic given their size. Consequently, each facility should actively monitor their zone 3 and 4 environments for the pathogen of greatest likelihood or else for appropriate indicator organisms. As previously stated, zone 2 sites (those in close proximity to zone 1) are frequently the source of directly measureable product contamination.

7.4.2.5   Degree of exposure of the product stream to the environment

Product not biocidally treated in their end-use container will be at risk from environmental contamination. However, this risk usually will be reduced if the product has minimal exposure to the environment (e.g., produced in sanitary enclosed equipment).

7.5   Critical control points and validation of critical limits

In any recall or pathogen-related crises, a review of the CCPs and procedures involved with validation of the CLs should be undertaken, among others. There are a number of ways in which CLs for CCPs are validated. The research related to the efficacy of any CCP should be challenged in the face of a product pathogen contamination event. Approaches observed in food production facilities include (1) tradition based on a presumption that no one has gotten ill to date (not recommended); (2) extrapolation from disparate matrices (not recommended; e.g., pasteurization times and temperatures for milk that destroy vegetative pathogens would not apply to thermal treatments given to cereal or pie fillings, etc.); (3) regulatory documents that define the process or safe harbor (e.g., thermal treatment for USDA commodity cheese, 2007); (4) published peer-reviewed scientific literature using the product matrix; (5) laboratory-based challenge studies with pathogens, wherein the process or processing conditions are simulated (often difficult or costly to do with highly dynamic processes); (6) predictive models; and (7) in-plant validation of processes with suitable nonpathogenic surrogate organisms (Kornacki, 2010a). Some have also used time-temperature indicators if the thermal resistance of the microbe is known or can be established in a laboratory with the food matrix.

A common misconception is that a microbe will always be destroyed by a single defined treatment irrespective of the product matrix. However, dry product matrices such as dry cereal will require far greater thermal treatments than products with high water activity like ground beef or fluid milk (Kornacki, 2010a). It is also important to recognize that for lethal CCPs, CLs are designed to destroy a predetermined microbial *population*, not an individual cell per se. Microbial populations comprised of individual cells, usually at varied metabolic and physiological states, have a varied degree of resilience to externally applied stresses. Therefore, death rates of microbial populations are dependent on many factors, notably physiology of individual cells represented as subpopulations, the composition of the food matrix, and the heat penetration into that matrix. (Kornacki, 2010a). Hence, the destruction of higher populations will take longer than lower populations, all other things being equal. It also important to recognize that differences in microbial inactivation can occur with the same treatment applied to different products, even under the same conditions of pH and water activity. These product differences may be related to bulk density differences (as may influence sterilant gas penetration), heat penetration into the product, type of low molecular weight solutes, fat, and so on as can impact thermal processing, among others (Kornacki, 2010a).

7.6   Testing considerations in light of changing knowledge

Despite great advances in the past 150 years in our knowledge of microorganisms, since Pasteur's "germ theory," all that can be known about the specific organisms that may contaminate food resulting in foodborne illness is not known. The Centers for Disease Control and Prevention (CDC) has reported that about 80% of foodborne illness in the United States is the result of "etiology unknown" (Mead et al., 1999; Scallan, Griffin, et al., 2011). Previously unrecognized pathogens will likely be discovered in the future, if past experience is any guide (see Table 7.4).

Microbes presently on the horizon include, among others, non-O157:H7 enterohemorrhagic *E. coli; Clostridium difficile; Helicobacter pylori; Cronobacter* spp. (*Enterobacter sakazakii;* in noninfant formula foods for high-risk populations?); other genera in the family Enterobacteriaceae (non-*Salmonella,* non-*Shigella,* non-*Yersinia*); *Arcobacter butzleri* (and other related

Table 7.4 Pathogens Previously Unrecognized by Date

Organism	First recognized in food
Escherichia coli O157:H7	1982
Listeria monocytogenes in foods	1981
E. coli foodborne gastroenteritis	1971
Vibrio parahemolyticus food poisoning	1951
C. perfringens from food	1945
Yersinia enterocolita gastroenteritis	1939
B. cereus food poisoning	1906
Staphylococcus aureus and food poisoning	1894
C. botulinum	1896
Salmonella	1888
Any microbe	1860s

Source: Kornacki, J. L. 2012. Sleuthing for Listeria and Salmonella in dairy processing plants—why, where, when, and how to look. Presentation for Oregon Dairy Industries Annual Conference, April 13, Salem, OR.

species); methicillin-resistant *Staphylococcus aureus*; and *Mycobacterium avium* subsp. *paratuberculosis*, among others, not to mention foodborne viruses (beyond the scope of this chapter). There is also greater recognition of the role of new and emerging pathogens in chronic disease (as illustrated by the association of *M. paratuberculosis* with Crohn's disease; Behling et al., 2010). Consequently, monitoring the product, ingredients, and environment for quality and safety indicators (e.g., via APCs, coliforms/enterobacteriaceae, yeasts and molds, anaerobic plate counts, HTEB, etc.) may provide key indicators to control presently unrecognized pathogens that may one day become sources of future food safety emergencies. The principles of statistical process control (ICMSF, 2002, Chapter 13; Evans and Lindsay, 1996) as applied to finished product with appropriate corrective actions and investigations in the light of out-of-control situations are also recommended.

References

AOAC International. 2009. AOAC RI launches emergency response validation program: *Salmonella* in peanut butter. http://www.aoac.org/News/ri_press_rel.htm (accessed November 29, 2010).

AOAC International. 2005. Appendix D: guidelines for collaborative study procedures to validate characteristics of a method analysis. http://www.eoma.aoac.org/app_d.pdf (accessed November 29, 2010).

Behling, R. G., J. Eifert, M. C. Erickson, et al. 2010. Selected pathogens of concern to industrial food processors: infectious, toxigenic, toxico-infectious, selected emerging pathogenic bacteria. In *Principles of microbiological troubleshooting in the industrial food processing environment*, ed. J. L. Kornacki. New York: Springer, Chapter 2.

Brenner, F. W., R. G. Villar, F. J. Angulo, et al. 2000. *Salmonella* nomenclature. *Journal of Clinical Microbiology* 38 (7):2465–2467.

Curiale, M. 2000. Validation of the use of composite sampling for *Listeria monocytogenes* in ready-to-eat meat and poultry products. August 15. http://www.fsis.usda.gov/PDF/Seminar_Listeria_Sample_Composite_Validation.pdf (accessed September 24, 2010).

Cutter, C. C., and W. R. Henning. 2001. Controlling *Listeria monocytogenes* in small and very small meat and poultry plants. http://www.fsis.usda.gov/OPPDE/Nis/Outreach/Listeria.htm#Top (accessed September 24, 2010).

Downes, F. P., and K. Ito (eds.). 2001. *Compendium of methods for the microbiological examination of foods*. 4th ed. Washington: American Public Health Association.

Eifert, J. D., and F. M. Arritt. 2002. Evaluating environmental sampling data and sampling plans. *Dairy, Food and Environmental Sanitation* 22(5):333–339.

Evancho, G., W. Sperber, L. J. Moberg, et al. 2001. Microbiological monitoring of the food processing environment. In *Compendium of methods for the microbiological examination of foods*, 4th ed., ed. K. Ito and F. P. Downes. Washington, DC: American Public Health Association, 25–35.

Evans, J. R., and W. M. Lindsay. 1996. Fundamentals of statistical process control. In *The management and control of quality*, 3rd ed., ed. J. R. Evans and W. M. Lindsay. New York: West, 639–698.

Gabis, D. A., and R. E. Faust. 1988. Controlling microbial growth in the food-processing environment. *Food Technology* December 81–82, 89.

Grocery Manufacturers of America (GMA). 2009. Control of *Salmonella* in low-moisture foods. February 4. www.gmaonline.org/science/SalmonellaControlGuidance.pdf.

International Commission on the Microbiological Specifications of Foods (ICMSF). (2002). *Microorganisms in foods 7: Microbiological testing in food safety management*. New York: Kluwer Academic Plenum.

International Organization for Standardization (ISO)/Conformity Assessment Policy Development Committee. 2007. IEC 17025:2005. General requirements for the competence of testing and calibration laboratories. AOAC International, Gaithersburg, MD.

Johnson, S. A., P. A. Goddard, C. Iliffe, B. Timmins, A. H. Rickard, G. Robson, and P. S. Handley. 2002. Comparative susceptibility of resident and transient hand bacteria to para-chloro-meta-xylenol and triclosan. *Journal of Applied Microbiology* 93(2):336–344.

Kornacki, J. L. 2009a. The missing element in microbiological food safety inspection approaches, part I. Process control section. February/March. http://www.foodsafetymagazine.com/article.asp?id=2800&sub=sub1.

Kornacki, J. L. 2009b. The missing element in microbiological food safety inspection approaches, part II. Process control section. April/May. http://www.foodsafetymagazine.com/article.asp?id=2914&sub=sub1.

Kornacki, J. L. (ed.). 2010a. *Principles of microbiological troubleshooting in the industrial food processing environment*. New York: Springer.

Kornacki, J. L. 2010b. IAFP symposium proposal (organizer and co-convener). An Indicator Approach to Enteric Contamination of At Risk Foods. Anaheim, CA, August 1–4.

Lunden, J. M., T. J. Autio, and H. J. Korkeala. 2002. Transfer of persistent *Listeria monocytogenes* contamination between food-processing plants associated with a dicing machine. *Journal of Food Protection* 65(7):1129–1133.

Martinez-Urtaza, J., and E. Liebana. 2005. Use of pulsed-field gel electrophoresis to characterize the genetic diversity and clonal persistence of *Salmonella senftenberg* in mussel processing facilities. *International Journal of Food Microbiology* 105:153–163.

McMahon, W. A., A. M. McNamara, and A. M. Schultz. 2005. Validation and use of composite sampling for the detection of *Listeria monocytogenes* in a variety of food products. Poster presented at the International Association for Food Protection, Baltimore, MD, August 17.

Mead, P., L. Slutsker, V. Dietz, et al. 1999. Food-related illness and death in the United States. *Emerging Infectious Diseases* 5(5):607–625.

Miettinen, M. K., K. J. Bjorkroth, and H. J. Korkeala. 1999. Characterization of *Listeria monocytogenes* from an ice cream plant by serotyping and pulsed-field gel electrophoresis. *International Journal of Food Microbiology* 46:187–192.

Moorman, M., P. Pruett, and M. Weidman. 2010. Value and methods for molecular subtyping of bacteria. In *Principles of microbiological troubleshooting in the industrial food processing environment,* ed. J. L. Kornacki. New York: Springer, Chapter 10.

Scallan, E., P. M. Griffin, F. J. Angulo, R. V. Tauxe, and R. M. Hoekstra. 2011. Foodborne illness acquired in the United States—unspecified agents. *Emerging Infectious Diseases* 17:16–22.

Scallan, E., R. M. Hoekstra, F. J. Angulo, R. V. Tauxe, M. A. Widdowson, S. L. Roy, J. L. Jones, and P. M. Griffin. 2011. Foodborne illness acquired in the United States—major pathogens. *Emerging Infectious Diseases* 17:7–15.

Silliker, J. H., and D. A. Gabis. 1973. ICMSF methods studies. I. Comparison of analytical schemes for detection of *Salmonella* in dried foods. *Canadian Journal of Microbiology* 19:475–479.

Silliker, J. H., and D. A. Gabis. 1974. ICMSF methods studies. II. Comparison of analytical schemes for detection of *Salmonella* in high-moisture foods. *Canadian Journal of Microbiology* 20:663–669.

Tompkin, R. B. 2002. Control of *Listeria* in the food-processing environment. *Journal of Food Protection* 65(4):706–723.

U.S. Department of Agriculture (USDA). 2007. PCD 6, USDA commodity cheese (11.05/07). http://www.fsa.usda.gov/Internet/FSA_File/pcd6.pdf.

U.S. Department of Agriculture/Food Safety Inspection Service. 2012 (May, last modification). *Microbiology laboratory guidebook,* 3rd ed. http://www.fsis.gov/Science/Microbiological_Lab_Guidebook/index.asp

U.S. Food and Drug Administration (FDA). 2003. Food sampling and preparation of sample homogenate. In *FDA Bacteriological Analytical Manual,* Chapter 1. April. http://www.fda.gov/Food/ScienceResearch/LaboratoryMethods/BacteriologicalAnalyticalManualBAM/UCM063335 (accessed September 24, 2010).

U.S. Food and Drug Administration (FDA). 2009. Peanut butter and other peanut containing products recall (list information current as of 12 PM October 28, 2009): 3918 entries in list. http://www.accessdata.fda.gov/scripts/peanutbutterrecall/index.cfm (accessed December 23, 2010).

U.S. Food and Drug Administration Center for Food Safety (USFDA/CFSAN) 2012 (March, last revision). Bacteriological Analytical Manual—Online. AOAC International, Gaithersburg, MD. http://www.fda/gov/Food/ScienceResearch/LaboratoryMethods/BacterialAnalyticalManual/BAM/default.htm

chapter eight

Predictive modeling
Principles and practice

Peter Wareing and Evangelia Komitopoulou

Contents

8.1   Introduction

Predictive microbial modeling has been an active area of study within food microbiology since 2000. Baranyi (Baranyi 2004) reported that the number of papers with modeling or predictive microbiology as keywords rose from about 20 in 1992 to approximately 80 in 2000. Public funding in this area has given rise to strong research and extension programs at several universities in the area of predictive modeling of foodborne pathogens. While large multinational food companies and food research institutes have used predictive microbiology to characterize risk of survival and growth of foodborne pathogens and spoilage organisms for several years, the recent advancements in this field have made pertinent microbial models more accessible to smaller food companies without considerable research and development resources. As such, these models have become integral tools for safety and quality assessments as part of food process development and food product design.

Predictive modeling to describe biological kinetics could be said to have started with Gompertz in 1825 (Betts and Everis 2005) and microbial modeling with the work of Monod in the 1940s (Baranyi 2004). There are complex predictive models that can apply to microbiology, such as models that describe the growth, survival, or destruction of pathogens or spoilage microbes. The Food MicroModel was one such model (Jones 1993; McClure et al. 1994). The frequency and duration of microbial contamination events in food-processing environments has also been modeled (Powell 2006). Other nonmicrobial models have been developed for modeling food process design or operations (Hills 2001) and the distribution chain (Sloof 2001). Models have also been developed for the heat and mass transfer dynamics that occur during the processing of foods or the kinetics of D and z values (Nicolai, Verboven, and Scheerlinck 2001; Van Boekel and Tijskens 2001). Still others have modeled cooling and freezing processes (Pham 2001). Finally, quality, sensory considerations, and organoleptic shelf life are the subjects of another group of models (Wilkinson and Tijskens 2001; Hills 2001).

Microbial models that predict the death or growth of microorganisms in foods are complex and are affected by many interacting parameters. These factors can be broadly divided into intrinsic and extrinsic factors; for example, intrinsic factors include

- pH and poising capacity
- Water activity (a_w)
- Nutrient content
- Redox potential
- Antimicrobials, including organic acids and barriers

Extrinsic factors include

- Temperature
- Humidity
- Packaging and gas atmosphere

The predictability of growth, survival, or death of microorganisms in different types of foods is essential to the ability of food microbiologists to make quick decisions about food process and food product risks or to make broad and encompassing decisions about categories of food fitting within defined thresholds of these parameters. Predictive modeling is a discipline within food microbiology that aims to predict outcomes based on parameters of numerous intrinsic and extrinsic factors occurring in and around a food matrix. Food microbiology researchers use models to describe and predict microbial growth, survival, death, toxin production, or spore production/germination/destruction under prescribed conditions. The beauty and utility of modeling is that these parameters can be modeled as if they occur simultaneously, sequentially, statistically, or dynamically depending on the objectives of the researcher (i.e., modeler or model user). Models allow questions to be answered with some degree of statistical confidence much more quickly and at reduced cost versus bench-level research. Of course, the utility and validity of any given predictive microbial model will depend on the quality of bench research, mathematical aptitude, and validation that went into its development.

To develop effective models, the disciplines of microbiology, mathematics, and statistics are combined (Whiting 1995; Forsythe 2002). As such, effective microbial models have usually been the outcome of years of multidisciplinary research with well-established long-term funding. Much of this research has occurred in the public sector at university and government research centers. However, the food industry has also taken part in microbial modeling research with groups at Unilever, Kraft-Oscar Mayer, Purac, and Leatherhead Food International having each produced notably effective models.

Models can be broadly grouped into kinetic and probabilistic types (Forsythe 2000, 2002). The extent and rate of microbial growth is the subject of kinetic models, whereas probabilistic models aim to predict the likelihood of an event occurring, such as a growth/no growth scenario in an acid-preserved product. Legan (Legan 2007) described still further categories, related to these issues: kinetic growth and death, growth boundary, probabilistic growth, time to inactivation, and survival models. There are advantages and disadvantages of each type; kinetic models are used to develop growth or death curves for microbes; the latter type is used to develop D and z values, for example. Growth boundary and time-to-inactivation models are useful for predicting endpoints—such as death of *Salmonella* in a high-pressure-processed product—but these cannot be used to construct growth curves.

8.2   Types of predictive microbiological models

Models can be categorized as primary, secondary, or tertiary types. Primary models typically quantify increases or decreases in microbial biomass with time under one set of growth conditions and are measured as colony-forming units (CFU) per milliliter or per gram, increase in turbidity, changes to a component of the growth medium, or time to toxin production. Gompertz and Baranyi equations have been used to describe primary growth models (Forsythe 2000; Baranyi and Roberts 1994). Secondary models further expand on the predictions of primary models to describe how the microbial response varies with changes in environmental parameters. This is where intrinsic and extrinsic factors come into play: pH, a_w, temperature, preservative concentration, and so on. Secondary models use Bêlehrádek square root or Arrhenius equations. Tertiary models then combine primary and secondary models to determine the response to changing environmental conditions and to compare the effects of the different conditions. These are usually presented as much more user-friendly computer-based models, either freely available over the Internet as Web-based software or a free download to be installed on the user's computer or a commercial package.

Two typical examples of tertiary models that can determine the growth or death of pathogens are the Pathogen Modeling Program (PMP) from the U.S. Department of Agriculture Agricultural Research Service (USDA-ARS) (Tamplin, Baranyi, and Paoli 2004), and Growth Predictor from the Institute of Food Research (IFR) in the United Kingdom. The latter can only handle growth predictions at present since it does not have a dataset for thermal death. The Food MicroModel was another example, but it is no longer available; the dataset has been subsumed into Growth Predictor and a much larger relational database called ComBase. ComBase is a large database of growth, death, and survival rate parameters (Baranyi and Tamplin 2004). An Internet-based version was launched in June 2003 and now contains over 50,474 records that can be extracted for study and model development. The database is now split into the ComBase Browser and ComBase Predictor; the latter works in much the same way as PMP and Growth Predictor.[*]

8.2.1 Simple models

A range of models has been developed, from simple calculations that predict the theoretical safety and stability of foods, to complex computer-based models. Examples of the former include a code of practice of the Federation of the Condiment Sauce Industries of Mustard and of Fruit and Vegetables Prepared in Oil and Vinegar of the European Union and the Preservation Index (Anonymous 1993; Federation of the Condiment Sauce Industries of Mustard and of Fruit and Vegetables Prepared in Oil and Vinegar of the European Union 2006). Both use simple models to determine microbial stability for acetic acid-based sauces and dressings. Both

[*] Internet links to these models and related databases can be found at the end of this chapter.

have limitations; they can only be used if acetic acid is the acidulant. The simple equation for the Preservation Index is

$$\frac{\text{Total acidity} \times 100}{(100 - \text{Total solids})} \geq 3.6\%$$

Another simple calculation is involved in the mold-free shelf life of flour-based confectionery and bakery products, developed by Seiler of Campden BRI in the United Kingdom.

8.2.2 Pathogen models

As noted, Growth Predictor and PMP are packages that are available free for download. Both are user-friendly models. The Pathogen Modeling Program is available in two formats, a downloadable version and an Internet version. Download version 6 is suitable for older Windows 95™ computers, version 7 for more recent computers. ComBase is also available as a downloadable or an Internet version. PMP consists of 10 pathogens in 38 models. These include growth, thermal inactivation, cooling, survival, irradiation, and time to turbidity or toxin. An example of the output of the PMP 7.0 model is shown in Figure 8.1.

ComBase is now administered by the ComBase Initiative, which is a collaboration between the research agencies in the United Kingdom, United States, and Australia. The ComBase Browser contains the ComBase database, and the ComBase Predictor (similar to PMP). There is also a Modeling Toolbox, which contains a collection of software tools based on ComBase data to predict the growth or inactivation of microorganisms. Currently available predictive tools include two online applications. The first application is ComBase Predictor. This is a set of 23 growth models and 6 thermal death models used to predict the response of the major foodborne pathogens and key spoilage microorganisms to key environmental factors. The second application is one that was developed by the IFR called Perfringens Predictor. This application is specifically designed to predict the growth of *Clostridium perfringens* during the cooling of cooked meats. The input is temperature over time, and the output is a viable count of *C. perfringens*. In addition, ComBase Toolbox has two packages that allow the user to compare ComBase records with prediction, or to fit the data, and a Web-based version of the IFR package, DMFit. This application is designed to fit log counts versus time data and extract parameters such as lag time/shoulder. MicroFit from the IFR is a stand-alone program that analyzes microbial growth data from sources like ComBase. The former MicroModel program contained growth models for all of the main pathogens, models for spoilage bacteria and yeasts,

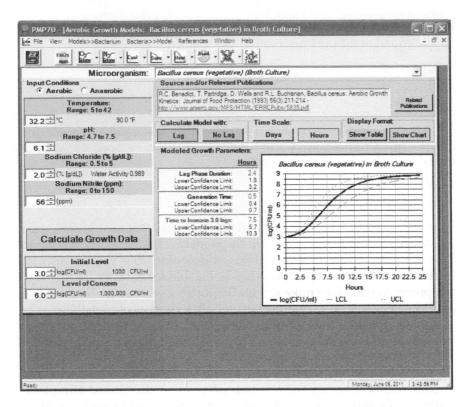

Figure 8.1 Tertiary model output of the Pathogen Modeling Program (PMP 7.0, version number 1.1.1433.15425; USDA-ARS [U.S. Department of Agriculture-Agricultural Research Service], NAA [north Atlantic area], ERRC [Eastern Region Research Center], MFSRU [Microbial Food Safety Research Unit, USDA]) showing predicted growth of *Bacillus cereus* vegetative cells in broth culture under aerobic conditions.

thermal death models for pathogens and yeasts, and a survival model for *Campylobacter jejuni* (McClure et al. 1994).

Growth Predictor is a downloadable package that, as the name suggests, is concerned purely with microbial growth. The majority of the 11 models are for pathogens, with *Brochothrix thermosphacta* an example of one of the spoilage models included in the program. Controlling factors in Growth Predictor are pH, salt, and temperature. The user can switch between salt and a_w, which is useful especially if the a_w of a food system is affected by ingredients other than salt. Additional factors include nitrite or CO_2 for some models within Growth Predictor. No survival or death models are included in the current version of Growth Predictor. A stand-alone version of Perfringens Predictor is associated with Growth Predictor. Table 8.1 shows the range of organisms for these models.

Table 8.1 Comparison of Available Models in Three Typical Tertiary Models

Microorganism	Growth predictor	Pathogen modeling program[a]				Forecast
	Growth	Growth	Survival	Thermal death	Cooling	Growth
Aeromonas hydrophila	Yes	Yes				
Bacillus spp.						Yes
Bacillus cereus	Yes	Yes				
Bacillus subtilis	Yes					
Bacillus licheniformis	Yes					
Brochothrix thermosphacta	Yes					
Clostridium botulinum (np)[b]	Yes			Yes		
Clostridium botulinum (p)[c]	Yes				Yes	
Clostridium perfringens	Yes	Yes			Yes	
Enterobacteriaceae						Yes
Escherichia coli O157	Yes	Yes	Yes	Yes		
Lactic acid bacteria			Yes			Yes
Listeria monocytogenes[d]	Yes	Yes	Yes	Yes		
Pseudomonas						Yes
Salmonella[d]		Yes	Yes			
Salmonella Typhimurium		Yes				
Shigella flexneri		Yes	Yes			
Staphylococcus aureus	Yes	Yes				
Yeasts (chilled)						Yes
Yersinia enterocolitica[d]		Yes				

[a] Version 7.0.
[b] Nonproteolytic.
[c] Proteolytic.
[d] Various parameters.

Risk Ranger is a simple food safety risk calculation tool intended as an aid for determining relative risks from different product, pathogen, and processing combinations. Risk Ranger provides a simple and quick means to develop a first estimate of relative risk, which means it can offer food safety managers a quick read about potential safety of new products and processes. It is a generic but robust model that uses information about all elements of food safety to make risk calculations. In addition to ranking risks, Risk Ranger helps to focus the attention of the users on the interplay of factors that contribute to foodborne disease. The model can be used to explore the effect of different risk-reduction strategies or the extent of change required to bring about a desired reduction in risk. The model can be used by risk managers and others without extensive experience in risk modeling. It is a useful tool for helping to train food safety managers to think in terms of risk and to communicate risk to the businesses they serve. The user is required to answer 11 questions related to the severity of the hazard, likelihood of a disease-causing dose of the hazard being present in a meal and the probability of exposure to the hazard.*

The Opti.Form® *Listeria* Control Model 2007 (Purac, the Netherlands) is a seventh-generation tertiary model specifically designed to predict the growth rate of *L. monocytogenes* in ready-to-eat meats as affected by levels of sodium or potassium lactate, sodium diacetate, NaCl, and moisture. Preceding iterations of the model were developed through cooperation with industry scientists at Kraft Foods (Oscar Mayer) (Seman et al. 2002; Legan et al. 2004), and the latest version incorporates their research as well as nearly 30 peer-reviewed articles and internal Purac data. The latest version also enables prediction of growth based on levels of other proprietary growth inhibitors. The model enables users to enter in NaCl ranges of 0 to 4.3 % w/w of finished product, as demonstrated in Figure 8.2.

However, model predictions in general should be interpreted with caution (Black and Davidson 2008) and should not replace individual product assessments, especially with substantial formulation changes. For instance, a tertiary model with a 95% confidence interval prediction may neglect the error in original experimental data and the error in the primary model fit, which can actually be greater than the secondary model uncertainty (Marks 2008).

For those dedicated to research, FARE Microbial™ (Exponent, Inc. [formerly Novigen Sciences, Inc.], Washington, DC) modular software is

---

* Editor's Note: This and other risk assessment tools can be found at http://www.foodrisk. org, an online repository of food safety-related models and data summaries developed by the Joint Institute of Food Safety and Applied Nutrition, a collaboration between the University of Maryland and the U.S. Food and Drug Administration. The repository contains models related to incidence and behavior of foodborne pathogens, as well as other risk-modeling tools.

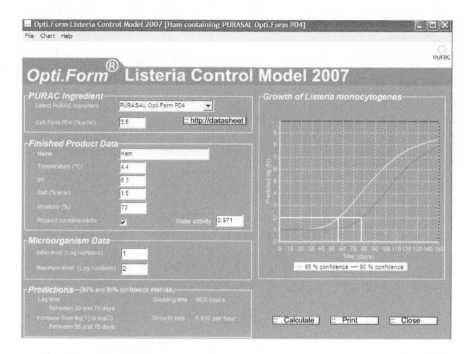

Figure 8.2 Tertiary model output of the Opti.Form Listeria Control Model 2007 (Purac Biochem bv, Gorinchem, the Netherlands) depicting predicted growth at 4.4°C of *Listeria monocytogenes* on a cured ham with 3.5% of an antimicrobial solution with a pH of 6.3, a_w of 0.971.

for performing probabilistic microbial risk assessment for a wide variety of foodborne pathogens. FARE Microbial consists of two modules: the Contamination and Growth Module and the Exposure Module. The program is available for free download (http://www.foodrisk.org). The Contamination and Growth Module (Figure 8.3.) must be run prior to running the Exposure Module. The output of the Exposure Module is a distribution representing colony-forming units of the microorganism of interest present in units of food at the time of consumption. This information can be used for prediction of a number of illnesses caused by a particular pathogen through the specific food vehicle modeled in the Contamination and Growth Module.

8.2.3 Spoilage models

In addition to models for the growth or death of pathogens, there are several models that predict the growth of spoilage organisms. Simple models include a mold spoilage model developed using *Penicillium roqueforti* (Valík, Baranyi, and Görner 1999). Such models can be useful for

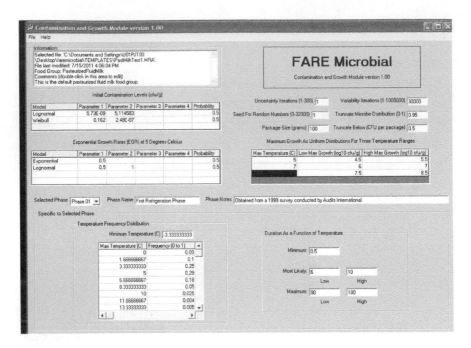

Figure 8.3 The Contamination and Growth Module of the FARE Microbial model (Exponent, Inc., Washington, DC). Users must input model choices for initial contamination and exponential growth rates. The output of this model can be used to run the Microbial Exposure Module.

determining the time for mold growth in intermediate- and low-moisture foods, for instance.

The Forecast model developed by Campden BRI (formerly the Campden and Chorleywood Food Research Association [CCFRA]) was developed to model various spoilage organisms or groups, including *Pseudomonas* and *Bacillus* (Walker and Betts, 2000; Betts and Everis, 2005). It is similar to the original Food MicroModel system, which was a phone-based subscriber service, before being licensed to users' sites. Most models deal with kinetic growth and can include lag time. Table 8.1 shows the organisms available for the Forecast model.

The Seafood Spoilage and Safety Predictor (SSSP) was developed by the Danish Institute for Fisheries Research at Lyngby. It was first developed as a spoilage model but now has been expanded to include safety models based on the growth of *L. monocytogenes*. The purpose of the model is to predict the shelf life and growth of bacteria in fresh and lightly preserved seafoods. Temperatures of products during storage and distribution are recorded via data loggers and used in one of two models: Relative Rate of Spoilage and Specific Spoilage Organism (SSO). The model is able to cope with temperature fluctuations in the logging data, as would happen in a live situation. There are growth and growth boundary model SSO models for *Listeria monocytogenes* (temperature, salt [NaCl/a_w], pH, CO_2, smoke intensity, nitrite, and organic acids). This model has been validated for both seafood and meat products. It also contains models to predict histamine formation by *Morganella psychrotolerans* and *Morganella morganii* and a model to predict the simultaneous growth of *Listeria monocytogenes* and lactic acid bacteria. Version 3.1 (2009) is available as a free download.

Food Spoilage Predictor was developed by McMeekin and Ross (1996) at the Department of Agricultural Science, University of Tasmania, originally as *Pseudomonas* Predictor. This used temperature function integration software to model the effects of temperature, a_w, and pH on the growth of psychrotrophic spoilage *Pseudomonas* spp. in proteinaceous foods. This was a commercial package that used datalogger results to determine the likely growth of psychrotolerant pseudomonads in proteinaceous foods.

ERH CALC is a program designed to calculate the equilibrium relative humidity (ERH) of foods and thus enable predictions of the mold-free shelf life (MFSL) for bakery products at either 21 or 27°C (Jones 1994; Blackburn 2000). It was developed by the Flour Milling and Baking Research Association and is now available from Campden BRI.

CoolVan, developed by the Food Refrigeration and Process Engineering Research Center (FRPERC), University of Bristol, is a program that predicts food temperatures in refrigerated transport, allowing potential adverse changes to be noted. This is a commercial package. It can calculate the effects of weather, delivery pattern, vehicle and refrigeration type, and food on onboard temperatures.

Other packages that have been developed include Chefcad, from Flanders Center for Postharvest Technology. This was a food recipe design and analysis package. It is no longer available.

8.3   Development of predictive models

A number of discrete phases to model development have been outlined (Walker and Betts 2000; Adams and Moss 2008):

- Planning: selection of target
- Data collection: inoculation of target into growth medium, incubation of test matrix, sampling of matrix at prescribed intervals
- Model fitting: construction of a model based on the target response to the variables
- Model validation: ideally in a typical food
- Refine model: use the model over time and determine if it is always a good fit for the data collected

Some questions that should be addressed during the planning stage include the following: What is the main concern: spoilage or safety, a particular organism or a group of organisms? Which is the important variable that needs to be considered: toxin production, growth or death rate, or time to spoilage? For spoilage, what is the independent variable that will be measured (time to microbial number, or rancidity level, measurable odor/slime/color change, visible growth—of mold, for example)? Will the model be based on existing models or start from scratch? Models include Arrhenius, nonlinear Arrhenius, square root (Ratkowsky or Bêlehrádek), polynomial or mechanistic, or dynamic models.

At least some basic background knowledge of food microbiology is required to know the relevant variables to be measured and the frequency of measurement. For example, how many strains should be studied? Are there any data on the most and least-resistant strains for the variables under study? Risk assessment from a food safety or spoilage perspective can guide selection of these factors.

- Food businesses need models that cover these aspects of microbial growth:
  - Lag phase
  - Growth
  - Death and inactivation
  - Survival
  - Probability of die-off
  - Effects of intrinsic, extrinsic factors & implicit factors

It is useful to be able to switch off any lag phase determinations within a model if the microbe is likely to have a limited (or no) lag phase, for example, modeling the behavior of a contaminant of different products with similar intrinsic properties such as pH or a_w. Another example would be modeling behavior of enteric pathogens from the slaughter process, as bacteria would likely be in log phase in animal host intestinal tracts. Models should include basic parameters: temperature, pH, and a_w. It is also helpful if models can examine the effects of specific factors for certain microbes, such as levels of nitrite, lactic acid, acetic acid, or CO_2.

It is better if the pH is adjusted with hydrochloric acid since weak organic acids are themselves antimicrobial, with each organic acid having its effects on different metabolic functions or structural processes. Again, it is helpful if models permit choice between sugar and salt—usually as a salt/a_w option in the model; in this way, products in which salt is not the main a_w-influencing factor can be examined.

8.3.1   Experimental design: Principles of challenge testing

The design of the modeling experiments is a vital phase in the development of models (van Boekel and Zwietering 2007). It is an iterative process, starting with hypothesis generation, leading to experimental design, carrying out the experiments, and results analysis. This leads back to a review of the hypothesis and changes where required. It is vital that experiments are well designed; poor design will lead to poor data collection, and no amount of statistical analysis can correct for experiments that were improperly designed, conducted, or interpreted. Guidance on proper procedures for challenge testing of foods has been published (Uyttendaele et al. 2004; Scott et al. 2005; National Advisory Committee on Microbiological Criteria for Foods 2010) and is covered in detail in Chapters 3 and 4. Challenge testing is covered here also with respect to building datasets from which to construct predictive models.

Challenge testing is distinct from shelf life testing, although often is confused with it (Betts and Everis 2000). Shelf life determinations using microbial assays answer this question: Is the product safe and stable under normal processing and storage conditions, given normal microbial loading? Shelf life therefore assumes good manufacturing conditions under a hazard analysis and critical control point (HACCP) plan. Shelf life testing targets microflora that are naturally present and could potentially cause spoilage during storage under stipulated conditions, whereas challenge testing seeks to answer this question: Would the product still be safe to consume and stable if it was contaminated with specific spoilage organisms or food pathogens? In other words, shelf life determinations do not involve the inoculation of the product with extraneous microbes, whereas challenge testing does. It may be that in the former case, the natural background flora may have a significant effect on shelf life, including pathogens if present. For example, psychrotrophic *Clostridium botulinum*, which could survive mild heating processes that kill spoilage microflora, may be permitted to grow if the storage temperature were above 3°C.

The same question could be asked during challenge testing. For example, suppose that contamination with psychrotrophic *C. botulinum* was infrequent but possible, and the thermal process was not expected to destroy it. A challenge test will then determine if the product is safe with respect to growth of *C. botulinum* under normal storage conditions and those of mild temperature abuse (e.g., a refrigerator running at above 8°C). Challenge testing can help to establish a product's food safety objectives and shelf life, as well as aid in product formulation or reformulation (intrinsic control factors: pH and a_w). It therefore simulates what could happen to a potential contaminating microorganism during production, processing, distribution, or handling from production to consumption.

An important and often-overlooked parameter is the question of which organisms should be used in a challenge test. A knowledge of the intrinsic properties of the product and previous history of incidences of foodborne illness associated with the product type is required in selecting inocula for challenge testing. These and other decisions should be left to trained and skilled individuals who would have an overall awareness of major product risks and familiarity with risk assessment procedures (i.e., based on critical product and process factors, which organisms would most likely grow/survive?).

Should natural isolates that are similar to the food matrix strains be used? Naturally occurring surrogate organisms may be necessary when attempting to determine behavior of hazardous microorganisms that cannot be used in the processing plant (e.g., *Clostridium sporogenes* instead of *Clostridium botulinum*; *Listeria innocua* instead of *Listeria monocytogenes*). In which instances is a surrogate acceptable, and will it perform similarly to the pathogen in tests? A disadvantage of some natural strains is that they can be slow growing or slow to recover from environmental stresses. In this case, strains from a culture collection could be adapted to specific environmental stresses. They may adapt to the new conditions and grow more rapidly; this can give faster growth rates and shorter lag times, which would give a fail-safe result. This is not ideal for measuring the effect of stress application; in that case, prolonged adaptation time in product would be advisable.

When there is a requirement for challenge testing with one species, the ideal challenge inoculum is a cocktail of three or four strains of the same species. This accounts for at least some degree of potential variability between strains within a species, provided there is a lack of antagonism. The same applies to the use of a cocktail of different species, which may re-create realistic scenarios but can be difficult to control. Some of the limitations of using cocktails are that strains in a cocktail may either grow faster (they can utilize a nutrient produced by other strains in the cocktail) or grow more slowly (such as by growth competition from strains). If different species are used in a cocktail, the use of selective media to distinguish between different species can potentially jeopardize recovery depending on the challenge matrix. This can be overcome in many cases by use of a thin agar layer method; the sample is plated on a thin nonselective nutrient agar layer, allowed to resuscitate for a brief period, and then overlaid with molten selective or differential agar medium. For example, this type of method has been shown to resuscitate stressed cells of *E. coli* O157:H7 successfully (Hajmeer et al. 2001).

Are the challenge tests for thermal death/survival or growth? In the former case, a higher inoculum is used (10^5–10^8), in the latter, lower (10^2–10^3). Note that less than 100 cells per gram or per milliliter is below detectable limits for some methods (it gives a "safe" product when it is not). Alternatively, inoculating with more than 1000 cells per sample could overcome intrinsic preservation properties of the food matrix (the product appears not to be safe when it is) (Betts and Everis 2000).

The growth/no growth interface can be determined experimentally in different food systems as affected by various parameters. This is appropriate and necessary when any increase in population of the microorganism under study cannot be tolerated in a given food system. If sufficient data are collected from enough parameters, the growth/no growth interface can be modeled (Boziaris et al. 2007; Skandamis et al. 2007; Hajmeer and Basheer 2003; McMeekin et al. 2000). For example, no increase in population of pathogens such as *C. botulinum* in sous vide food products would be tolerable, and a growth/no growth determination would be most fitting.

Inocula can be cultivated under optimum growth conditions for many types of experiments. However, in many cases a process of adaptation or "stress hardening" must be undertaken to avoid shocking cells when they are added to food matrices or to mimic worst-case conditions in which pathogens could become introduced into the food. In any case, it is often appropriate to centrifuge pelletize and wash cells to avoid transfer of nutrients from the growth medium to the food matrix. Pellets are resuspended in a suitable carrier matrix, such as buffer, saline, or liquid product (e.g., juice). Different methods may be employed, such as lyophilization of bacterial culture and product inoculation by use of a dry, dust-like vector (De Roin et al. 2003). Dry carriers should particularly be considered for inoculation of dehydrated products such as flour, dry spice, and other dry ingredients. With the liquid inoculum method, typically the product is inoculated at minimum volume (e.g., 0.1 mL per 100 g sample) to avoid affecting the properties of the matrix. If spores are used, they are washed and stored in distilled water to prevent germination and then heat shocked immediately prior to inoculation to represent processing conditions. For vegetative cells, alternative cell cultivation using lawn plates (24–48 h) can be used. Lawn-collected cells show enhanced survival compared to broth-collected cells and can represent the worst-case scenario in challenge testing (Komitopoulou and Penaloza 2009). The inoculation method should not substantially alter any of the product's intrinsic or extrinsic properties and should be reproducible. Inoculation methods should be properly validated as such by pre- and postinoculation analysis of the product's critical characteristics. Except for special circumstances when a non-homogeneous distribution of the inoculum is called for, even distribution of the inoculum within the product matrix or on the appropriate surface can maximize microbial exposure to the product's environment and minimize sampling and enumeration errors.

If doing a challenge study throughout a product shelf life, trial duration should be at least for the desired/intended shelf life of the product, preferably including a margin beyond this time to account for consumers storing product beyond its recommended life (such as 1.25 times the shelf life). Storage temperature needs to replicate conditions the product is likely to be exposed to under normal storage conditions, and usually slightly abusive storage conditions (during transport or at retail) should be included. Simple models to simulate the temperature changes in the manufacturing and retail chain may prove useful in researching temperature cycling effects on inocula in food.

Sample analysis times should include time zero analysis immediately after product inoculation to verify inocula levels. Subsequent data points depend on the duration of trials, but a minimum of seven points during storage is a good general guideline. At each point, a minimum of two independent samples should be taken. Positive and negative controls are required to demonstrate that the inoculum behaved as expected during the study irrespective of the variables being tested. It is wise to test the background flora and the intrinsic factors of the uninoculated matrix concurrent to analysis of the inoculated samples.

Interpretation should involve the use of all sets of data: microbial growth, any changes in physicochemical and other intrinsic/extrinsic properties of the product, or the growth or otherwise of positive and negative controls. Trend analysis of data by graphical plotting of mean log counts against time or mean survivor curves is a useful technique. It is important to make sure that the pass and fail criteria for the product are defined.

Challenge testing experimentation does have some limitations. It is time consuming, requiring much physical effort undertaken in the laboratory or pilot plant. An appropriately designed and maintained lab is required with correct containment facilities and with qualified microbiologists on hand in the lab. Also, a considerable amount of time must elapse before results are obtained, particularly for challenge studies that span the anticipated shelf life of a food product. The heavy labor requirement means that it is an expensive endeavor not only for conducting specific experiments but also for the maintenance of a laboratory space for this purpose (see Chapter 2). It can be cost effective if bulk trials are carried out and the amount of work and impact to the food business are sufficient to justify continued investment. It does require microbiological and statistical skills to design, set up, monitor, and interpret the results. Finally, results are only valid for the specific product formulations challenged under controlled and often-limited conditions for the purposes of validating specific products or processes. This means that variations to either product or process may negate prior validation work, leading to the need to repeat the expensive and time-consuming challenge testing. This reinforces the need for development of predictive models, which can rapidly model outcomes due to variations. Although at least some challenge testing is a necessary part of robust model development, experimental designs such as response surface modeling can facilitate this process by selecting representative variables and limiting the scope of challenge studies.

8.3.2   Data collection and modeling

Once the experiments have been designed and set up, the relatively easy and fun process of data collection can occur. However, data collection can also be an error-strewn process since the experimenter is trying to take measurements from a biological system that is inherently variable, sometimes unpredictably (van Boekel and Zwietering 2007). The models that are developed are an attempt to best fit the data to the model, sometimes better (closer) than others. Ascribing relevant time intervals and the right parameters to monitor are vital: too few, and critical processes will be missed; too many, and there is a danger that there are too many data to analyze in the time required, and that as the number of parameters increases, the uncertainty in the data also increases (van Boekel and Zwietering 2007). Some understanding of experimental errors is required to be able to separate out the noise from the data.

After data collection, the next phase is data processing; the data are fitted to the model. A goodness-of-fit estimate is made on the data to the model. Regression techniques (linear, nonlinear, which may be weighted) are carried out on the model parameters like growth rate. Weighting is often applied because the variance of microbial growth data is not constant. Transformations can also be applied to stabilize the variance (log, square root or both, depending on the parameter; these were reviewed by van Boekel and Zwietering 2007). Comparisons can be made between the model and other existing models to discriminate between models in terms of effectiveness.

8.3.3   Model validation

There are several phases to model validation. These include internal validation against the data from which the model was derived, and external validation, using one of several methods (Alzamora et al. 2005). The phases are approached as follows:

- By testing against new experimental data under a range of circumstances
- By testing in a variety of food matrices
- By validating against existing challenge test data collected from the literature
- By using naturally contaminated products in the lab or from industry, under controlled conditions

Models need to be validated on a continual basis to make sure that they are current for strains of the species under test. Graphical and statistical methods can be used for validation (observed vs. predicted values, or mean square error, goodness of fit, bias, and accuracy factors).

8.4   Applications of predictive models

Some of the parameters that a good model can estimate are lag time, generation time, growth curves, and time to event x. These parameters are useful in a food-manufacturing context (Betts and Everis 2005). For example, lag time of a bacterial pathogen like *Salmonella* at refrigeration would be helpful in determining a temperature critical limit in production of raw meat.

8.4.1   Product development and modification

Models can be used at the new product development (NPD) stage to answer questions such as: How stable is this product under normal storage conditions? What will happen if temperature abuse occurs? Modeling can be used to develop specifications; putting different levels of spoilage organisms into a model can show the maximum starting inoculum that can be tolerated before spoilage occurs. This can be used to set food safety objectives for products.

The NPD trend today is for low levels of preservatives, including traditional "chemical" preservatives such as sugar and salt. A model can show the effects of reduced salt or acid levels (increased pH) on microbial growth: When will *C. botulinum* move out of the lag phase and start to grow? Can we achieve the required 25-day shelf life with low-salt chilled fresh pasta, particularly if organic ingredients are used with higher levels of spoilage yeasts? Models can help to modify the product formulation until the required shelf life is achieved. For example, using the Public Health Laboratory Service (PHLS, now Health Protection Agency, HPA) microbiological criteria guidelines for ready-to-eat food (Gilbert et al. 2000) and noting what the microbiological specifications should be at the start of shelf life, modeling can work out what the specifications should be at the start.

8.4.2   Process design

Modeling can also be used to determine the effects of changing processing parameters on the microbial loading for safety or spoilage issues. For instance, several attempts have been made to model the high hydrostatic pressure inactivation of *L. monocytogenes* (Youart et al. 2010; Saucedo-Reyes et al. 2009) and spoilage organisms in processed meats (Slongo et al.

2009). Process design modeling can be used during risk assessment studies, to determine worst-case scenarios, and to identify the hazards in the food or in the process and then to design the process to remove the hazards. For example, the presence of *Salmonella* in eggs should be taken into account for any egg-based product, including mayonnaise. For how long should the finished product be stored, and at what temperature, to cause sufficient death of *Salmonella*? The formulation and storage conditions can be designed to reduce the likelihood of pathogen growth or toxin production, and modeling can help determine what level of risk reduction occurs with respect to alteration of parameters (e.g., decrease the pH below 4.5, increase the salt level by 0.5%).

8.4.3   Shelf life determination

Similar to initial NPD studies, the final formulation and storage conditions can be factored into the model to determine the expected shelf life under normal and abuse conditions. This can be done to simulate the distribution chain, retail display, and the consumer home. Such models can answer: What is the effect of opening a multiuse product on shelf life with respect to spoilage organisms—lactic acid bacteria, yeasts or molds? Does this mean the consumer has to store the opened pack in the refrigerator, or is normal ambient temperature sufficient? It may be that by looking at the effects of the formulation and storage conditions on the growth of different spoilage organisms it becomes apparent that lactics are the biggest problem, in which case the specification needs to reflect this.

8.4.4   Quality assurance

Models are a useful training tool, showing staff how subtle changes in formulation can have major spoilage or safety implications. Wilkinson and Tijskens (Wilkinson and Tijskens 2001) discussed the use of models to help companies to understand the relationship between perceived quality and product attributes, and from that knowledge to predict and control quality through the production process and throughout the distribution chain, and by that means to optimize product quality. Processes can be product or consumer centered; the former approach is science based, and the latter uses consumers to either assess appreciation (like/dislike) or evaluate a parameter (saltiness, toughness). Quality assessment models examine the relationship between quality attributes and perceived quality, whereas dynamic product models describe the effects of external parameters (process or storage temperature, light, gas atmosphere). Models developed from the parameters described can then be integrated into quality control plans. Gathering these data and then running them through the model will show the effect on perceived product quality.

8.5   Challenges and limitations

Sometimes, populations of microbes do not behave as predicted. For example, they may not die as quickly as expected, or one subpopulation may die at a different rate than another. Diels et al. (2007) reviewed this for thermal and high-pressure treatments, noting that non-first-order decay would not always occur. Indeed, linear inactivation curves are easiest to analyze, but in practice they are often sigmoidal with "shoulders" and "tails." Conversely, growth curves do not always follow the expected sigmoidal shape and may deviate from distinct lag, exponential, stationary, and death phases. Sometimes, there are biological reasons like a second-order decay mechanism or overlapping populations of the same organism, each with different reactions to growth substrates or heating matrices. Sometimes, these confounding phenomena are merely due to experimental artifacts like poor experimental design or variability in data collection techniques. Even cellular adaptations can disrupt first-order reaction kinetics if the process examined is mild enough and carries on for long enough. The development of resistance to adverse environmental factors can be modeled, such as the acid tolerance response in *Salmonella typhimurium* and *L. monocytogenes* (Greenacre et al. 2006; Greenacre and Brocklehurst 2006; Greenacre et al. 2003).

One of the biggest problems with any model is how well it works in a real food matrix, as opposed to a broth-based system. It may be that bacteria are more resistant to heat or pressure in a complex matrix (for example, meat or a ready meal). Reyns et al. (2000) noted that the simple model developed for *Zygosaccharomyces bailii* in buffer at pH 6.5 worked in low-pH buffer but not apple or orange juice. Heat inactivation models developed in broth systems may have much different outcomes in food systems that offer heat protection due to solute or lipid.

Extrapolation of models beyond their validated parameters can be treacherous due to the issues mentioned. Conversely, interpolation may also be inaccurate if a combination of parameters beyond the experimental data set is chosen. For example, a model might have data for different parameters from pH 3.5 to pH 8.5 and a_w 0.999–0.90 and from 5 to 35°C, but not for every combination of factors. Although response surface modeling can overcome some of these data gaps, the validity and robustness of model use will depend on the prior selection of variables for experiments that were used to develop the model originally. Food microbiologists should make interpretations of model outcomes for these extrapolated or interpolated parameters. The source of the data used in models can be laboratory media based, food matrices, or data from published literature. Researchers should note these data sources when applying models to specific food microbiology questions.

8.6   Future trends

At the time of writing, the food industry was firmly in the grip of the "clean label" movement. As such, there is a particular need to predict the effects of the removal of preservatives while maintaining shelf life (never easy). There is a demand for natural ingredients and product reformulations, for the replacement of chemical preservatives, and for salt, sugar, and fat reduction. For these reasons, the use of relevant predictive models will continue to be part of the first stage in any shelf life and safety assessment.

The continued growth of online databases like ComBase will increase the access to and usefulness of modeling (Legan 2007). ComBase data will be used by individuals to develop their own models that relate to their particular product or processing situation by the use of curve-fitting programs such as DMFit. Tailor-made models will help in product reformulation, and this may be a growing area of specialized research. It is unlikely that modeling will ever completely replace challenge testing or other bench science. What is likely is that predictive microbial modeling is here to stay as a necessary research and development tool. Integration of microbial predictive models into food process monitoring and food safety management is possibly one way modeling will become even more prevalent.

8.7   Sources of further information

CoolVan was available for purchase from http://www.frperc.bris.ac.uk/; now, any enquiries regarding modeling should be directed to http://www.rdandt.co.uk/.

> ComBase: http://www.combase.cc/index.php/en
>
> ComBase Toolbox, including ComBase Predictor, Perfringens Predictor, and DMFit: http://www.combase.cc/index.php/en/downloads/category/11-dmfit
>
> ERH CALC: available for purchase from http://www.campden.co.uk/content.htm
>
> FoodRisk.org, Joint Institute for Food Safety and Applied Nutrition: http://foodrisk.org
>
> Food Spoilage Predictor: http://www.ars.usda.gov/SP2UserFiles/ad_hoc/19353000CenterforExcellence/FSPsoftware.pdf
>
> Forecast Bureau-based (paid for) system: http://www.campden.co.uk/content.htm
>
> Growth Predictor and Perfringens Predictor download: http://www.ifr.ac.uk/Safety/GrowthPredictor/
>
> Listeria Growth Model: http://www.purac.com
>
> Pathogen Modeling Program: http://www.ars.usda.gov/services/docs.htm?docid=11550

Predictive Microbiology Information Portal: http://portal.arserrc.gov/overview.aspx

Risk Ranger: http://www.foodsafetycentre.com.au/riskranger.php

Seafood Spoilage Predictor: http://sssp.dtuaqua.dk/Download.aspx

References

Adams, M. R., and M.O. Moss. 2008. *Food microbiology*. Cambridge, UK: RSC.

Alzamora, S. M., S. Guerrero, P. E. Voilaz, and J. Welti-Chanes. 2005. Experimental protocols for modeling the response of microbial populations exposed to emerging technologies: some points of concern. In *Novel Food Processing Technologies*, ed. G. V. Barbosa-Canovas, M. S. Tapia, and M. P. Cano, pp. 591–607. Boca Raton, FL: CRC Press.

Anonymous. 1993. Code for the production of microbiologically safe and stable emulsified and non-emulsified sauces containing acetic acid. The CIMSCEE code. In *Shelf life of foods—guidelines for its determination and prediction*. London: Institute of Food Science and Technology, pp. 49–51.

Baranyi, J. 2004. Predictive modeling to control microbial hazards in the food processing industry. In *Food safety assurance and veterinary public health— safety assurance during food processing*, ed. F. J. M. Smulders and J. D. Collins. Wageningen, the Netherlands: Wageningen Academic, pp. 145–156.

Baranyi, J., and T. A. Roberts. 1994. A dynamic approach to predicting bacterial growth in food. *International Journal of Food Microbiology* 23:277–294.

Baranyi, J., and M. L. Tamplin. 2004. ComBase: a common database on microbial responses to food environments. *Journal of Food Protection* 67 (9):1967–1971.

Betts, G., and L. Everis. 2000. Shelf-life determination and challenge testing. In *Chilled foods. A comprehensive guide*, ed. M. Stringer and C. Dennis. Boca Raton, FL: CRC Press, pp. 259–285.

Betts, G., and L. Everis. 2005. Modeling systems and impact on food microbiology. In *Novel food processing technologies*, ed. G. V. Barbosa-Canovas, M. S. Tapia, and M. P. Cano. Boca Raton, FL: CRC Press, pp. 555–578.

Black, D. G., and P. M. Davidson. 2008. Use of modeling to enhance the microbiological safety of the food system. *Comprehensive Reviews in Food Science and Food Safety* 7:159–167.

Blackburn, C. D. W. 2000. Modeling shelf-life. In *The stability and shelf-life of food*, ed. D. Kilcast and P. Subramanian. Boca Raton, FL: CRC Press, pp. 55–78.

Boziaris, I. S., P. N. Skandamis, M. Anastasiadi, and G. J. E. Nychas. 2007. Effect of NaCl and KCl on fate and growth/no growth interfaces of *Listeria monocytogenes* Scott A at different pH and nisin concentrations. *Journal of Applied Microbiology* 102 (3):796–805.

De Roin, M. A., S. C. Foong, P. M. Dixon, and J. S. Dickson. 2003. Survival and recovery of *Listeria monocytogenes* on ready-to-eat meats inoculated with a desiccated and nutritionally depleted dustlike vector. *Journal of Food Protection* 66 (6):962–969.

Diels, A., I. V. Opstal, B. Masschalck, and C. W. Michiels. 2007. Modeling of high-pressure inactivation of microorganisms in foods. In *Modeling microorganisms in food*, ed. S. Brul, S. van Gerwen, and M. Zwietering. Boca Raton, FL: CRC Press, pp. 161–197.

Federation of the Condiment Sauce Industries of Mustard and of Fruit and Vegetables Prepared in Oil and Vinegar of the European Union. 2006. Code of practice. Available from http://fic-webrelease.agepnet.com/FICEUROPE_Code_of_Practice_EN.pdf; accessed October 24, 2006.

Forsythe, S. J. 2000. *The microbiology of safe food*. Oxford, UK: Blackwell.

Forsythe, S. J. 2002. *The microbial risk assessment of foods*. Oxford, UK: Blackwell.

Gilbert, R. J., J. de Louvois, T. Donovan, C. Little, K. Nye, C. D. Ribeiro, J. Richards, D. Roberts, and F. J. Bolton. 2000. Guidelines for the microbiological quality of some ready-to-eat foods sampled at the point of sale. PHLS Advisory Committee for Food and Dairy Products. *Communicable Diseases and Public Health* 3 (3):163–167.

Greenacre, E. J., and T. F. Brocklehurst. 2006. The acetic acid tolerance response induces cross-protection to salt stress in *Salmonella* Typhimurium. *International Journal of Food Microbiology* 112:62–65.

Greenacre E. J., T. F. Brocklehurst, C. R. Waspe, D. R. Wilson, and P. D. G. Wilson. 2003. *Salmonella enterica* serovar typhimurium and *Listeria monocytogenes* acid tolerance response induced by organic acids at 20 degrees C: optimization and modeling. *Applied and Environmental Microbiology* 69:3945–3951.

Greenacre, E. J., S. Lucchini, J. C. D. Hinton, and T. F. Brocklehurst. 2006. The lactic acid-induced acid tolerance response in *Salmonella enterica* serovar Typhimurium induces sensitivity to hydrogen peroxide. *Applied and Environmental Microbiology* 72:5623–5625.

Hajmeer, M. N., and I. A. Basheer. 2003. A hybrid Bayesian-neural network approach for probabilistic modeling of bacterial growth/no-growth interface. *International Journal of Food Microbiology* 82 (3):233–243.

Hajmeer, M. N., D. Y. Fung, J. L. Marsden, and G. A. Milliken. 2001. Effects of preparation method, age, and plating technique of thin agar layer media on recovery of *Escherichia coli* O157:H7 injured by sodium chloride. *Journal of Microbiological Methods* 47 (2):249–253.

Hills, B. P. 2001. The powers and pitfalls of deductive modeling In *Food process modeling*, ed. L. M. M. Tijskens, M. L. A. T. M. Hertog, and B. M. Nicolaï. Cambridge, UK: Woodhead, pp. 3–18.

Jones, H. P. 1994. Ambient packaged cakes. In *Shelf life evaluation of foods*, ed. C. M. D. Man and J. A. Jones. London: Blackie Academic and Professional, pp. 179–201.

Jones, J. E. 1993. A real-time database/models base/expert system in predictive microbiology *Journal of Industrial Microbiology* 12:268–272.

Komitopoulou, E., and W. Penaloza. 2009. Fate of *Salmonella* in dry confectionery raw materials. *Journal of Applied Microbiology* 106 (6):1892–1900.

Legan, J. D. 2007. Application of models and other quantitative microbiology tools in predictive microbiology. In *Modeling microorganisms in food*, ed. S. Brul, S. van Gerwen, and M. Zwietering. Boca Raton, FL: CRC Press, pp. 82–109.

Legan, J. D., D. L. Seman, A. L. Milkowski, J. A. Hirschey, and M. H. Vandeven. 2004. Modeling the growth boundary of *Listeria monocytogenes* in ready-to-eat cooked meat products as a function of the product salt, moisture, potassium lactate, and sodium diacetate concentrations. *Journal of Food Protection* 67 (10):2195–2204.

Marks, B. P. 2008. Status of microbial modeling in food process models. *Comprehensive Reviews in Food Science and Food Safety* 7:137–143.

McClure, P. J., C. W. Blackburn, M. B. Cole, P. S. Curtis, J. E. Jones, J. D. Legan, I. D. Ogden, M. W. Peck, T. A. Roberts, J. P. Sutherland, and S. J. Walker. 1994. Modeling the growth, survival and death of microorganisms in foods: the UK Food Micromodel approach. *International Journal of Food Microbiology* 23:265–275.

McMeekin, T. A., K. Presser, D. Ratkowsky, T. Ross, M. Salter, and S. Tienungoon. 2000. Quantifying the hurdle concept by modeling the bacterial growth/no growth interface. *International Journal of Food Microbiology* 55 (1–3):93–98.

McMeekin, T. A., and T. Ross. 1996. Shelf-life prediction: status and future possibilities. *International Journal of Food Microbiology* (31):65–84.

National Advisory Committee on Microbiological Criteria for Foods. 2010. Parameters for determining inoculated pack/challenge study protocols. *Journal of Food Protection* 73 (1):140–202.

Nicolai, B. M., P. Verboven, and N. Scheerlinck. 2001. The modeling of heat and mass transfer. In *Food process modeling*, ed. L. M. M. Tijskens, M. L. A. T. M. Hertog, and B. M. Nicolaï. Cambridge, UK: Woodhead, pp. 60–86.

Pham, Q. T. 2001. Modeling thermal processes: cooling and freezing. In *Food process modeling*, ed. L. M. M. Tijskens, M. L. A. T. M. Hertog, and B. M. Nicolaï. Cambridge, UK: Woodhead, pp. 312–339.

Powell, M. R. 2006. Modeling the frequency and duration of microbial contamination events. *International Journal of Food Microbiology* 110 (1):93–99.

Reyns, K. M. A., C. C. F. Soontjens, K. Cornelius, C. A. Weemaes, M. E. Hendrickx, and C. W. Michiels. 2000. Kinetic analysis and modeling of combined high pressure-temperature inactivation of the yeast *Zygosaccharomyces bailii*. *International Journal of Food Microbiology* 56:199–210.

Saucedo-Reyes, D., A. Marco-Celdrán, M. C. Pina-Pérez, D. Rodrigo, and A. Martínez-López. 2009. Modeling survival of high hydrostatic pressure treated stationary- and exponential-phase *Listeria innocua* cells. *Innovative Food Science and Emerging Technologies* 10 (2):135–141.

Scott, V. N., K. M. J. Swanson, T. A. Freier, W. P. Pruett, W. H. Sveum, P. A. Hall, L. A. Smoot, and D. G. Brown. 2005. Guidelines for conducting *Listeria monocytogenes* challenge testing of foods. *Food Protection Trends* 25 (11):818–825.

Seman, D. L., A. C. Borger, J. D. Meyer, P. A. Hall, and A. L. Milkowski. 2002. Modeling the growth of *Listeria monocytogenes* in cured ready-to-eat processed meat products by manipulation of sodium chloride, sodium diacetate, potassium lactate, and product moisture content. *Journal of Food Protection* 65 (4):651–658.

Skandamis, P. N., J. D. Stopforth, Y. Yoon, P. A. Kendall, and J. N. Sofos. 2007. Modeling the effect of storage atmosphere on growth-no growth interface of *Listeria monocytogenes* as a function of temperature, sodium lactate, sodium diacetate, and NaCl. *Journal of Food Protection* 70 (10):2329–2338.

Slongo, A. P., A. Rosenthal, L. M. Quaresma Camargo, R. Deliza, S. P. Mathias, and G. M. Falcão de Aragão. 2009. Modeling the growth of lactic acid bacteria in sliced ham processed by high hydrostatic pressure. *LWT—Food Science and Technology* 42 (1):303–306.

Sloof, M. 2001. Problem decomposition. In *Food process modeling,* ed. L. M. M. Tijskens, M. L. A. T. M. Hertog, and B. M. Nicolaï. Cambridge, UK: Woodhead, pp. 19–34.

Tamplin, M., J. Baranyi, and G. Paoli. 2004. Software programs to increase the utility of predictive microbiology information. In *Modeling microbial responses in food,* ed. R. C. McKellar and X. Lu. Boca Raton, FL: CRC Press, pp. 223–242.

Uyttendaele, M., A. Rajkovic, G. Benos, K. Francois, F. Devlieghere, and J. Debevere. 2004. Evaluation of a challenge testing protocol to assess the stability of ready-to-eat cooked meat products against growth of *Listeria monocytogenes. International Journal of Food Microbiology* 90 (2):219–236.

Valík, L., J. Baranyi, and F. Görner. 1999. Predicting fungal growth: the effect of water activity on *Penicillium roqueforti. International Journal of Food Microbiology* 47:141–146.

Van Boekel, M. A. J. S., and L. M. M. Tijskens. 2001. Kinetic modeling. In *Food process modeling,* ed. L. M. M. Tijskens, M. L. A. T. M. Hertog, and B. M. Nicolaï. Cambridge, UK: Woodhead.

van Boekel, M. A. J. S., and M. H. Zwietering. 2007. Experimental design, data processing and model fitting in predictive microbiology. In *Modeling microorganisms in food,* ed. S. Brul, S. van Gerwen, and M. Zwietering. Boca Raton, FL: CRC Press, pp. 35–59.

Walker, S. J., and G. Betts. 2000. Chilled foods microbiology. In *Chilled foods. A comprehensive guide,* ed. M. Stringer and C. Dennis. Cambridge, UK: Woodhead, pp. 153–186.

Whiting, R. C. 1995. Microbial modeling in foods. *Critical Reviews in Food Science and Nutrition* 35:467–494.

Wilkinson, E. C., and L. M. M. Tijskens. 2001. Modeling food quality. In *Food process modeling,* ed. L. M. M. Tijskens, M. L. A. T. M. Hertog, and B. M. Nicolaï. Cambridge, UK: Woodhead, pp. 366–382.

Youart, A. M., Y. Huang, C. M. Stewart, R. M. Kalinowski, and J. D. Legan. 2010. Modeling time to inactivation of *Listeria monocytogenes* in response to high pressure, sodium chloride, and sodium lactate. *Journal of Food Protection* 73 (10):1793–1802.

chapter nine

*Detection and identification
of bacterial pathogens
in food using biochemical
and immunological assays*

Hari P. Dwivedi, Patricia Rule, and John C. Mills

Contents

9.1 Introduction

Over the last century, various microbiological methods have been reported for their ability to detect foodborne pathogens directly from contaminated foods. Traditionally, culture-based methods have been used to detect and confirm foodborne pathogens. These culture-based methods, which could detect a single target cell in the various sizes of samples (10 to 375 g or more), are considered the "gold standard." Many of the newer pathogen detection methods elicit presumptive-positive results and still rely on post hoc biochemical or immunological confirmation. The general approach for testing consists of the sequential steps of cultural enrichment, selective and differential plating, final confirmation, and subtyping of isolates. Traditional culture-based methods usually require 4–6 days for the confirmation of bacterial identity (Dwivedi and Jaykus, 2011). Biochemical and immunological assays are routinely used for the detection, identification, and confirmation of pathogens in foods. These assays have been constantly modified to make them more sensitive, specific, less labor intensive, and rapid (Feng, 2001). The advancements in instrumentation have resulted in the integration of multiple steps to perform the assay (including sample inoculation, reagent mixing, incubation, assay reading, and report generation) into a single automated system. This automation has resulted not only in the reduction of labor required but also in the reduction of hands-on manipulation, resulting in more time efficient processes with a decrease in laboratory errors. Besides various advancements, the success of pathogen detection and identification using biochemical and immunological approaches still heavily relies on upstream sample preparation, including cultural enrichments. The cultural enrichment facilitates the amplification of target pathogen by severalfold, which enhances the sensitivity of the assay. Further, selective and differential plating is used to enumerate and obtain isolated colonies of suspected pathogens. Isolated colonies are required for proper identification using biochemical and immunological assays. Immunological assays resulting in presumptive positives require further confirmation using so-called gold standard biochemical techniques.

This chapter describes biochemical and immunological assays for the detection and identification of foodborne pathogens and considerations that might be addressed during the development of improvements to these methods. The beginning of this chapter describes approaches for the enrichment of food samples that are essential for the amplification of bacteria before detection and identification using biochemical and immunological assays that are described in further parts of the chapter.

9.2 Enrichment of food samples

9.2.1 Primary enrichment

Cultural enrichment required for the detection of pathogens in foods usually consists of two steps: (1) primary enrichment using a nonselective broth and (2) secondary enrichment using a selective broth medium. Primary enrichment is typically performed for 16–24 h to resuscitate the bacterial cells that are stressed or sublethally damaged during food processing, packaging, transport, and storage. The resuscitation of injured cells is particularly important for microbial analysis as these cells either may be unable to grow or may take more than the normal time to grow to a detectable level for the different detection methods. Primary enrichment assists in neutralizing or diluting the effects of various inhibitory substances (i.e., preservatives, other antimicrobials) present in food matrices, which could hinder the growth of bacterial cells. Furthermore, primary enrichment helps in rehydrating cells sampled from foods, such as dried or processed foods. Approaches such as filtration, centrifugation, and immunoconcentration could be applied for the concentration of microbial population from foods prior to primary enrichment. This could help in the reduction of the duration of the primary enrichment. For example, environmental aqueous chilling solutions and surface rinse solutions could be concentrated using filtration through a glass-fiber filter and a 0.45-µm hydrophobic grid membrane filter in a vacuum filter system for the detection of *Listeria monocytogenes* (U.S. Department of Agriculture [USDA], 2008). However, care must be taken to avoid stress to cells when using these concentration approaches. If filters are used for the concentration of target pathogens, excessive drying of the filter must be avoided and must immediately be added to appropriate enrichment broths once the filtration process is completed. Similarly, it must be noted that immunoconcentration methods could be beneficial in concentrating the target agent for further enrichment, but they are not 100% efficient. An ideal concentration method must be able to recover all target cells directly from complex food matrices.

For the enrichment of samples, the choice of enrichment broth depends on various factors, such as target pathogen and type of food sample. For example, for *Salmonella* detection, primary enrichment of foods is typically performed using buffered peptone water (BPW) at a 1:10 dilution ratio. Other broth media, such as lactose broth and tryptic soy broth (TSB), could also be used for specific food categories (Food and Drug Administration [FDA], 1998). Food items such as spices, cinnamon, cloves, and the like may be diluted 1:100 (spices, cinnamon) or 1:1000 (cloves) to minimize the inherent antimicrobial activity of these foods. Similarly, for products containing amounts of sugar and salt, enrichment at higher dilution (achieving a final concentration approximately less than 2%) could be performed to reduce the adverse effect of osmotic imbalance on bacterial growth due to high sugar/salt concentrations.

For the enrichment of food products rich in organosulfur compounds such as onion and garlic, neutralizing agents such as potassium sulfite may be used to reduce their antimicrobial effect (Mansfield and Forsythe, 1995). Food products such as cocoa powder and chocolate confectionery could contain sublethally injured cells. Nonfat dry milk (10%) or nonacidic casein (5%) could be added to recover stressed/injured cells from these products (Jasson et al., 2011). Alternatively, universal preenrichment broth could be used to support the recovery and growth of injured cells in raw and processed food with high background microflora (Jiang et al., 1998). Dried food products can be allowed to stand at room temperature for a brief period before incubation as this might help in loosening microbes from food surfaces. If the examination of large-size samples is performed, it is advisable to use prewarmed media for the primary enrichment, which could help in achieving a suitable temperature for faster bacterial multiplication.

It is important that the pH of enrichment broths with special category foods such as fermented products, mayonnaise, fruits, and vinegar-containing products is adjusted to pH 6.8–7.0 before incubation. Double-strength BPW may be used for the enrichment of highly acidic food products (juices, berries, etc.) to avoid reduction of pH during enrichment. When examining foods with high fat contents such as cheese, surfactants (such as Tergitol 7, Triton™ X-100, TWEEN® 80) could be used to disperse the food contents in broth. Antimicrobials used for carcass rinse such as trisodium phosphate (TSP) has been reported to decrease the recovery of salmonellae from processed poultry carcasses (Kim et al., 1994). Thus, proper neutralization of antimicrobials must be performed to recover the bacteria from carcass rinse during preenrichment. If sponges and gauzes are used for the surface swabbing to collect environmental samples, proper neutralization of such samples should be performed using neutralizing buffer before the cultural enrichment. The sponges and gauzes could have anionic detergents that may inhibit the recovery of target bacteria (Downes and Ito, 2001).

The components of media composition, particularly source and type, and concentration could affect the recovery of injured cells from preenrichment medium prior to selective enrichment (Clark and Ordal, 1969; Edel and Kampelmacher, 1973). Media preparation, particularly autoclaving and overheating, could further affect media performance. For example, autoclaving may result in autooxidation of phosphate buffers and sugars of BPW (Baumgartner, 1938), generating toxic oxygen by-products such as hydrogen peroxide (Mackey and Derrick, 1986). The buffering capacity of media also affects the recovery of stressed and injured cells from food products. Media with a higher buffering capacity such as BPW could counteract conditions such as extreme pH changes caused by either food (such as milk) or metabolic activity of microbial populations during enrichment (Nojoumi et al., 1995).

If a high bacterial background is present in food matrices, the effect of a dominant population on a minority population could be expected (Jameson, 1962). In mixed cultures, the growth of target bacteria could be terminated prematurely if the competitor organisms reached the maximum cellular concentration that the enrichment broth can support, resulting in reduced numbers of target bacteria after enrichment (Bail, 1929). Primary enrichment at higher temperatures, for example, incubation at 42°C for *Salmonella* and *E. coli* O157:H7 enrichment using nonselective broth, could be helpful to reduce the background flora.

Heat stress during food processing to bacterial cells such as *E. coli* O157:H7 may lead to blebbing and vesiculation of cell surfaces, increasing the permeability of the cell (Katsui et al., 1982). The selective enrichment of such injured cells would lead to further stress to these cells, therefore detrimental to the repair of injury. In such cases, nonselective broth generally supports better repair and colony formation by heat-stressed cells (Rochelle et al., 1996). Bacterial growth enhancer such as sodium pyruvate could help in recovery of target cells in enrichment broths. For example, modified BPW with pyruvates enhances the growth of enterohemorrhagic *E. coli* (EHEC). Similarly, the addition of exogenous pyruvate to repair media could stimulate recovery of heat-stressed *L. monocytogenes* (Busch and Donnelly, 1992). The decreased catalase and superoxide dismutase activities in injured cells could make them susceptible to lethal effects of hydrogen peroxide and the superoxide radical (Dallimer and Martin, 1988). Divalent cations, specifically iron, could also enhance the growth of *L. monocytogenes* as it may supplement iron needed for essential redox reactions for the enzymes, such as catalase, peroxidase, and cytochromes (Busch and Donnelly, 1992). Cell injury may lead to the loss of essential compounds through damaged cell membrane; thus, these compounds must be supplied in the recovery medium. Failure to employ a proper nonselective repair-enrichment step when attempting to recover injured cells could lead to missed detection.

9.2.2 Secondary enrichment

Secondary enrichment is performed on preenriched samples using selective broth medium to enhance the growth of target bacteria and simultaneously minimize the background microbial population. The length of secondary enrichment depends on factors such as the numbers of target cells expected in the primary enrichment during the transfer to secondary enrichment and the growth rate of target bacteria in the selective broth. It is also important to consider the lag phase in bacterial growth that could be the result of the transfer from nonselective to selective broth. Usually, a 1:10 to 1:100 dilution of primary enrichment is performed during the secondary enrichment. Selective enrichment could help in reducing the interference by competing flora during recovery of target organism (Arroyo and Arroyo, 1995). Further, dilution of preenriched samples during the secondary enrichment helps in achieving a cleaner sample as it helps to get rid of excessive food components present in the preenriched sample. The incubation temperature for selective enrichment depends on factors such as the organism targeted, type of broth medium, and background microbial load in the sample. For example, the enrichment of foods for commonly found *Salmonella enterica enterica* subspp. serovars could be performed at either 35 or 42°C using broth such as Rappaport-Vassiliadis (RV) medium or tetrathionate (TT) broth, depending on the expected background microbial loads. Specialized selective broth medium could be used for targeting a specific pathogen. For example, selenite cystine (SC) broth could be used for food enrichments targeting *S*. Typhi and *S*. Paratyphi (FDA, 1998).

The matrices and expected background play a significant role in the need to "differentiate" or spread apart the enhancement of the target bacteria and background flora. A cooked product or a dried product will not contain the same background potential as a raw product. This is why so much attention should be paid to the matrices from which the target bacteria are being enriched.

Various broth media for selective enrichment of food samples are listed in Table 9.1. Many of the enrichment broths are modified using supplements to make the broth more selective and suitable for the growth of the target pathogen. For example, for the recovery of microaerophilic bacteria *Campylobacter* spp., oxygen-quenching agents such as lysed or defibrinated blood, or charcoal; a combination of ferrous sulfate, sodium metabisulfite, and sodium pyruvate (FBP); and hemin or hematin could be added in the enrichment broth to protect campylobacters from the toxic effect of oxygen derivatives (Corry et al., 1995; FDA, 1998). Sodium pyruvate, sodium metabisulfite, and sodium carbonate increase the aerotolerance of *Campylobacter* spp. by acting as oxygen scavengers (George et al., 1978). Supplements such as novobiocin and acid digest of casein could be used to enhance the growth of *E. coli* O157: H7 selectively (Okrend and Rose, 1989; USDA, 2008). Similarly, the selective agents consisting of antibiotics and other antimicrobials could be used for selective suppression of a nontarget bacterial population. For example, a combination of acriflavin, sodium nalidixate, and the antifungal agent cycloheximide can be added

Table 9.1 Selected Cultural Media for the Enrichment and Presumptive
Identification of Foodborne Pathogens

Broth/Agar media	Target pathogen	Type of use
Modified EC broth with novobiocin	*E. coli* O157 including H7	Selective enrichment
Buffered peptone water plus SOC	*E. coli* O157 including H7	Selective enrichment
EHEC enrichment broth (EEB)	*E. coli* O157 including H7	Selective enrichment
Modified buffered peptone water with supplements (pyruvate and acriflavin-cefsulodin-vancomycin supplement)	*E. coli* O157 including H7	Selective enrichment
Tryptic soy broth modified with novobiocin and acid digest of casein	*E. coli* O157:H7	Selective enrichment
CT-sorbitol MacConkey agar (CT-SMAC)	*E. coli* O157:H7	Selective/differential isolation
Sorbitol MacConkey agar	*E. coli* O157:H7	Selective/differential isolation
Buffered *Listeria* enrichment broth with supplements (sodium pyruvate, cycloheximide, natamycin, etc.)	*Listeria* spp.	Selective enrichment
UVM broth with nalidixic acid and acriflavine hydrochloride supplements	*Listeria* spp.	Selective enrichment
Demi-Fraser broth (with ferric ammonium citrate and reduced nalidixic acid and acriflavine concentration)	*Listeria* spp.	Selective enrichment
Fraser broth (ferric ammonium citrate with supplement, including lithium chloride, nalidixic acid, and acriflavin)	*Listeria* spp.	Selective enrichment
Esculin selective agars with supplements (Oxford medium [OXA], PALCAM, modified Oxford [MOX], lithium chloride-phenylethanol-moxalactam agar [LPM])	*Listeria* spp. including *L. monocytogenes*	Selective/differential isolation
Campylobacter enrichment broth/ Bolton broth with lysed horse blood and antibiotic supplements (sodium cefoperazone, rifampicin, amphotericin)	*Campylobacter* spp.	Selective enrichment

continued

Table 9.1 *(continued)* Selected Cultural Media for the Enrichment
and Presumptive Identification of Foodborne Pathogens

Broth/Agar media	Target pathogen	Type of use
Abeyta-Hunt bark agar with lysed horse blood and antibiotic supplement (sodium cefoperazone, rifampicin, amphotericin)	*Campylobacter* spp.	Selective isolation
Modified campy blood-free agar (mCCDA) with antibiotic supplement (sodium cefoperazone, rifampicin, amphotericin)	*Campylobacter* spp.	Selective isolation
Skirrow's medium with vancomycin, polymixin-B, and trimethoprim	*Campylobacter* spp.	Selective isolation
Campy-Cefex agar (with supplements cycloheximide, cefoperazone)	*Campylobacter* spp.	Selective isolation
Buffered peptone water	*Salmonella* spp.	Enrichment
Lactose broth	*Salmonella* spp.	Enrichment
Selenite cystine broth (SC)	*Salmonella* spp.	Enrichment
Trypticase (tryptic) soy broth	*Salmonella* spp.	Enrichment
Tetrathionate broth (TT) (or variants such as TT broth-Hajna)	*Salmonella* spp.	Selective enrichment
Rappaport-Vassiliadis (RV) medium (or variants such as Rappaport-Vassiliadis R10 broth, modified Rappaport-Vassiliadis broth [mRV], Rappaport-Vassiliadis soya peptone Broth [RVS])	*Salmonella* spp.	Selective enrichment
Xylose lysine desoxycholate agar (XLD)	*Salmonella* spp.	Selective/differential isolation
Xylose lysine Tergitol 4 agar (XLT4)	*Salmonella* spp.	Selective/differential isolation
Bismuth sulfite agar (BS)	*Salmonella* spp.	Selective/differential isolation
Hektoen enteric agar (HE)	*Salmonella* spp.	Selective/differential isolation
Brilliant green sulfa agar (with 0.1% sodium sulfapyridine)	*Salmonella* spp.	Selective/differential isolation
Double modified lysine iron agar (DMLIA)	*Salmonella* spp.	Selective/differential isolation

to buffered *Listeria* enrichment broth (BLEB) after 4 h of primary enrichment to make it a selective enrichment for total incubation of 48 h at 30°C.

If possible, freshly prepared selective media must be used for the recovery of microaerophilic and anaerobic pathogens as media could absorb oxygen during storage, which may lead to stress injury to pathogens such as *Campylobacter* spp. and *Clostridium* spp. If preparing fresh media, supplements must be added to the media after the broth is cooled to 45–55°C as supplements could be heat sensitive and lose their potency at higher temperature. Specific instructions, if provided, must be followed for the preparation and storage of certain media to retain their proper activity (D'Aoust, 1977). Deionized or distilled water is recommended for all media.

Many methodologies combine primary and secondary enrichments into a single enrichment step to reduce the time to detection and cost of performing an assay. These single media enrichments employ specific growth promoters for the target species, thus combining resuscitation and growth stages into a single enrichment. However, a single enrichment step could affect the resuscitation of injured or stressed cells as direct enrichment using broth with selective supplements could further stress the injured cell. Strategies such as a brief enrichment without selective supplements followed by selective enrichment could be helpful when using a single enrichment step. Addition of selective supplements after 4 h of *Listeria* spp. enrichment in BLEB media is a good example of this notion. In some cases, it might help to reduce the concentration of the selective antimicrobial agents used during the enrichments, but in doing so, the background organisms could be resuscitated as well. In this case, optimizing the temperature for the target analyte in combination with the reduced antimicrobial agents could reintroduce the desired selectivity along with target enrichment. The use of Demi Fraser medium for the recovery of *Listeria* spp. is a good example of how the reduction of the antimicrobial quantity in preenrichment could enhance the recovery of stressed *Listeria* cells.

Overall, the enrichment process results in exponential amplification of the target pathogen by as much as a millionfold; at this concentration, detection becomes much easier. The goal would be to create a single universal enrichment step that would be applicable to amplify the target bacteria from diverse food matrices. As we continue to work with food pathogens, which are physiologically and metabolically impacted in the environment from which they are derived, there will always be expectations to address the challenges to resuscitate and recover them. The enrichment becomes an exceedingly critical strategy regarding the scenario of a single cell suspected in food. The enrichment process would remain a critical step in assay development due to its importance and complexity. In addition, enrichment steps provide break points to store enriched samples if the testing process is to be briefly discontinued (i.e., refrigerated for some period of time) or stored until the completion of the identification process as the organism in enriched samples could remain at the levels detectable by identification methods.

9.3   Culture method for microbial isolation and presumptive identification

9.3.1   Selective and differential plating

Most culture methods for selective and differential identification of foodborne bacteria use a combination of selective agents (such as a combination of antimicrobials), which suppress the growth of competitive microorganisms, and differential agents, which allow the organism to be readily distinguished from other microorganisms present. The result of selective and differential plating is the isolation of one or more colonies that fulfill the presumptive-positive criteria. In the absence of typical colonies, the analysis is completed, and the results could be reported as negative. In the case of a presumptive positive, additional biochemical or immunological testing is performed to confirm that the isolate is indeed the target pathogen. Performed in the traditional manner, the combined enrichment and culture plating steps take 24–48 h each, which means that the presumptive detection of a pathogen can take about 4 days or longer for confirmation. Enrichment-based pathogen detection only provides qualitative (presence/absence) results; however, it could be made quantitative by using approaches such as most probable number (MPN) (Dwivedi and Jaykus, 2011).

Selective and differential media could be used for the differentiation and isolation of bacteria based on their biochemical properties. For examples, lactose fermenters (such as *E. coli*) and non-lactose-fermenting (such as *Shigella, Salmonella*) enteric pathogens could be differentiated using media such as eosin methylene blue (EMB) agar and MacConkey agar. Dye combinations such as eosin and methylene blue in EMB and pH indicator such as neutral red in MacConkey agar provide the characteristic differentiation capability to these media (Leininger et al., 2001; MacConkey, 1905). It must be noted that these media could not be used for differentiation within lactose fermenters and nonfermenters.

Selective and differential media are routinely used for presumptive identification and isolation of pathogens from food samples. Most of these media are based on utilization and fermentation of one or more carbohydrate sources by bacterial cells producing by-products that lower media pH and can then be detected using pH indicators. Additional reactions, such as the ability of pathogens to reduce sulfur to hydrogen sulfide, can also be applied to further differentiate them from background colonies. For example, Hektoen enteric (HE) agar differentiates enterics based on carbohydrate fermentation (lactose, sucrose, and salicin). *Salmonella* spp. cannot ferment them and remain colorless; however, *E. coli* produces yellow to salmon color colonies due to fermentation of these carbohydrates. Further, the production of hydrogen sulfide by *Salmonella* turns *Salmonella* colonies into black or blue-green colonies with a black center (Taylor and Schelhaut, 1971; Goo et al., 1973).

Similarly, phenol red pH indicator turns red media to yellow if xylose is fermented in xylose lysine deoxycholate (XLD) agar, another medium for the isolation of *Salmonella* spp. (Taylor and Schelhaut, 1971; Warburton et al., 1994). On prolonged incubation, the exhaustion of carbohydrate sources may lead to utilization of amino acids present in the medium, which could reverse media pH to alkaline, turning it back to red. Alkaline reversion by other lysine-positive organisms is prevented by excess acid production from fermentation of lactose and sucrose. The red color *Salmonella* colonies could be confused with *Shigella*, which also produces red colonies. However, H_2S production in the alkaline condition could aid further differentiation of *Salmonella* from *Shigella* in such a condition. For example, thiosulfate in XLD is hydrolyzed by *Salmonella*, producing hydrogen sulfide, which turns *Salmonella* colonies red with black centers but not *Shigella* colonies. However, H_2S negative *Salmonella* strains also do not produce black colonies. Some other enterics, such as *Proteus* and *Pseudomonas* genus, may also produce black colonies and thus could be tough to differentiate from *Salmonella*. Various other carbohydrates and H_2S-producing systems along with selective agents have been tried to enhance the selectivity and differentiation capability of the media for *Salmonella* isolation. For example, bismuth sulfite agar (BSA), another selective and differential medium for *Salmonella* (including *S.* Typhi), employs glucose as the primary carbon source. However, the medium accentuates the ability of *Salmonella* to utilize ferrous sulfate and convert it to hydrogen sulfide, which appears as a black precipitate. The combination of different selective and differential media is helpful for isolating and identifying target microorganisms from complex food matrices, often with background microflora. For example, XLD, HE, and BSA could be used in combination for the selective and differential isolation and presumptive identification of *Salmonella* from different food items (McCoy and Spain 1969; FDA, 1998).

Some of the selective and differential media could be theoretically applied for *Salmonella* isolation directly from primary enrichments due to their high selectivity. For example, Brilliant Green Sulfa (BGS) agar has brilliant green and sulfadiazine, which provide the media high selectivity against the majority of the Gram-positive and selected Gram-negative bacteria (*S.* Typhi, *Shigella* spp., *E. coli*, and *Proteus* spp.). However, bacteria such as *Proteus, Pseudomonas,* and *E. coli* could still grow in this medium, although not frequently (Warburton et al., 1994; Downes and Ito, 2001). Xylose-Lysine-Tergitol 4 (XLT4) is another medium based on the fermentation of xylose, lactose, and sucrose; decarboxylation of lysine; and production of hydrogen sulfide (Miller et al., 1991). Phenol red indicator in XLT4 depicts pH changes due to fermentation and decarboxylation reactions, while hydrogen sulfide production is detected by the addition of ferric ions. Typical *Salmonella* colonies are black or black centered (H_2S positive) with a yellow periphery (carbohydrate fermentation) after 18–24 h of incubation. The colonies may turn either entirely black or pink to red with black centers on continued incubation. Although XLT4 is reported to be highly selective due to Tergitol 4, which inhibits growth of non-*Salmonella* organisms, *Citrobacter* spp., *Shigella* spp., *Enterobacter aerogenes,* and *E. coli* may still grow on this medium. XLT4 and BGS media are used in combination for selective isolation and presumptive identification of *Salmonella* from various foods, including meat and carcass rinses after enrichments (USDA, 2011).

The selective and differential media commonly used for the isolation of *Listeria* spp. are mostly based on esculin hydrolysis (esculinase activity). These media include Oxford agar, polymyxin, acriflavin, lithium chloride, ceftazidime, aesculine, mannitol (PALCAM) agar, and modified Oxford (MOX) agar. Ferric ammonium citrate in the medium such as Oxford agar helps in the differentiation of esculin-hydrolyzing *Listeria* spp. Esculin-hydrolyzing colonies and surrounding zones turn black due to formation of 6,7-dihydroxycoumarin on reacting with the ferric ions (Fraser and Sperber, 1988). Selectivity is attributed to high salt tolerance of *Listeria* spp. along with added supplements including a combination of antibiotics and lithium chloride. MOX agar contains the Oxford base medium plus antibiotics (moxalactam and colistin sulfate) to achieve the enhanced selectivity for detection of *L. monocytogenes* (Lee and McClain, 1989). As esculinase activity is present in all *Listeria* spp., it cannot be used as the sole criterion for differentiation of *L. monocytogenes* from other nonpathogenic species of *Listeria.* Therefore, growth on this medium is often described as *Listeria*-like organisms. Enzymatic activity exclusive to pathogenic *Listeria* species, principally phosphatidylinositol-specific phospholipase C (PI-PLC), has been targeted in some specialty media (Goodridge and Bisha, 2011). Differentiation of *L. monocytogenes* on media called PALCAM agar is based on esculin hydrolysis and mannitol fermentation. Selected *Enterococci* and *Staphylococci* strains ferment mannitol and could be differentiated using pH indicators from nonfermenting *Listeria* spp. Selectivity to the media is provided by a combination of antibiotics, including polymyixin B, acriflavin, ceftazidime, and lithium chloride (Grau and Vanderlinde, 1992).

Media base formulas can be supplemented using various compounds that provide additional selectivity against nontarget bacteria while differentiating a target group of bacteria. Various bacterial growth stimulators are also added. For example, Campy-Cefex media for the detection of campylobacters constitutes supplements such as sterile laked horse blood to provide essential growth factors. Blood-free selective media containing charcoal are also available for *Campylobacter* isolation (such as Campy blood-free selective medium, Karmali agar). Antimicrobial agents such as cycloheximide or amphotericin B are applied in these *Campylobacter* media to ensure selectivity against fungal growth. Similarly, cefoperazone is used as a selective agent to inhibit enteric flora. Basically, Campy-Cefex agar is derived from brucella agar, which has additional selective agents cycloheximide (antifungal) and cefoperazone to inhibit Gram-positive and Gram-negative bacteria (Goodridge and Bisha, 2011). The enhanced selectivity of this medium could pose trouble in recovering cephalothin-sensitive *Campylobacter* spp. such as *C. fetus* and *C. upsaliensis* (Murray et al., 2003). To enhance the aerotolerance of campylobacters, which are microaerophilic, oxygen-scavenging compounds such as sodium pyruvate, sodium metabisulfite, and sodium carbonate could be added to the medium. The incubation at specific environmental conditions and temperature is needed for appropriate growth and isolation of microaerophilic and anaerobic bacteria. For example, thermotolerant campylobacters could be grown at 42°C in an atmosphere composed of 5–6% oxygen, 3–10% carbon dioxide, and 84–85% nitrogen for 24–48 h.

Many of the existing selective and differential media have been modified to extend their application scope. MacConkey agar with sorbitol, a fermentable carbohydrate, is a good example. The addition of sorbitol instead of lactose in MacConkey agar (SMAC) could help in the differentiation of enteropathogenic *E. coli* serotypes as these are typically sorbitol negative. Selective agents bile salts and crystal violet are used to inhibit the growth of Gram-positive cocci (Rappaport and Henig, 1952). Sorbitol-fermenting organisms produce pink colonies. However, colonies that are sorbitol positive can revert and can be mistaken for sorbitol negative. It has been reported that *E. coli* O157:H7, on prolonged incubation, could also ferment sorbitol (Adams, 1991). Thus, care should be taken to interpret the result. Cefixime tellurite-sorbitol MacConkey (CT-SMAC) is another selective and differential medium that produces colorless colonies of *E. coli* O157 and is reportedly more sensitive than SMAC (Zadik et al., 1993).

Overall, selective and differential media play a major role in the isolation and presumptive identification of target pathogens. Many of these media utilize more than one reaction to indicate the presence of a suspected colony. Conversely, a combination of different media could be used to enhance the isolation and presumptive identification of target bacteria from a background population.

9.3.2 Chromogenic media

Various chromogenic culture media are commercially available for the selective or differential isolation and identification of target bacteria. Some of the early designed chromogenic media were based on extracellular enzymatic activity, with detection based on diffusion of the color into the surrounding media. These reactions were not specifically associated with the target colony. With the recent advancement in the understanding of bacterial gene expression, chromogenic media targeting expression of specific intracellular enzyme systems could be made. These media provide substrates that produce specific color associated with growth of target bacterial colonies. The substrates are usually sugar or amino acid conjugated to a chromogen or fluorogen (Manafi, 2000), which on utilization by bacterial enzyme produce a chromogenic or fluorogenic reaction. This reaction results in visual changes that could be easily observed for the identification of a colony of target bacteria. Multiple enzyme-substrate systems could be utilized to provide the detection of multiple pathogens using the same medium plate.

The chromogenic and fluorogenic selective media are particularly appealing as they could be potentially employed for the early identification of target pathogen directly after primary enrichment without requirement of further confirmation. For example, chromogenic media having substrates such as 5-bromo-4-chloro-3-indoxyl myo-inositol-1-phosphate (X-phos-inositol) and a selected lecithin mixture are commercially available for the differential and selective isolation of pathogenic *Listeria* spp. (Roczen et al., 2003). The detection in these media is based on the cleavage of X-phos-inositol forming turquoise colonies of pathogenic *Listeria* spp. with production of a white precipitate surrounding the colonies on addition of a selected lecithin mixture. The identification is based on the expression of two virulence genes (*plc*A [PI-PLC] and *plc*B [lecithinase]). Chromogenic media could be suitable for primary plating (Gaillot et al., 1999; Cooke et al., 1999; Jinneman et al., 2003) due to their better performance (sensitivity and specificity) than routine selective culture media. Overall, chromogenic culture media could be an attractive alternative to the conventional methods and could facilitate rapid detection and enumeration of target pathogens (i.e., quantitative pathogen detection). Further advancement in chromogenic media technology could provide species-specific diagnostics available on primary plating media. When combined with a single-step enrichment concept, chromogenic media could significantly reduce the time to detection. The acceptance and validation of performance of several chromogenic media by regulatory agencies such as the Association of Analytical Communities International (AOAC) and the International Organization for Standardization (ISO) have further broadened the future scope of chromogenic media in food pathogen isolation and detection.

9.4 Biochemical assays for microbial identification and confirmation

9.4.1 Biochemical assays

Biochemical assays are among the oldest methods used for microbial identification, but they have not lost their critical need in today's laboratories. They depend on the study of defined metabolic activities of different species of known microorganisms for identification purposes. A series of traditional biochemical assays can be performed to generate the identity of an unknown isolate. Traditional biochemical reactions could be categorized into various categories, including (1) pH-based reactions, in which a positive reaction is indicated by change in color of a pH indicator dye. For example, the fermentation of carbohydrate results in acidic pH, but utilization of proteins results in release of alkaline by-products. Also, there are (2) enzymatic reactions, in which hydrolysis/utilization of a colorless substrate by marker enzyme is detected by either chromogenic or fluorogenic by-products; (3) assimilation reactions that are based on utilization of carbon sources, which could be detected by colorimetric changes using a redox indicator dye such as tetrazolium violet; (4) detection of growth of the test organism (using turbidity) in the presence of a substrate, and the growth results could be determined by comparison to a test control; (5) detection based on metabolic end products such as fatty acid metabolism; and (6) other specific tests such as motility determination.

Biochemical assays for the identification of specific pathogens from foods require a pure bacterial culture having genetically identical cells. Because all the cells in a pure colony theoretically develop from a single cell, an isolated individual colony should be used to prepare a bacterial suspension to perform biochemical tests. Two successive streaks of a colony, each made with a new disposable sterile loop, on a nutrient medium could help in obtaining isolated colonies. Purification by streaking on a general-purpose medium is an important step because isolated colonies on selective agar media may still be in contact with invisible background cells of partially inhibited competing bacteria. The required concentration of bacterial cell suspensions should be prepared from an isolated colony as per the requirement of the assay. The use of a concentrated bacterial suspension may considerably reduce the incubation time for biochemical reaction (Hartman et al., 1992). More than one isolated colony should be analyzed for the identification of a pathogen. For microbes with several known pathogenic species, such as *Listeria* spp., at least five isolates should be analyzed (FDA, 1998).

To perform a biochemical test, the bacterial suspension is mixed with reaction medium and incubated at a suitable temperature for a certain period of time to facilitate the biochemical reaction. The reaction could be later examined to determine the changes in reaction outcome. The observed changes could be compared with a reference guide to determine if it indicates a positive or negative result. Many of the biochemical reactions are influenced by the inoculum size, incubation time, and temperature of the reaction (Gracias and McKillip, 2004). In general, 7–12 chemicals are needed for a general category of organism, such as Gram-positive or Gram-negative organisms or yeast. They are selected at a hierarchy that would not only differentiate at the genus level but also have to be specific enough to help speciate. It is usually cumbersome and costly and requires a range of media, reagents, and other consumable material to perform individual biochemical tests; thus, commercially available miniaturized kits or automated systems are preferred. The next section discusses a group of routine biochemical reactions that are used in the identification of foodborne pathogens. The selected biochemical reactions for the identification of foodborne pathogens are listed in Table 9.2.

9.4.2 Routine biochemical tests

The Oxidase Test: The oxidase test determines if bacteria produce certain cytochrome c oxidases. The oxidase-producing bacteria (such as *Pseudomonas* spp.) turn the oxidase test reagents (such as N,N,N′,N′-tetramethyl-p-phenylenediamine [TMPD] or N,N-dimethyl-p-phenylenediamine [DMPD]) into dark purple to maroon color within a few minutes (Stern et al., 2001). The members of the *Enterobacteriaceae* family are oxidase negative. The oxidase test is performed in the beginning of biochemical analysis, which helps in determining the presence/absence of *Enterobacteriaceae* members.

Catalase Enzyme Activity: The test determines if bacteria can produce catalase enzyme, which decomposes hydrogen peroxide into water and bubbles of oxygen (or froth) (Entis et al., 2001). As the enzyme is present only in viable cultures, cultures older than 24 h may give false-negative reactions. Some inoculating loops or wires can react with the hydrogen peroxide to produce false-positive reactions. Cultures of anaerobic bacteria should be exposed to air for 30 min prior to testing. Some bacteria, such as *Enterococcus*, may produce a pseudocatalase reaction (Gilmore, 2002). This test is important when identifying Gram-positive organisms, such as to differentiate *Staphylococcus* sp. from *Streptococcus* and *Enterococcus* species.

IMViC Test: IMViC, a combination of tests for indole, methyl red (MR), Voges-Proskauer (VP), and citrate, is performed for the identification of bacteria within the *Enterobacteriaceae* family (Newton et al., 1997; Huang et al., 1997). These tests could be performed using three separate test tubes having (1) tryptone broth for the indole test; (2) methyl red–Voges-Proskauer broth for the MR-VP test; and (3) citrate.

Table 9.2 Selected Typical Biochemical Reaction Demonstrated by Common Foodborne Pathogens

Biochemical reactions[a]	*Listeria monocytogenes*	*Campylobacter jejuni*	Enterotoxic *E. coli*	*Salmonella*
Catalase	Positive	Positive		
Gram strain	Positive[b]	Negative		
Hemolysis	Positive			
Nitrate reduction test	Negative	Positive		
Motility	Positive[c]	Motile on wet mount, corkscrew motion	Motile (H7) Nonmotile (NM)	
Indole	Negative		Positive	Negative
Oxidase	Negative	Positive		
Urease	Negative	Positive for rapid urease		Negative
H_2S production	Negative	Positive (lead acetate strip) Negative (TSI)		Positive[d]
Methyl red	Positive		Positive	Positive
Voges-Proskauer	Positive		Negative	Negative
Hippurate hydrolysis		Positive		
Citrate			Negative	Variable
Dextrose	Positive			
Glucose	Positive			
Esculin	Positive			
Rhamnose	Positive			
Mannitol	Negative			
Xylose	Negative			
Maltose	Positive			
Glucose	Positive	Negative		Positive
Sorbitol			Negative	
Cellobiose			Negative	
MUG[e]			Negative	
CAMP	Positive[f]			
Decarboxylase broth base			Negative (yellow)	

continued

Table 9.2 *(continued)* Selected Typical Biochemical Reaction Demonstrated
by Common Foodborne Pathogens

Biochemical reactions[a]	*Listeria monocytogenes*	*Campylobacter jejuni*	Enterotoxic *E. coli*	*Salmonella*
TSI	Positive[g]		Yellow butt/ yellow slant, no H$_2$S	Yellow butt with H$_2$S
Decarboxylase broth base			Negative (yellow)	
Lysine decarboxylase			Positive (purple)	
Ornithine decarboxylase			Positive (purple)	

Source: FDA, 1998; Downes and Ito, 2001.

[a] Typical reactions; strains with certain atypical reactions are known for pathogens.
[b] Older colonies could be Gram variable.
[c] Positive for both umbrella-like pattern on SIM and tumbling motility.
[d] H$_2$S negative strain are reported.
[e] MUG = 4-methylumbelliferyl-3D-glucuronide.
[f] Positive with *S. aureus* and negative with *Rhodococcus*.
[b] Acid slant/acid butt.

The indole test determines if bacteria produce tryptophanase enzyme. The tryptophanase-positive bacteria, when inoculated in tryptone broth, cleave tryptophan into indole and other products. Indole in broth could be detected using Kovac's reagent (*p*-dimethylaminobenzaldehyde), which produces a dark pink color on reacting with indole. Alternatively, Ehrlich's reagent could be used when performing the test on nonfermenters and anaerobes. The indole test should be performed in bacterial culture within 24–48 h of incubation as indole could be degraded during prolonged incubation. About 96% of *E. coli* is indole positive (Giammanco and Pignato, 1994), whereas other enterics such as *Shigella* and *Salmonella* are frequently indole negative.

The MR and VP tests are other components of the IMViC test. MR-VP tests are performed on unknown pure culture in MRVP broth (having glucose and peptone) for 24–28 h. The reagent used for the MR test is methyl red pH indicator. Bacteria such as *E. coli* that produce acids causes a drop in the pH of the broth below 4.4, turning the broth into a cherry red color indicative of a positive MR test. However, enterics such as *Klebsiella* and *Enterobacter* spp. produce neutral end products (such as ethyl alcohol, acetyl methyl carbinol) on glucose oxidation. As the neutral pH does not affect the bacterial growth, the bacteria start utilizing the peptones in MRVP broth, causing the pH to rise above 6.2. On higher pH, MR indicator turns broth into a yellow color, depicting a negative MR test.

The VP test is performed using Barritt's A (alpha-naphthol) and Barritt's B (potassium hydroxide) reagents. These reagents, when added to MRVP broth, react with acetyl methyl carbinol, produced as glucose oxidation by-products, producing a pink-burgundy color depicting a positive VP test. *Escherichia coli* does not produce acetyl methyl carbinol and is VP negative; however, *Enterobacter* and *Klebsiella* are VP positive.

The citrate test is the last component of the IMViC test. The citrate test determines if bacteria produce citritase enzyme and can grow utilizing the citrate as its sole carbon and energy source. The green color Simmon's citrate medium (pH 6.9) used to perform the citrate test contains sodium citrate as a carbon source, inorganic ammonium salts as the sole source of nitrogen, and bromothymol blue as a pH indicator. The growth of bacteria producing an acidic environment turns bromothymol blue in the medium to a yellow color. However, bacteria producing enzyme citritase break down citrate into oxaloacetate and acetate; finally, sodium bicarbonate and ammonia are produced, resulting in alkaline pH. The alkaline pH turns green-colored medium into a Prussian blue color, which depicts a positive citrate reaction. *Enterobacter* and *Klebsiella* are citrate positive. However, *E. coli* shows a negative citrate reaction. The routine use of the IMViC test for identification of lactose-fermenting Gram-negative rods and even *E. coli* is questionable; thus, additional specific assays should be performed for their detection (Kornacki and Johnson, 2001; Wolfe and Amsterdam, 1968).

H_2S Production in Sulfide-Indole-Motility (SIM) Media: This test differentiates members of the *Enterobacteriaceae* family for their ability to reduce sulfur into H_2S in a SIM media slant. H_2S produced during the growth of bacteria (such as *S. enterica* ser. Choleraesuis) reacts with ferrous sulfate in the media, producing a black precipitate. However, H_2S-negative bacteria (*E. coli* and *Klebsiella pneumoniae*) do not produce black precipitate (Tsai et al., 1988).

Urea Hydrolysis: The urea hydrolysis test detects the ability of bacteria to hydrolyze urea to generate ammonia using the enzyme urease. The assay is performed using the urea broth (yellow-orange color medium) that contains phenol red indicator, which turns into a bright pink color due to the alkaline pH resulting from ammonia accumulation. Enterics such as *Proteus* spp. and *Klebsiella* spp. are urease positive (Ferguson and Hook, 1943).

Motility Test: A motility test determines if an organism is motile in the motility medium slant that has a color indicator. The motility medium is stabbed with test organism and incubated aerobically.

The motile bacteria grow going out away from the stab line, as indicated by a color indicator. However, nonmotiles grow along the stab line. It must be noticed that many bacteria do not grow deep in the stab. *Enterobacter aerogenes, E. coli,* and *Proteus vulgaris* show motility in the medium. However, *Shigella flexneri* does not show motility in the medium.

Lactose Fermentation: This test determines the ability of an organism to ferment lactose to acid and gas. The test is typically performed using phenol red broth medium containing peptone and lactose. When organisms ferment lactose and generate an acidic environment, the medium turns yellow. However, the reddish pink color of broth does not change if lactose is not fermented. Some enterics produce gases during lactose fermentation, which can be detected by inserting Durham tubes in the broth before inoculation. *Escherichia coli* is a lactose fermenter and produces gases. However, *S.* Typhimurium does not ferment lactose.

The bacterium's ability to ferment other carbohydrates, such as glucose, sucrose, mannitol, dextrose, inositol, and sorbitol, and gas production can also be detected using phenol red broth containing respective carbohydrates. Simultaneously, Durham tubes may be placed in the broth to record gas production. Alternatively, broths such as bromocresol purple broth could be used for studying the fermentation of carbohydrates by *Enterobacteriaceae.*

Triple-Sugar Iron (TSI) Agar: A TSI slant, which has phenol red indicator, is used to differentiate enterics based on their ability to reduce sulfur and ferment carbohydrates (lactose, sucrose, glucose). The lactose fermenters typically produce a yellow slant and yellow butt due to an acid reaction. However, nonlactose fermenters may result in a pink or yellow slant based on the carbohydrate fermentation in the aerobic/anaerobic condition. The H_2S production due to some bacteria turns the butt a black color, masking the acid reaction (yellow) in the butt. *Salmonella* typically produce yellow color with blackening of the butt (Andrews et al., 2001).

Besides commonly performed biochemical assays for the detection of enteric pathogens, there are other specific biochemical assays that may be helpful in the identification of nonenteric pathogens. For example, hippurate hydrolysis tests could be performed to identify *C. jejuni.* The hippurate hydrolysis test is performed to determine if an organism produces the enzyme hippurate hydrolase, which hydrolyzes sodium hippurate into benzoic acid and glycine. A solution containing 1% sodium hippurate in distilled water is inoculated with test organism and incubated at 37°C for 2 h. The medium is later treated with ninhydrin solution and incubated further to detect the presence of glycine in the solution. The presence of glycine turns the solution into a blue-colored complex (FDA, 1998; Stern et al., 2001).

Similarly, the Christie-Atkins-Munch-Peterson (CAMP) test can be performed to determine *Listeria* spp. based on the enhancement of their hemolytic ability in the presence of a β-hemolytic *Staphylococcus aureus* and *Rhodococcus equi*. The assay is performed by (1) streaking *S. aureus* and *R. equi* culture in parallel and diametrically opposite to each other on a sheep blood agar; (2) streaking test cultures in between and at a right angle to the *S. aureus* and *R. equi* streaks; and (3) observation of plates for hemolytic enhancement after incubation at 35°C for 24–48 h. It must be made sure that the streak lines do not touch each other while performing the CAMP assay. The hemolytic reaction of *L. monocytogenes* is enhanced at the zone near the intersection with the *S. aureus* streak. Although most of the *L. monocytogenes* strains do not produce hemolytic enhancement at the zone near *R. equi*, some strains are known to do so (Ryser and Donnelly, 2001). The CAMP assay and test for acid production from sugars, including D-xylose, L-rhamnose, alpha-methyl-D-mannoside, and D-mannitol, could be performed for presumptive identification of *Listeria* spp. (FDA, 1998; Swaminathan et al., 2007).

9.4.3 Miniaturized and automated systems for biochemical-based identification

Concurrent advancements, such as customized miniaturized tests and automation, have enhanced the utility of biochemical tests for routine identification of commonly found foodborne pathogens. Miniaturized strips containing the reagents for a battery of biochemical tests are commercially available for the manual identification of pathogens at genus and species levels. Selected commercially available manual biochemical test kits to detect Gram-negative and Gram-positive foodborne pathogens are listed in Table 9.3. Many of these kits were initially introduced for clinical sample analysis and later became a part of microbial analysis in foods (Stager and Davis 1992). These kits are based on technologies such as pH indicator-based reactions (e.g., fermentation of a battery of carbohydrates, hydrolysis of specific amino acid), enzyme-based reaction (e.g., substrate utilization or hydrolysis, etc.), utilization of carbon sources, and detection of bacterial growth and their growth metabolic products.

These kits require a pure culture to set up the reactions and at least 16–24 h of incubation at a suitable temperature to obtain results. A suspension of required concentration using a pure colony of test organism is made turbidometrically in saline (McFarland, 1907). Some of these kits also require prior results from conventional biochemical reactions such as

Table 9.3 Selected Commercial Manual Microbial Identification Systems Based on Biochemical Reactions

Assay name and source	Select target foodborne pathogens
API 20E (bioMerieux SA)	Gram-negative rods
API Listeria (bioMerieux SA)	*Listeria* spp. including *L. monocytogenes*, without the CAMP test
API Campy (bioMerieux SA)	*Campylobacter* spp.
RapiD 20E (bioMerieux SA)	Enterobacteriaceae
Crystal E/NF (BD Diagnostics)	Aerobic glucose-fermenting and nonfermenting Gram-negative bacilli of the family Enterobacteriaceae
Enterotube II (BD Diagnostics)	Oxidase-negative, glucose nonfermenting Gram-negative bacilli and oxidase-negative, glucose-fermenting bacilli (Enterobacteriaceae)
Microgen® GN A + B – ID (Microgen Bioproducts Ltd.)	Enterobacteriaceae and oxidase-positive Gram-negative bacilli
MicroLog 2 and MicroLog 1 (Biolog, Inc.)	Gram-positive and Gram-negative bacteria
ID 32E (bioMerieux SA)	Enterobacteriaceae
Microbact™ TM 24E (Oxoid Ltd., Thermo Fisher Scientific, Inc.)	Gram-negative bacteria
Microbact™ TM12E (Oxoid Ltd., Thermo Fisher Scientific, Inc.)	Enteric pathogens
Microbact™ TM 12L (Oxoid Ltd., Thermo Fisher Scientific, Inc.)	*Listeria* spp.
RapID ONE (Remel, Thermo Fisher Scientific, Inc.)	Enterobacteriaceae, including *Salmonella* spp., *Shigella* spp., *Escherichia* spp.
RapID SS/u (Remel, Thermo Fisher Scientific, Inc.)	Gram-negative bacilli, including *E. coli*
RapID CB Plus (Remel, Thermo Fisher Scientific, Inc.)	Gram-positive *Coryneform* bacilli, including *Listeria* spp.
R/b Enteric Differential (Remel, Thermo Fisher Scientific, Inc.)	Enterobacteriaceae including H_2S-negative *Salmonella* strains
Oxi/Ferm II (BD Diagnostics)	Fermentative, oxidase positive, and nonfermentative Gram-negative bacteria, including *Vibrio* spp.

catalase, coagulase, and oxidase tests (where appropriate) and the results of Gram staining before inoculation of test strips with unknown culture.

The reactions in the test wells are examined for color and other observable changes following the appropriate incubation. The outcome patterns of the reactions in test wells are converted into a digit profile number, which is matched against the electronic reference database patterns provided online by proprietary Web server software for most of the commercial kit providers. The utility and evolution of many of these reactions in the form of miniaturized biochemical systems have been previously reported (Shi and Fung, 2000; O' Hara, 2005; Feng, 2001; Murtiningsih and Cox, 1997). Manual miniaturized kits are simpler to use, generate less waste, and are more economical than standard identification methods. The combination of tests in the kits is reported to show 90–99% accuracy in the identification of unknown pathogens (Cox et al., 1987; Hartman et al., 1992). Limitations to the miniaturization kits are that they still require analysis by the microbiologist and are at times subjective as not all organisms utilize/produce reactions at the same rate and efficiency. Most of these miniaturized tests require 4–48 h for results based on the inoculum level and the class of organisms.

Besides manual miniaturized kits, automated systems for microbial identification are also available (Table 9.4). These systems provide a single platform to perform all steps required for identification, including

Table 9.4 Selected Commercial Automatic Microbial Identification Systems Based on Biochemical Reactions

Assay name and source	Select target organism(s) of food microbiology significance
Phoenix™-NID Panel (BD Diagnostics)	Gram-negative bacteria
Vitek 2GN (bioMerieux, Inc.)	Fermenting and nonfermenting Gram-negative bacilli, including several *Salmonella* spp. and *E. coli* O157
Vitek 2 GP (bioMerieux, Inc.)	Gram-positive organisms, including *Listeria* and *Staphylococcus* spp.
MicroScan®	
Conventional or Rapid Panel	
(Siemens Health Care Diagnostics, Inc.)	Gram-negative organism
Sensititre ARIS 2X GNID (TREK Diagnostic Systems)	Common aerobic Gram negative
OmniLog® ID fully automated system (Biolog, Inc.)	Aerobic and anaerobic bacteria

(1) automated culture inoculation in test panels (plates/cards); (2) incubation at the temperature required for the reactions present in the test panel; and (3) real-time monitoring/reading of growth kinetics of microbes during biochemical reactions using novel technologies such as measuring light attenuation with an optical scanner to read changes automatically. The test panels for these systems, such as cards and plates, contain reaction reagents in several wells and negative control wells to assess growth and viability of the suspension.

Automated systems can handle large sample volumes with much less labor needed for only preparation of the bacterial suspensions and to load them with test panels (test cards or plates) in the systems. The inventory (test panels) in various systems is automatically tracked throughout the identification process using bar code information present on each panel. The time-temperature controlled incubation within these systems eliminates the requirement for separate incubation. The robotics within the system transports panels to the reading unit. The changes in reactions present on each test panel are read periodically for real-time monitoring of test reactions. Eventually, a phenotypic profile is generated by summarizing the biochemical changes and matched automatically against the reference library algorithm in the system's software to predict the identity of the pathogen (Dziezak, 1987; O'Hara, 2005; Shetty et al., 1998).

Sometimes, more than one organism is listed under the identification output of manual and automatic identification systems. In these situations, the correct organism can be chosen based on supplemental confirmatory tests provided by identification systems. Alternatively, the organism from the suitable wells in the test panels could be collected aseptically and streaked on nutrient media plates to check the purity of test culture if a purity test on bacterial suspension is not performed before the setup of the test panel. If more than one type of colony is present from the test inoculum, confirmatory testing should be performed. Confirmatory tests should also be performed if a rare organism is listed by the identification system with a higher confidence level or as the first identification choice while the suspected organism is listed as an alternate choice. Similarly, in the case of lack of identification or identification with low discrimination, identification using other specialized assays may be performed.

Many of these commercially available biochemical assays are certified by third-party certification agencies such as AOAC International, ISO, Association Française de Normalization (AFNOR), USDA-FSIS (Food Safety and Inspection Service), and so on. These methods are accepted and recognized as the standard methods for detection of pathogens in foods. Automated systems are reported to be cost effective and less labor intensive for the rapid identification of bacteria. Further, these systems preclude the need for manual reading and interpretation to obtain identification results. This eliminates human errors associated with the identification.

9.5 Immunological assays

Immunological assays are based on the reaction of target antigen with the specific antibody. Immunological assays could be applied during different steps of the detection of pathogens in food matrices. For example, antibodies tethered to solid surfaces could be applied for the selective capture and concentration of target bacteria from sample matrices. Thus, it could be helpful in preanalytical sample processing. Similarly, antibodies could be used as ligands to capture and detect target antigens/pathogens in different formats of enzyme-based colorimetric and fluorogenic assays. Serotyping is also a class of immunological assays mainly applied for definitive identification and subtyping of pathogens. The different immunological assay-based approaches that could be applied for the development of assays for detection of pathogens in foods are discussed in the next section.

9.5.1 Immunoconcentration

The antibody specific to target antigen tethered to solid surfaces could be used to capture and immobilize target bacteria on their surface. Immunoconcentration employing antibodies linked to plastic spurs could be applied to concentrate target pathogens selectively from food. Similarly, immunomagnetic separation (IMS) employing paramagnetic particles linked to target-specific antibodies could be employed in selective separation and concentration of target pathogens from complex sample matrices. Although immunoconcentration is not a detection method in itself, it could be used for sample preparation preceding detection assays. For example, IMS in combination with culture-based methods can successfully detect *S. enterica* in shell and liquid eggs (Hara-Kudo et al., 2001). IMS could be combined with a nucleic acid-based detection method such as polymerase chain reaction (PCR). For example, many foodborne pathogens such as *S. enterica* in milk, ground beef, and alfalfa sprouts (Mercanoglu and Griffiths 2005); *E. coli* O157:H7 in ground beef (Fu et al., 2005); and *L. monocytogenes* in milk (Amagliani et al. 2006) have been detected using the combination of IMS followed by PCR. IMS could help increase the sensitivity of detection assays such as PCR, which are usually affected by food samples (Hsih and Tsen, 2001). The capture efficiency of IMS methods for target recovery depends on the type of target pathogen and its antigenic expression, the binding affinity of the antibody for its antigens, and physicochemical properties of the food matrix from which capture of target pathogens has to be performed. Theoretically, IMS could be directly applied to samples in an effort to circumvent the need for cultural enrichment; however, cultural enrichment steps could still be necessary to detect low cell levels in foods. For example, cultural enrichment linked to IMS was used to detect low levels of target pathogens in foods such as 1.1 CFU (colony-forming unit) of *L. monocytogenes* in a 25-g ham sample (Hudson et al., 2001) and 2–3 CFU *E. coli* O157 in 25 g of raw beef (Chapman and Ashton, 2003).

Various commercial immunoconcentration platforms employing target-specific antibodies linked to a solid support such as spurs (VIDAS®) and paramagnetic particles (Pathatrix®) are commercially available (Table 9.5). Commercially available automated recapture systems (Table 9.5) could increase the contact between target cells and antibodies to enhance the target recovery from various sizes of sample volumes. These systems could process multiple samples in a single run.

9.5.2   Lateral flow immunoassay

The target antigens present in food samples can be visualized using specific antibody-coated test strips. The test strip for a lateral flow assay has capture antibodies impregnated in a solid support such as a nitrocellulose membrane at a defined distance from the sample application slot. The detection antibodies coupled to colloidal latex or gold particles are placed near the sample application slot. When a food sample is applied, the target antigens bind with the detection antibody, and the complex moves laterally toward the impregnated capture antibody due to capillary action. The antigen and detection antibody in the moving fluid segregate into two different capture zones, one specific for the target antigen and the other specific for the unbound detection antibody. A positive test result is visually evident (usually by two colored lines) when a target-specific reaction occurs. This differs from the visible signal caused by the detection of antibody-only control (such as a single line). Detection using these strips is rapid and takes around 5–10 min (Chapman and Ashton 2003). Lateral flow immunoassays (LIAs) have been reported for the detection of pathogens, including verotoxigenic *E. coli* (VTEC), *Salmonella enterica*, *C. jejuni*, and *L. monocytogenes* from a variety of food matrices, such as raw and processed meat, poultry products, raw milk, and juices (Aldus et al., 2003; Bohaychuk et al., 2005). Several LIAs for the detection of foodborne pathogens are commercially available (Table 9.5). Lateral flow assays are usually reported to have a high limit of detection; thus, sample enrichment prior to test is required (Banada and Bhunia, 2008). There are reports of a relatively higher number of false-positive results using LIAs as compared to more traditional microtiter plate enzyme-linked immunosorbent assay (ELISA) methods (Bohaychuk et al., 2005). Advancements such as automated readers to facilitate the integration of target quantification could further improve the usefulness of this platform (Leung et al., 2008). Overall, the ease and rapidity to perform the LIA makes it convenient to perform preliminary screening of pathogen contamination in foods and

Table 9.5 Selected Immunology-Based Commercial Products
for Identification of Food Pathogens

Assay name and source	Technique	Target organism(s)
VIDAS® Easy SLM (bioMerieux SA)	ELFA	*Salmonella* spp.
TECRA™ *Salmonella* VIA (3M)	ELISA	*Salmonella* spp.
TECRA™ *Salmonella* ULTIMA (3M)	ELISA	*Salmonella* spp.
TECRA™ *Listeria* visual immunoassay (3M)	ELISA	*Listeria* spp.
TECRA™ *Campylobacter* visual immunoassay (3M)	ELISA	*Campylobacter*
VIP® Gold™ for *Salmonella* (BioControl Systems)	Lateral flow immunoassay	*Salmonella* spp.
VIP® Gold™ for *Listeria* (BioControl Systems)	Lateral flow immunoassay	*Listeria* spp.
VIP® Gold™ for EHEC (BioControl Systems)	Lateral flow immunoassay	Enterohemorrhagic *E. coli* (EHEC)
Reveal® test kits for *Listeria* (Neogen Corp.)	Lateral flow immunoassay	*Listeria* spp.
Reveal® test kits *Salmonella* (Neogen Corp.)	Lateral flow immunoassay	*Salmonella* spp.
Reveal® test kits *E. coli* O157:H7 (Neogen Corp.)	Lateral flow immunoassay	*E. coli* O157:H7
Oxoid *Salmonella* Latex Test (Oxoid, Thermo Fisher Scientific, Inc.)	Latex agglutination	*Salmonella* spp.
Oxoid Dryspot *E. coli* O157 test (Oxoid, Thermo Fisher Scientific, Inc.)	Latex agglutination	*E. coli* O157
Assurance® EIA *Listeria* (BioControl Systems)	Enzyme immunoassay	*Listeria* spp.
Assurance® EIA *Salmonella* (BioControl Systems)	Enzyme immunoassay	*Salmonella* spp.
Assurance® EIA EHEC (BioControl Systems)	Enzyme immunoassay	EHEC
Oxoid *Listeria* Rapid Test (Oxoid, Thermo Fisher Scientific, Inc.)	Lateral immunoassay	*Listeria* spp.
Remel™ RIM[a] *E. coli* O157:H7 Latex Test (Remel, Thermo Fisher Scientific, Inc.)	Latex agglutination	*E. coli* O157:H7

continued

Table 9.5 (continued) Selected Immunology-Based Commercial Products for Identification of Food Pathogens

Assay name and source	Technique	Target organism(s)
Premier™ EHEC (Meridian Bioscience, Inc.)	ELISA	Shiga toxins of EHEC
RapidChek® *Salmonella* (SDIX)	Lateral flow immunoassay	*Salmonella* spp.
RapidChek® *Listeria* spp. (SDIX)	Lateral flow immunoassay	*Listeria* spp.
RapidChek® SELECT™ *Salmonella* Enteritidis (SDIX)	Lateral flow immunoassay	*Salmonella* Enteritidis
RapidChek® *E. coli* O157 (SDIX)	Lateral flow immunoassay	*E. coli* O157
Dynabeads® anti-*Salmonella* antibody (Life Technologies)	Immunomagnetic beads	*Salmonella*
Dynabeads® anti-*E. coli* O157 antibodies (Life Technologies)	Immunomagnetic beads	*E. coli* O157
BeadRetriever™ system (Life Technologies)	IMS system	Foodborne pathogens
Pathatrix (Life Technologies)	IMS system	Foodborne pathogens
Microgen *Salmonella* Latex Kit (KeyDiagnostics)	Latex slide agglutination test	*Salmonella*
Microgen *E. coli* 0157 Latex Test (KeyDiagnostics)	Latex slide agglutination test	*E. coli* O157
Microgen *Listeria* Latex (KeyDiagnostics)	Latex slide agglutination test	*Listeria* spp.
Wellcolex® Color *Salmonella* (Remel, Thermo Fisher Scientific, Inc.)	Latex agglutination test (detection and serogrouping)	*Salmonella*
Wellcolex® *E. coli* O157:H7 (Remel, Thermo Fisher Scientific, Inc.)	Latex agglutination test	*E. coli* O157:H7
Premier™ EHEC (Meridian Bioscience)	Enzyme immunoassay	Shiga toxin-producing *E. coli* (STEC)
ImmunoCard STAT!® EHEC (Meridian Bioscience)	Lateral flow immunoassay	STEC
ImmunoCard STAT!® CAMPY (Meridian Bioscience)	Lateral flow immunoassay	*Campylobacter* spp.
Premier™ CAMPY (Meridian Bioscience)	Enzyme immunoassay	*Campylobacter* spp.
VIDAS® ECO (bioMerieux SA)	ELFA	*E. coli* O157

Table 9.5 *(continued)* Selected Immunology-Based Commercial Products
for Identification of Food Pathogens

Assay name and source	Technique	Target organism(s)
VIDAS® ICE (bioMerieux SA)	Immunoconcentration	*E. coli* O157
VIDAS® SLM (bioMerieux SA)	ELFA	*Salmonella* spp.
VIDAS® ICS (bioMerieux SA)	Immunoconcentration	*Salmonella* spp.
VIDAS® Easy SLM (bioMerieux SA)	ELFA	*Salmonella* spp.
VIDAS® LSX (bioMerieux SA)	ELFA	*Listeria* spp.
VIDAS® LMO2 (bioMerieux SA)	ELFA	*Listeria monocytogenes*
VIDAS® CAM (bioMerieux SA)	ELFA	*Campylobacter* spp.
VIDAS® LIS (bioMerieux SA)	ELFA	*Listeria* spp.
VIDAS® UP *Salmonella* (bioMerieux SA)	ELFA[a]	*Salmonella* spp.

[a] Combine specific phage capture technology.

environmental samples despite its comparatively lower specificity and sensitivity (Sharma et al., 2005; Dwivedi and Jaykus, 2011).

Latex agglutination is another immunoassay that is mainly performed on the isolated colonies as a means of confirming a presumptive pathogen identification. Visible clumping indicative of a positive reaction can be compared with positive and negative control samples (Gracias and McKillip, 2004). Latex agglutination assays are rapid and easy to perform, providing results within a few minutes. Many latex agglutination assays are commercially available for the detection of foodborne pathogens (Table 9.5).

9.5.3 Enzyme immunoassays

The ELISA is an immunoassay for detection of target antigens in food samples using specific antibodies tethered to solid supports such as microtiter plates. The enriched food samples are applied to the primary antibody-coated microtiter plate. The primary antibodies bind specifically with the target antigen if present in the food sample or enrichment. The detection of specific antigen-antibody complexes is facilitated by enzymatic conjugate (such as alkaline phosphatase) labeled with secondary antibody, which develop detectable color when appropriate substrate (such as *p*-nitrophenyl phosphate [PNP]) is applied to conjugate (Shekarchi et al., 1998). Various components of pathogens such as somatic antigens, flagellar antigens, or bacterial products such as enterotoxin may act as antigen targets for ELISA (Notermans and Wernars, 1991). Washing steps remove unbound antibody to ensure specificity of the assay. Relative quantification of antigen in sample could be performed by application of serially diluted antigen (food samples) corresponding to related spectrophotometric readings. The direct application of food samples on microtiter plates may result in nonspecific binding of matrix components, which in turn may interfere with antigen-antibody reactions or lead to nonspecific positively interpreted results (i.e., false positives) (Weeratna and Doyle, 1991). Several ELISA assays have been developed to detect antigens of foodborne pathogens, for example, *Salmonella* spp., *E. coli* O157:H7, *Campylobacter* spp., and *L. monocytogenes* (Fratamico and Strobaugh, 1998; Bohaychuk et al., 2005; Valdivieso-Garcia et al., 2001; de Paula et al., 2002; Bolton et al., 2002; Shim et al., 2008). Selected commercially available ELISA assays for detection of foodborne pathogens are also listed in Table 9.5.

Although the sensitivity of ELISA assays depends on the food matrices being analyzed (among other factors), the typical limit of detection is around 10^4 CFU/mL (Cox et al., 1987). The enrichment of the food sample is required to achieve the limit of detection by enzyme-based immunoassays. For application in several immunoassays, boiling of enriched samples is performed to release the antigens from bacteria attached to the food matrix. Many modified versions of ELISA assays have been reported that require relatively smaller sample volumes and have improved detection limits with fewer reports of false-positive results (Milley and Sekla, 1993; Bohaychuk et al., 2005). Further modifications have been reported to enhance the sensitivity and multiplexing ability of the assay and to make it more quantitative (Leng et al., 2008). In this effort, fluorogenic and electrochemiluminescent reporters have been employed in place of traditional chromogenic reporter substrates. Automated ELISA formats are amenable to performing large numbers of sample analyses with relative ease and rapidity.

Enzyme-linked fluorescent antibody (ELFA) assays are based on enzymatic reactions similar to ELISA but instead use fluorogenic substrates like 4-methylumbelliferyl phosphate (4MUP) as the reporter (Shekarchi et al., 1998). 4MUP can detect much lower levels of alkaline phosphatase (10^{-9} M) by converting 4MUP into the fluorescent product 4-methylumbelliferone (Ishikawa and Kato 1978). However, colorimetric substrates such as PNP can detect only 10^{-5} M of alkaline phosphatase as it converts PNP into the yellow-pigmented p-nitrophenol. ELFA is reported as more sensitive and faster than ELISA and traditional culture-based methods (Ali and Ali, 1983; Yolken and Stopa, 1979; Johnson and Jechorek, 2011). The automation, rapidity, and ease of use of the automated ELFA formats have further increased the popularity of ELFA. Various commercially available ELFA assays are listed in Table 9.5. It must be noted that all samples analyzed using enzyme immunoassays must be confirmed using culture-based procedures (see Section 9.4) to conclude the test results. Until cultural confirmation procedures are completed, a positive immunoassay should be considered presumptive. Immunoassays for detection are not to be confused with definitive serological methods, such as serotyping, which can be confirmatory.

9.5.4 Serotyping

Serotyping is mainly used for the typing of foodborne pathogens. Serotyping is the process of identifying the specific cell surface antigens, including somatic (O) and flagellar (H) antigens (Voogt et al., 2002). The serological reaction between target antigens and a panel of known antisera specific to various O and H antigen epitopes determines isolate serotype. This is a particularly important tool for *Salmonella* phenotyping as there are more than 2500 *Salmonella enterica* serotypes. Similarly, over 700 antigenic types (serotypes) of *E. coli* have been recognized based on O, H, and K (capsular) antigens. The other major foodborne pathogens for which serotyping has been widely applied include *Campylobacter* and *L. monocytogenes* (Cortes et al., 2010; Frost et al., 1999; Gianfranceschi et al., 2009).

Commercial sera are available for serotyping of only limited foodborne pathogens. Further, clear interpretation of the serotyping reaction is not always straightforward, limiting the universal applicability of serotyping of all foodborne bacterial pathogens.

9.6 Considerations for assay validation

Like all other assays, biochemical and immunological assays must include a number of essential attributes: sensitivity (limit of detection), inclusivity (ability to detect the complete range of species or serovars of the pathogen), and specificity (ability to exclude nontarget organisms).* Assay sensitivity is determined by the detection of low levels of the target pathogen in a variety of sample matrices. Ideally, assays would be sensitive to as few as 1 cell per volume or weight sampled. Sensitivity analysis should be performed on as many serovars/strains as possible for the target organism; therefore, inclusivity can be evaluated simultaneously. To validate inclusivity, at least 50 pure strains of the target organism should be analyzed. For organisms such as *Salmonella enterica*, for which more than 2500 serotypes are known, at least 100 strains representing various serovars should be analyzed (Feldsine et al., 2002). Sensitivity of the assay is important to prevent false-negative results as it ensures that a contaminated sample is identified accurately.

---

* For more discussion on assay validation as it relates to sensitivity, inclusivity, and specificity (exclusivity), see Chapter 7, Section 7.2.2.

Another important feature is assay specificity, also referred to as exclusivity. This speaks to the ability to classify a sample as negative if the pathogen is absent. Exclusivity ensures the absence of interference from nontarget strains and limits false-positive test results. As such, a specific assay helps save additional time and resources from being unnecessarily spent confirming test results on products that do not represent a risk to consumer health. Exclusivity validation should be performed on at least 30 nontarget strains (Feldsine et al., 2002), which should include organisms commonly recognized as potentially cross-reacting with the target organism.

Several other assay features should also be evaluated for assay performance. Ruggedness should be determined to assess the assay performance due to variation in operational and environmental conditions. Is the assay appropriate for the laboratory in which it is to be used? For example, a LIA would be appropriate for laboratories housed within processing plants, while an ELISA or ELFA may not. Similarly, stability (shelf life) of the assay components and lot-to-lot variation should be analyzed. Assay validation should be performed by third-party certification agencies such as the AOAC International, AFNOR, USDA-FSIS, and the like. The guidelines of certification agencies should be followed for the validation. The validation process could include many analyses, such as ruggedness testing, inclusivity and exclusivity studies, an interlaboratory collaborative study, final data collection, and statistical analyses (Feldsine et al., 2002). The performance of an assay for the detection of a target organism such as *Salmonella* spp. can be evaluated in a variety of food matrices for the recovery of target pathogens. However, an assay for target organisms such as *Campylobacter* spp. should be evaluated and validated for specific food matrices that are commonly associated with contamination from *Campylobacter* spp. The background flora of the test matrix should be quantified to specify its relative impact on the performance of assay, if any. Besides artificially inoculated samples with different levels of inocula including fractional recovery levels, naturally contaminated samples must also be included in the study if possible. The type of artificial inocula (dry or wet) and conditions (temperature and duration) to stress equilibrate the inoculum in foods for validation studies depend on the physical state of the food matrices (liquid or dry) and storage conditions (room temperature/refrigeration/freezing); thus, samples must be prepared strictly as per the guidelines of the method validation agencies (e.g., AOAC International methods committee guidelines). For examples, for assay validation in ground beef, wet inoculum must be used, and inoculated beef must be stored at refrigeration temperature for 48–72 h to stress equilibrate the inoculum before analysis. However, for dry food matrix such as ground pepper, dry inoculum must be used and stored at room temperature for about 7 days to stress equilibrate the inoculum before analysis. For more rigorous performance analysis, a competitor microorganism at higher contamination levels (at least one logarithm) than the target microorganism could also be included in some food matrices artificially inoculated for assay performance analysis (Feldsine et al., 2002).

The performance of assays should be compared to a well-established reference method such as AOAC-Official Methods of Analysis (OMA), FDA-BAM (*Bacteriological Analytical Manual*), or USDA-FSIS. The comparison of performance data should be done using suitable statistical analysis, and it must be established that there is no statistically significant difference between the performance of the assay under validation and the reference method. However, statistically significant differences due to better performance of the test assay under validation could occur.

Finally, it should be mentioned that validation efforts might reveal that novel assays perform better than the gold standard culture-based detection methodologies. In many instances, the collaborative studies for validation of new methods have shown greater sensitivity, inclusivity, or specificity of new methods compared to standard methods even when controlling for the influences of sample matrix and sample preparation techniques. This combined with rapidity and cost can and should influence choices of assays for pathogen detection.

References

Adams, S. 1991. Screening for verotoxin-producing *Escherichia coli*. *Clinical Laboratory Science* 4:19–20.

Aldus, C. F., A. Van Amerongen, R. M. Ariens, M. W. Peck, J. H. Wichers, and G. M. Wyatt. 2003. Principles of some novel rapid dipstick methods for detection and characterization of verotoxigenic *Escherichia coli*. *Journal of Applied Microbiology* 95 (2):380–389.

Ali, A., and R. Ali. 1983. Enzyme-linked immunosorbent assay for anti-DNA antibodies using fluorogenic and colorigenic substrates. *Journal of Immunological Methods* 56:341–346.

Amagliani, G., E. Omiccioli, A. Campo, I. J. Bruce, G. Brandi, and M. Magnani. 2006. Development of a magnetic capture hybridization-PCR assay for *Listeria monocytogenes* direct detection in milk samples. *Journal of Applied Microbiology* 100 (2):375–383.

Andrews, W. H., R. S. Flowers, J. Silliker, and J. S. Bailey. 2001. *Salmonella*. In *Compendium of methods for the microbiological examination of foods*, 4th ed., ed. F. P. Downes and K. Ito. Washington, DC: American Public Health Association, 358.

Arroyo, G., and J. A. Arroyo. 1995. Efficiency of different enrichment and isolation procedures for the detection of *Salmonella* serotypes in edible offal. *Journal of Applied Bacteriology*. 79 (4):360–367.

Bail, O. 1929 Ergebnisse experimenteller populationsforschung. *Zeitschrift für Immunitatsforschung* 60:1–22.

Banada, P. P., and A. K. Bhunia. 2008. Antibodies and immunoassays for detection of bacterial pathogens. In *Principles of bacterial detection: biosensors, recognition receptors and microsystems*, ed. M. Zourob, S. Elwary, and A. Turner. New York: Springer, 567–602.

Baumgartner, J. G. 1938. Heat sterilised reducing sugars and their effects on the thermal resistance of bacteria. *Journal of Bacteriology* 36:369–382.

Bohaychuk, V. M., G. E. Gensler, R. K. King, J. T. Wu, and L. M. McMullen. 2005. Evaluation of detection methods for screening meat and poultry products for the presence of food borne pathogens. *Journal of Food Protection* 68 (12):2637–2647.

Bolton, F. J., A. D. Sails, A. J. Fox, D. R. A. Wareing, and Greenway, D. L. A. 2002. Detection of *Campylobacter jejuni* in foods by enrichment culture and polymerase chain reaction enzyme-linked immunosorbent assay. *Journal of Food Protection* 65:760–767.

Busch, S. V., and C. W. Donnelly. 1992. Development of a repair-enrichment broth for resuscitation of heat-injured *Listeria monocytogenes* and *Listeria innocua*. *Applied and Environmental Microbiology* 58 (1):14–20.

Chapman, P. A., and R. Ashton. 2003. An evaluation of rapid methods for detecting *Escherichia coli* O157 on beef carcasses. *International Journal of Food Microbiology* 87 (3):279–285.

Clark, C. W., and Z. J. Ordal. 1969. Thermal injury and recovery of *Salmonella* Typhimurium and its effect on enumeration procedures. *Applied Microbiology* 18:332–336.

Cooke, V. M., R. J. Miles, R. G. Price, and A. C. Richardson. 1999. A novel chromogenic ester agar medium for detection of salmonellae. *Applied and Environmental Microbiology* 65 (2):807–812.

Corry, J. E., D. E. Post, P. Colin, and M. J. Laisney. 1995. Culture media for the isolation of campylobacters. *International Journal of Food Microbiology* 26 (1):43–76.

Cortes, P., V. Blanc, A. Mora, G. Dahbi, J. E. Blanco, M. Blanco, C. Lopez, A. Andreu, F. Navarro, M. P. Alonso, G. Bou, J. Blanco, and M. Llagostera. 2010. Isolation and characterization of potentially pathogenic antimicrobial-resistant *Escherichia coli* strains from chicken and pig farms in Spain. *Applied and Environmental Microbiology* 76 (9):2799–2805.

Cox, N. A., D. Y. C. Fung, J. S. Bailey, P. A. Hartman, and P. C. Vasavada. 1987. Miniaturized kits, immunoassays and DNA hybridization for recognition and identification of food borne bacteria. *Dairy, Food and Environmental Sanitation* 7:628–631.

Dallimer, A. W., and S. E. Martin. 1988. Catalase and superoxide dismutase activities after heat-injury of *Listeria monocytogenes*. *Applied and Environmental Microbiology* 54:581–582.

D'Aoust, J. Y. 1977. Effect of storage conditions on the performance of bismuth sulfite agar. *Journal of Clinical Microbiology* 5:122–124.

de Paula, A. M. R., D. S. Gelli, M. Landgraf, M. T. Destro, and B. Franco. 2002. Detection of *Salmonella* in foods using Tecra *Salmonella* VIA and Tecra *Salmonella* UNIQUE rapid immunoassays and a cultural procedure. *Journal of Food Protection* 65:552–555.

Downes, F. P., and K. Ito (eds.). 2001. *Compendium of methods for the microbiological examination of foods*. 4th ed. Washington, DC: American Public Health Association.

Dwivedi, H. P., and L. A. Jaykus. 2011. Detection of pathogens in foods: the current state-of-the-art and future directions. *Critical Reviews in Microbiology* 37 (1):40–63.

Dziezak, J. D. 1987. Rapid methods for microbiological analysis of foods. *Food Technology* 41 (7):56–73.

Edel, W., and Kampelmacher, E. W. 1973. Comparative studies on isolation methods of 'sub-lethally' injured salmonellae in nine European laboratories. *Bulletin of the World Health Organization* 48:167.

Entis, P., D. Y. C. Fung, M. W. Griffiths, et al. 2001. Rapid methods for detection, identification, and enumeration. In *Compendium of methods for the microbiological examination of foods*, 4th ed., ed. F. P. Downes and K. Ito. Washington, DC: American Public Health Association, 100.

Feldsine, P., C. Abeyta, and W. H. Andrews. 2002. AOAC International Methods Committee Guidelines for Validation of Qualitative and Quantitative Food Microbiological Official Methods of Analysis. *Journal of AOAC International* 85 (5):1187–1200.

Feng, P. 2001. Archived BAM method: rapid methods for detecting foodborne pathogens. In *Bacteriological analytical manual*, 8th ed. http://www.fda.gov/food/scienceresearch/laboratorymethods/bacteriologicalanalyticalmanualbam/ucm109652.htm#authors

Ferguson, W. W., and A. E. Hook. 1943. Urease activity of *Proteus* and *Salmonella* organisms. *Journal of Laboratory and Clinical Medicine* 28:1715–1720.

Food and Drug Administration (FDA). 1998. *Bacteriological analytical manual*. 8th ed. http://www.fda.gov/Food/ScienceResearch/LaboratoryMethods/BacteriologicalAnalyticalManualBAM/default.htm

Fraser, J., and W. Sperber. 1988. Rapid detection of *Listeria* in food and environmental samples by esculin hydrolysis. *Journal of Food Protection* 51:762–765.

Fratamico, P. M., and T. P. Strobaugh. 1998. Evaluation of an enzyme-linked immunosorbent assay, direct immunofluorescent filter technique, and multiplex polymerase chain reaction for detection of *Escherichia coli* O157:H7 seeded in beef carcass wash water. *Journal of Food Protection* 61 (8):934–938.

Frost, J.A., J.M. Kramer, S.A. Gillanders. 1999. Phage typing of *Campylobacter jejuni* and *Campylobacter coli* and its use as an adjunct to serotyping. *Epidemiology and Infection* 123 (1):47–55.

Fu, Z., S. Rogelj, and T. L. Kieft. 2005. Rapid detection of *Escherichia coli* O157:H7 by immunomagnetic separation and real-time PCR. *International Journal of Food Microbiology* 99 (1):47–57.

Gaillot, O., P. di Camillo, P. Berche, R. Courcol, and C. Savage 1999. Comparison of CHROMagar *Salmonella* medium and Hektoen enteric agar for isolation of salmonellae from stool samples. *Journal of Clinical Microbiology* 37 (3):762–765.

George, H. A., P. S. Hoffman, and N. R. Krieg. 1978. Media for growth and aerotolerance of *Campylobacter fetus*. *Journal of Clinical Microbiology* 8:36–41.

Giammanco, G., and S. Pignato. 1994. Rapid identification of microorganisms from urinary tract infections by beta-glucuronidase, phenylalanine deaminase, cytochrome oxidase and indole tests on isolation media. *Journal of Medical Microbiology* 41 (6):389–392.

Gianfranceschi, M. V., M. C. D'Ottavio, A. Gattuso, A. Bella, and P. Aureli. 2009. Distribution of serotypes and pulsotypes of *Listeria monocytogenes* from human, food and environmental isolates (Italy 2002–2005). *Food Microbiology* 26 (5):520–526.

Gilmore, M. S. 2002. *The enterococci: pathogenesis, molecular biology, and antibiotic resistance*. Washington, DC: ASM Press.

Goo, V. Y., G. Q. Ching, and J. M. Gooch. 1973. Comparison of brilliant green agar and Hektoen enteric agar media in the isolation of salmonellae from food products. *Applied Microbiology* 26 (3):288–292.

Goodridge, L. D., and B. Bisha. 2011. Chromogenic and accelerated culture methods: In *Rapid detection, characterization and enumeration of foodborne pathogens*, ed. J. Hoorfar. Washington, DC: ASM Press, 47–61.

Gracias, K. S., and J. L. McKillip. 2004. A review of conventional detection and enumeration methods for pathogenic bacteria in food. *Canadian Journal of Microbiology* 50:883–890.

Grau, F. H., and P. B. Vanderlinde. 1992. Occurrence, numbers, and growth of *Listeria monocytogenes* on some vacuum-packaged processed meats. *Journal of Food Protection* 55:4–7.

Hara-Kudo, Y., S. Kumagai, T. Masuda, K. Goto, K. Ohtsuka, H. Masaki, H. Tanaka, K. Tanno, M. Miyahara, and H. Konuma. 2001. Detection of *Salmonella* Enteritidis in shell and liquid eggs using enrichment and plating. *International Journal of Food Microbiology* 64 (3):395–399.

Hartman, P. A., B. Swaminathan, M. S. Curiale, R. Firstenberg-Eden, A. N. Sharpe, N. A. Cox, D. Y. C. Fung, and M. C. Goldschmidt. 1992. Rapid methods and automation. In *Compendium of methods for the microbiological examination of foods*, 3rd ed., ed. C. Vanderzant and D. F. Splittstoesser. Washington, DC: American Public Health Association, 665–746.

Hsih, H. Y., and H. Y. Tsen. 2001. Combination of immunomagnetic separation and polymerase chain reaction for the simultaneous detection of *Listeria monocytogenes* and *Salmonella* spp. in food samples. *Journal of Food Protection* 64 (11):1744–1750.

Huang, S. W., C. H. Chang, T. F. Tai, and T. C. Chang. 1997. Comparison of the beta-glucuronidase assay and the conventional method for identification of *Escherichia coli* on eosin-methylene blue agar. *Journal of Food Protection* 60 (1):6–9.

Hudson, J. A., R. J. Lake, M. G. Savill, P. Scholes, and R. E. McCormick. 2001. Rapid detection of *Listeria monocytogenes* in ham samples using immunomagnetic separation followed by polymerase chain reaction. *Journal of Applied Microbiology* 90 (4):614–621.

Ishikawa, E., and K. Kato. 1978. Ultrasensitive enzyme immunoassay. *Scandinavian Journal Immunology* 8:43–55.

Jameson, J. E. 1962. A discussion of the dynamics of salmonella enrichment. *Journal of Hygiene Cambridge* 60:193–207.

Jasson, V., L. Baert, and M. Uyttendaele. 2011. Detection of low numbers of healthy and sub-lethally injured *Salmonella enterica* in chocolate. *International Journal of Food Microbiology* 145 (2–3):488–491.

Jiang, J., C. Larkin, M. Steele, C. Poppe, and J. A. Odumeru. 1998. Evaluation of universal preenrichment broth for the recovery of foodborne pathogens from milk and cheese. *Journal of Dairy Science* 81 (11):2798–2803.

Jinneman, K. C., J. M. Hunt, C. A. Eklund, J. S. Wernberg, P. N. Sado, J. M. Johnson, R. S. Richter, S. T. Torres, E. Ayotte, S. J. Eliasberg, P. Istafanos, D. Bass, N. Kexel-Calabresa, W. Lin, and C. N. Barton. 2003. Evaluation and interlaboratory validation of a selective agar for phosphatidylinositol-specific phospholipase C activity using a chromogenic substrate to detect *Listeria monocytogenes* from foods. *Journal of Food Protection* 66:441.

Johnson, R. L., and R. P. Jechorek. 2011. Evaluation of VIDAS *Listeria* species Xpress (LSX) immunoassay method for the detection of *Listeria* species in foods: collaborative study. *Journal of AOAC International* 94 (1):159–171.

Katsui, N., T. Tsuchido, R. Hiramatsu, S. Fujikawa, M. Takano, and I. Shibasaki. 1982. Heat-induced blebbing and vesiculation of the outer membrane of *Escherichia coli. Journal of Bacteriology* 151:1523–1531.

Kim, J. W., M. F. Slavik, M. D. Pharr, D. P. Raben, C. M. Lobsinger, and S. Tsai. 1994. Reduction of *Salmonella* on postchill chicken carcasses by trisodium phosphate (NA_3PO_4) treatment. *Journal of Food Safety* 14:9–17.

Kornacki, J. L., and J. L. Johnson. 2001. Enterobacteriaceae, coliforms, and *Escherichia coli* as quality and safety indicators. In *Compendium of methods for the microbiological examination of foods*, 4th ed., ed. F. P. Downes and K. Ito. Washington, DC: American Public Health Association, 69–82.

Lee, W. H., and D. McClain. 1989. *Laboratory communication no. 57* (revised May 24, 1989). Beltsville, MD: USDA, FSIS Microbiology Division.

Leininger, D. J., J. R. Roberson, and F. Elvinger. 2001. Use of eosin methylene blue agar to differentiate *Escherichia coli* from other gram-negative mastitis pathogens. *Journal of Veterinary Diagnostic Investigation* 13:273–275.

Leng, S., J. McElhaney, J. Walston, D. Xie, N. Fedarko, and G. Kuchel. 2008. ELISA and multiplex technologies for cytokine measurement in inflammation and aging research. *Journal of Gerontology—Series A: Biological Sciences and Medical Sciences* 63 (8):879–884.

Leung, W., C. P. Chan, T. H. Rainer, M. Ip, G. W. H. Cautherley, and R. Renneberg. 2008. InfectCheck CRP barcode-style lateral flow assay for semi-quantitative detection of C reactive protein in distinguishing between bacterial and viral infections. *Journal of Immunological Methods* 336 (1): 30–36.

MacConkey, A.T. 1905. Lactose-fermenting bacteria in faeces. *Journal of Hygiene (London)* 5 (3):333–379.

Mackey, B. M., and C. M. Derrick. 1986. Peroxide sensitivity of cold-shocked *Salmonella* Typhimurium and *Escherichia coli* and its relationship to minimal medium recovery. *Journal of Applied Bacteriology* 60:501–511.

Manafi, M. 2000. New developments in chromogenic and fluorogenic culture media. *International Journal of Food Microbiology* 60:205–218.

Mansfield, L., and S. Forsythe. 1995. Collaborative ring-trial of Dynabeads® anti-*Salmonella* for immunomagnetic separation of stressed *Salmonella* cells from herbs and spices. *International Journal of Food Microbiology* 29:41–47.

McCoy, J. M., and G. E. Spain. 1969. Bismuth sulphite media in the isolation of salmonellae. In *Isolation methods for microbiologists*, ed. D. A. Shapton and G. W. Gould. London: Academic Press, 20.

McFarland, J. 1907. The nephelometer: an instrument for estimating the number of bacteria is suspensions used for calculating the opsonic index and for vaccine. *Journal of the American Medical Association* 49:1176–1178.

Mercanoglu, B., and M. W. Griffiths. 2005. Combination of immunomagnetic separation with real-time PCR for rapid detection of *Salmonella* in milk, ground beef, and alfalfa sprouts. *Journal of Food Protection* 68 (3):557–561.

Miller, R. G., C. R. Tate, E. T. Mallinson, and J. A. Scherrer. 1991. Xylose-lysine-tergitol 4: an improved selective agar medium for the isolation of Salmonella. *Poultry Science* 70 (12):2429–2432.

Milley, D. G., and L. H. Sekla. 1993. An enzyme-linked immunosorbent assay-based isolation procedure for verotoxigenic *Escherichia coli*. *Applied and Environmental Microbiology* 59 (12):4223–4229.

Murray, P., E. J. Baron, J. H. Jorgensen, M. A. Pfaller, and R. H. Yolken (eds.). 2003. *Manual of clinical microbiology*. 8th ed. Washington, DC: ASM Press.

Murtiningsih, S., and Cox, J. M. 1997. Evaluation of the Serobact and Microbact systems for the detection and identification of *Listeria* spp. *Food Control* 8:205–210.

Newton, K. G., J. C. Harrison, and K. M. Smith. 1977. Coliforms from hides and meat. *Applied and Environmental Microbiology* 33 (1):199–200.

Nojoumi, S. A., D. G. Smith, and R. J. Rowbury. 1995. Tolerance to acid in pH 5.0-grown organisms of potentially pathogenic gram-negative bacteria. *Letters in Applied Microbiology* 21:359–363.

Notermans, S., and K. Wernars. 1991. Immunological methods for detection of food-borne pathogens and their toxins. *International Journal of Food Microbiology* 12:91–102.

O'Hara, C. M. 2005. Manual and automated instrumentation for identification of Enterobacteriaceae and other aerobic gram-negative bacilli. *Clinical Microbiology Reviews* 18 (1):147–162.

Okrend, A. J. G., and B. E. Rose. 1989. *USDA communication no. 38, rev. 4.* Washington, DC: USDA.

Rappaport, F., and E. Henig. 1952. Media for the isolation and differentiation of pathogenic *Escherichia coli* (serotypes 0111 and 055). *Journal of Clinical Pathology* 5:361–362.

Roczen, D., M. Knödlseder, K. Friedrich, G. Schabert, H. Spitz, R. Müller, and R. Reissbrodt. 2003. Development of a new chromogenic *Listeria monocytogenes* plating medium and comparison with three other chromogenic plating media. Poster O-48 presented at the ASM 103rd General Meeting, May 18–20, Washington, DC.

Ryser, E. T., and C. W. Donnelly. 2001. *Listeria*. In *Compendium of methods for the microbiological examination of foods*, 4th ed., ed. F. P. Downes and K. Ito. Washington, DC: American Public Health Association, 351.

Sharma, S. K., B. S. Eblen, R. L. Bull, D. H. Burr, and R. C. Whiting. 2005. Evaluation of lateral-flow *Clostridium botulinum* neurotoxin detection kits for food analysis. *Applied and Environmental Microbiology* 71:3935–3941.

Shekarchi, I. C., J. L. Sever, L. Nerurkar, and D. Fuccillo. 1985. Comparison of enzyme-linked immunosorbent assay with enzyme-linked fluorescence assay with automated readers for detection of rubella virus antibody and herpes simplex virus. *Journal of Clinical Microbiology* 21 (1):92–96.

Shetty, N., G. Hill, and G. L. Ridgway. 1998. The Vitek analyser for routine bacterial identification and susceptibility testing: protocols, problems, and pitfalls. *Clinical Pathology* 51:316–323.

Shi, X., and D. Y. C. Fung. 2000. Control of foodborne pathogens during sufu fermentation and aging. *Critical Reviews in Food Science and Nutrition* 40:399–425.

Shim, W. B., J. G. Choi, J. Y. Kim, Z. Y. Yang, K. H. Lee, M. G. Kim, S. D. Ha, K. S. Kim, K. Y. Kim, C. H. Kim, S. A. Eremin, and D. H. Chung. 2008. Enhanced rapidity for qualitative detection of *Listeria monocytogenes* using an enzyme-linked immunosorbent assay and immunochromatography strip test combined with immunomagnetic bead separation. *Journal of Food Protection* 71 (4):781–789.

Stager, C. E., and J. R. Davis. 1992. Automated systems for identification of micro-organisms. *Clinical Microbiology Reviews* 5:302–327.

Stern, N. J., J. E. Line, and H. Chen. 2001. *Campylobacter*: In *Compendium of methods for the microbiological examination of foods*, 4th ed., ed. F. P. Downes and K. Ito. Washington, DC: American Public Health Association, 305.

Swaminathan, B., D. Cabanes, W. Zhang, and P. Cossart. 2007. *Listeria monocytogenes*: In *Food microbiology—fundamentals and frontier*, 3rd ed., ed. M. P. Doyle and L. R. Beuchat. Washington, DC: ASM Press, 457–491.

Taylor, W. I., and D. Schelhaut. 1971. Comparison of xylose lysine desoxycholate agar, Hektoën enteric agar, *Salmonella-Shigella* agar, and eosin methylene blue agar with stool specimens. *Applied Microbiology* 21:32–37.

Tsai, W. C., C. M. Tarng, J. King, C. D. Ko, and H. J. Wu. 1988. Evaluation on the capability of indole pyruvic acid production by members of Proteus on different brands of sulfide-indole-motility *Chinese Journal of Microbiology and Immunology* 21 (2):117–124.

U.S. Department of Agriculture (USDA). 2008. *Food Safety and Inspection Service, detection, isolation and identification of* Escherichia coli O157:H7 *from meat products.* MLG 5.04. Washington, DC: USDA/FSIS Microbiology Laboratory Guidebook.

U.S. Department of Agriculture (USDA). 2011. *Food Safety and Inspection Service. Isolation and identification of* Salmonella *from meat, poultry, pasteurized egg and catfish products.* MLG 4.05. Washington, DC: USDA/FSIS Microbiology Laboratory Guidebook.

Valdivieso-Garcia, A., E. Riche, O. Abubakar, T.E. Waddell, and B. W. Brooks. 2001. A double antibody sandwich enzyme-linked immunosorbent assay for the detection of *Salmonella* using biotinylated monoclonal antibodies. *Journal of Food Protection* 64:1166–1171.

Voogt, N., W. J. Wannet, N. J. Nagelkerke, and A. M. Henken. 2002. Differences between national reference laboratories of the European community in their ability to serotype *Salmonella* species. *European Journal of Clinical Microbiology and Infectious Diseases* 21:204–208.

Warburton, D. W., B. Bowen, A. Konkle, C. Crawford, S. Durzi, R. Foster, C. Fox, L. Gour, G. Krohn, P. LaCasse, et al. 1994. A comparison of six different plating media used in the isolation of Salmonella. *International Journal of Food Microbiology* 22 (4):277–289.

Weeratna, R. D., and M. P. Doyle. 1991. Detection and production of verotoxin 1 of *Escherichia coli* O157:H7 in food. *Applied and Environmental Microbiology* 57 (10):2951–2955.

Wolfe, M. W., and D. Amsterdam. 1968. New diagnostic system for the identification of lactose-fermenting gram-negative rods. *Applied Microbiology* 16 (10):1528–1531.

Yolken, R. H., and P. J. Stopa. 1979. Enzyme-linked fluorescence assay: ultrasensitive solid-phase assay for detection of human rotavirus. *Journal of Clinical Microbiology* 10:317–321.

Zadik, P. M., P. A. Chapman, and C. A. Siddons 1993. Use of tellurite for the selection of verocytotoxigenic *Escherichia coli* O157. *Journal of Medical Microbiology* 39:155–158.

chapter ten

Microbiological growth-based and luminescence methods of food analysis

Ruth Eden and Gerard Ruth

Contents

10.1 Introduction

This chapter discusses microbiological assays designed to detect metabolic activity. Such assays detect changes in impedance or optical signal changes and relate these changes to microbial populations. Also covered in the chapter are luminescence assays (i.e., adenosine triphosphate [ATP] bioluminescence assays). Principles of these technologies, areas of research and development, and commercial applications are covered.

10.2 Principles of growth-based assays

10.2.1 Effect of microbial metabolism on measured signal

All growth-based methods (metabolic methods) rely on an electrical or an optical signal change as a result of microbial metabolism. Most of the work with growth-based methods has been applied to bacteria, yeast, and mold. These microorganisms become metabolically active and grow in a broth medium, consequently changing the chemical composition of the medium. The two main growth-based methods operate on the principles of impedance and optical signal changes. The impedance method relies on the generation of small molecules or ions from metabolism of the larger molecules in the medium. These changes in the composition of the medium increase conductivity, thereby decreasing the impedance signal. The optical method relies on optical changes resulting from pH change (using pH indicators such as bromocresol purple or phenol red) or cleavage of unique molecules.

Figure 10.1 demonstrates the relationship between the microbial growth and the generated signal in a medium. The upper graph shows microbial growth curves on a logarithmic scale starting from three initial concentrations of bacteria (curves A, B, and C). The corresponding optical or electrical signal is shown in the curves in the bottom graph. In the example shown, 5.0×10^6 cells are required to change the signal and produce a detectable acceleration on the curve. This point in time is called the "detection time" (DT). The DT is a function of the initial concentration of bacteria, the generation times (growth rates) of the organisms in the medium, the incubation temperature, and the type of signal monitored. The figure shows that the DT occurs earlier in samples that contain higher initial numbers of organisms.

A lower initial number of organisms in a sample causes delayed DT. Therefore, under certain conditions a calibration curve relating log colony-forming units (CFU)/gram to DT can be generated as shown in

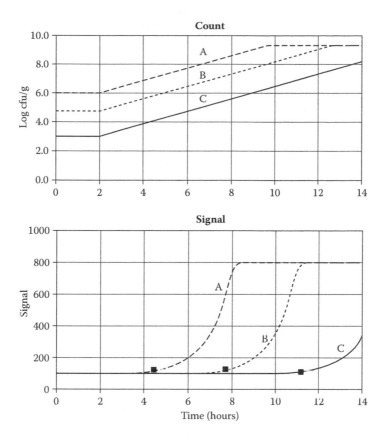

Figure 10.1 Relationship between microbial count and instrument signal.

Figure 10.2. This calibration curve defines the relationship between DT and log colony-forming units/gram obtained by standard methods such as plate counting. The slope of the line equation generated equals log 2/Tg, where Tg is the generation time of the dominant flora. The scatter along the line is due to errors in both methods.

There are several key requirements for the generation of a good calibration curve:

- Number of data points: It is important to have at least 50 data points, and there must be enough data points to represent the available microflora.
- Samples along the range of counts: It is important to include samples for the calibration that will span levels of organisms distributed over at least four or five log cycles.
- Effect of generation time: Consistent generation times among the dominant flora will result in a good calibration curve.

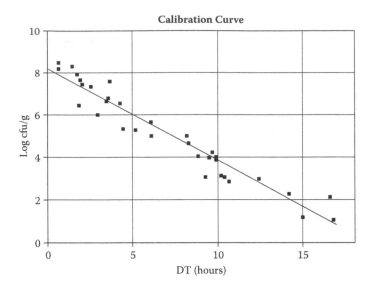

Figure 10.2 Correlation between bacterial counts and instrument detection times.

- Natural microflora: It is important the samples obtained are naturally contaminated and not inoculated with stock cultures. The metabolic state of the organisms in the sample should be similar to those in the natural product.

Sometimes, when low initial numbers of bacteria are present, such samples can create a "tail" in the calibration curve. In many cases, low counts are the reason for the lack of accuracy at the low end, with both the standard method and the metabolic method.

10.2.2   Components of growth-based systems

All growth-based methods use instruments that serve as incubators and monitor the changes in signals over time. All systems on the market use a 6-min time interval of monitoring samples since this sampling rate is compatible with the rate of growth for most relevant microorganisms at incubation temperatures of 20 to 65°C. The second component of each system is a disposable container or vial in which the sample is mixed with the appropriate growth medium. Selective and nonselective media are utilized to test for total aerobic count, yeast and mold, coliforms, Enterobacteriaceae, lactic acid bacteria, and unique species such as *Escherichia coli, Pseudomonas aeruginosa, Staphylococcus aureus, Salmonella,* and so on. The third element of growth-based systems is a software package that allows for data analysis and reporting. The systems include a detection algorithm that automatically determines the detection time for each sample and displays the result in real time. Most software packages use a "traffic light" color code approach to report on the screen the status of the samples. This approach was first used with the impedance-based systems (Firstenberg-Eden and Eden 1984) and was consequently adopted by other growth-based systems. With this approach, samples that have detection times indicating that they are above the specified level (unacceptable samples) will appear in red to alert the operator that they require immediate action. Samples that have microbial levels lower than the allowed level appear in green, while marginal samples appear in yellow.

10.2.3   Dilute-to-specification approach

Since every combination of product type and assay requires an individual calibration curve, the calibration approach is feasible to use only in cases with few combinations of product types and assays. An alternative approach used by many of the growth-based methods is the "dilute-to-specification approach" or "dilute to spec" for short. The dilute-to-spec protocol requires diluting the sample to the specification limit required for product action or release. For example, if the specified level for yeast and molds is less than 100 CFU/g, an initial 1:10 dilution of the sample is carried out followed by the addition of 0.1 mL of the diluted sample to the appropriate growth medium. If there is growth, the sample fails; if there is no detection, the sample passes since the counts are below the specification limit.

10.2.4   Applications

10.2.4.1   Estimation of total aerobic plate count

To estimate the total count in a sample using a growth-based method, special attention must be given to controlling differences in generation times of the major microbial populations in the product. This can be done by the selection of the most appropriate medium and temperature of incubation (Ogden 1986), which could require some research. By minimizing the differences in generation times between various genera in a product, the organisms will multiply at a similar rate and will achieve similar DT for similar concentrations. High correlations between the plate count method and DT were obtained for a variety of food products, such as meat, poultry, milk, fish, and so on.

10.2.4.2   Estimation of numbers of selected groups of organisms

Unique media have been developed for the detection and estimation of numbers of selected groups of organisms by growth-based methods in a variety of products, such as dairy products, poultry and meat, and wastewater, including coliforms, fecal coliforms, Enterobacteriaceae, *E. coli*, and psychrotrophs. As with the estimation of total count, special attention is required to ensure that the generation times of the organisms are similar. In most sample matrices, the use of selective media targeting defined groups of microorganisms can result in consistent generation times.

10.2.4.3   Detection of specific bacteria

Salmonella: Several impedance media are described as a screen for the detection of salmonellae. One impedance method has been given AOAC first action approval following a collaborative study by 17 laboratories (Gibson et al. 1992). Implementation of this method could require research and validation studies, especially for food, beverage, ingredient, and environmental samples that may not have been tested in the collaborative study.

Listeria: *Listeria*-specific media have been developed for impedance and optical systems using various chemicals that are inhibitory to other bacteria. *Listeria* can be differentiated from other bacteria in the impedance system by the distinct shape of the curve. For optical systems of detection, the hydrolysis of esculin during growth leads to the formation of a black precipitate that decreases the optical signal.

Staphylococcus aureus: Color change due to mannitol fermentation in conjunction with inhibitory biochemicals in the substrate is the base for the detection of *S. aureus* in color-based systems.

Specific media for the optical detection of other species of bacteria have also been developed for *Pseudomonas aeruginosa, E. coli,* and other bacteria. There are opportunities to perfect these specific media to aid in specificity and sensitivity without compromising signal detection.

10.2.4.4   Shelf life estimation

Methods have been developed for the estimation of shelf life of fresh food products by detecting the spoilage organisms responsible for the production of spoilage defects (odor, slime, and other objectionable properties). Similar methods for the estimation of shelf life of milk are described for impedance and optical-based systems (Bishop et al. 1984; White et al. 2006). The shelf life of fresh fish was evaluated with trimethylamine oxide nitrogen medium using impedance (Jørgenson et al. 1988). Using this method, H_2S-producing bacteria were detected, and their presence was highly correlated with shelf life. The shelf life of fresh poultry was predicted in 18 h with high correlation to storage shelf life (Russell 1997).

10.2.4.5   Sterility of UHT (ultra-high-temperature processing) products

The growth methods are perfectly suited to screen for absence of viable microorganisms or spores after processing. In most cases, the food samples are preincubated with or without growth medium to allow viable microorganisms, including dormant spores, to proliferate. Samples are then placed in one or more growth media in the instrument. In some cases, the samples are analyzed at different temperatures (e.g., 55 and 35°C) to allow thermophiles and mesophiles to grow.

10.2.5 Impedance assays

The impedance technology was the first growth-based method on the market. The observation that microbial growth causes impedance changes was reported by Steward in 1899. Research and development on a commercial system began in the mid-1970s (Cady 1975; Ur and Brown 1974), and in the early 1980s, the first impedance growth-based systems appeared on the market.

10.2.5.1 Principle of operation

Impedance can be defined as the resistance to flow of an alternating current as it passes through a conducting material. Impedance Z has two elements: a resistive element R and a capacitive element C that depends on the applied current frequency f. The formula for impedance is calculated as follows:

$$Z = \sqrt{\left(R^2 + \left(1/\left(2\pi f C\right)\right)^2\right)}$$

Conductance G is the inverse of resistance, $G = 1/R$, and depends on the ionic strength of the medium. Some impedance systems measure admittance Y, $Y = 1/Z$. As microorganisms metabolize, they transfer uncharged or weakly charged compounds into smaller, higher-charged end products. As a result, microbial metabolism causes an increase in capacitance and conductance. To measure impedance, the measuring container needs to have a pair of electrodes. Most systems on the market use stainless steel electrodes.

10.2.5.2 Indirect method

The indirect technique provides an impedance method that monitors the amount of carbon dioxide produced by growing organisms. This technique is particularly suitable for detecting organisms that do not produce highly charged metabolites (Timms et al. 1966). It is a useful technique for organisms that favor fermentative metabolism. As microorganisms metabolize, they produce CO_2. The measuring device contains an insert with electrodes and an alkaline solution. The production of CO_2 causes the conductance of the alkaline solution to change. The method is advantageous for measuring the growth of yeast because the conductance signal is greatly magnified (Deak and Buchat 1993). In addition, the indirect method can work well in media containing high salt or other compounds that irritate the electrodes.

10.2.5.3   Commercial products

The Bactometer (bioMerieux, Marcy l'Etoile, France) was the first commercial impedance-based system, which is now discontinued from production. It measures impedance, capacitance, and conductance. The Bactometer Processing Unit (BPU) is capable of operating in the temperature range of 10 to 55°C. One BPU can monitor from 1 to 128 samples simultaneously, and each BPU contains two separate incubators. One computer can drive up to 512 samples simultaneously. The system monitors signals for conductance G and capacitance C components, and it calculates total impedance Z from the C and G components.

The Bactometer utilizes disposable sterile plastic "modules," each containing 16 measuring wells. The module well has a capacity of 2.0 mL. Each well contains a pair of stainless steel electrodes attached to a lead frame molded into the module's plastic base. Either broth or agar media can be utilized in the module well.

The BacTrac system is produced by Sy-Lab (Geräte GmbH, Purkersdorf, Austria). Each BacTrac 4300 unit contains two 32-sample dry-block incubators that are independently controlled. Up to 12 units (768 samples, 24 incubators) can be controlled by one personal computer (PC). The system uses either disposable or reusable measurement cells. The temperature range of each incubator can be adjusted by its software from 0 to +65°C and maintained with an accuracy of ±0.1°C. For incubator temperatures from 0 to 7°C above room temperature, an optional water circulation cooler is required. Direct impedance measurements are made using either 20-mL disposable plastic containers or 10-mL reusable glass cells. Glass cells and their electrode assemblies are autoclavable. Custom-formulated impedance media are available for reconstitution and filling by the user, or the cells can be ordered prefilled. Indirect impedance measurements of yeasts and molds can be made using nested cells: a 20-mL reusable outer cell with integral electrodes used with a disposable plastic (7-mL) mating inner vial.

The RABIT (Rapid Automated Bacterial Impedance Technique) (Don Whitley Scientific Ltd., Shipley, West Yorkshire, England) measures conductance or utilizes the indirect assay. The base unit is a single 32-channel dry-block incubator module. The computerized system can be expanded to provide a total of 512 channels by adding more incubator modules. The test cells are reusable and contain an electrode assembly at the bottom of the tube. The tube and electrode assembly can be washed and reused. The cells have a working volume of 2–10 mL of medium.

10.2.6   Advantages and disadvantages of impedance

10.2.6.1   Advantages

Impedance was the first method offering automation and rapidity for the detection of indicator organisms. The time to results is shorter than with the classical plate count methodology. It also offers labor savings due to simplification and automation of the methodology.

10.2.6.2   Disadvantages

The growth media used for the various impedance signals greatly influence the electrical signals interpreted by the instrument. Media components must be optimized for the electrical signal during microbial growth to result in good impedance curves. Ions from food systems or other samples may interfere with the impedance signal and cause excessive drift by reacting with components in the medium to produce compounds that mimic microbial metabolism. Chlorine (Cl^-) ions cause excessive drift and destroy the oxide layer of electrodes. Drift in the baseline of curves can eliminate change of direction in the growth curves, especially near the DT. When developing media for impedance applications, attention must be given to eliminate weak accelerating curves, biphasic curves that are difficult to interpret, and erratic curves such as those caused by gas production by the bacteria or interaction of media chemicals with the electrodes. Also, the initial impedance values must be within the instrument dynamic range to avoid baseline drift. The media must be carefully developed to optimize the signal by minimizing ingredients that contribute to erratic curves. As such, there are areas of further research for new and emerging pathogens and spoilage microorganisms in various food systems.

10.3   Color and fluorescent assays

10.3.1   Principle of operation

10.3.1.1   Differential media

Color and fluorescence reactions are widely used in microbiology to distinguish one microorganism type from another, both growing on the same media. This type of media uses the biochemical characteristics of a microorganism growing in the presence of specific nutrients or indicators (such as pH indicators phenol red, bromocresol purple, bromocresol green, etc.) or other chromogenic or fluorogenic compounds that can be cleaved by the target organism or that can react with enzymes secreted by the microorganisms to indicate their defining characteristics. In general, fluorogenic and chromogenic substrates are effective at beaconing specific enzymatic activities of certain microorganisms.

In the early 1990s, many new media and diagnostic tests on unique color changes or fluorescence changes have been introduced. By incorporation of fluorogenic or chromogenic substrates into primary selective media, the detection of specific groups of organisms can be performed (Manafi et al. 1991; Manafi 2000). The introduction of many of these media and identification tests has led to improved accuracy and faster detection of target organisms, often reducing the need for isolation of pure cultures and confirmatory tests.

A few examples of unique biochemical reactions that can be used include

1. A variety of media are based on a pH change in the media due to the production of acid from various carbohydrates causing change of pH (e.g., mannitol salt agar, MacConkey agar) or decarboxylation of amino acids (lysine decarboxylase broth).
2. A number of media utilize 4-methylumbelliferone glucuronide (MUG), which is hydrolyzed by glucuronidase to yield a fluorogenic product in combination with selective agents for the detection of *E. coli*.
3. A chromogenic agar was developed for isolating presumptive colonies of *Enterobacter sakazakii* (now *Cronobacter sakazakii*) from foods and environmental sources. The medium contains two chromogenic substrates (5-bromo-4-chloro-3-indoxyl-α-D-glucopyranoside and 5-bromo-4-chloro-3-indoxyl-β-D-cellobioside); three sugars (sorbitol, D-arabitol, and adonitol); a pH indicator; and inhibitors (bile salts, vancomycin, and cefsulodin), which all contribute to its selectivity and differential properties (Restaino et al. 2006).
4. A selective medium for the detection of *Salmonella* spp. includes two substrates: 3,4-cyclohexenoesculetin-β-D-galactoside and 5-bromo-4-chloro-3-indolyl-α-D-galactopyranoside. This medium exploits the fact that *Salmonella* spp. may be distinguished from other members of the family Enterobacteriaceae by the presence of α-galactosidase activity in the absence of β-galactosidase activity (Perry et al. 1999).

Optical systems utilize a variety of such reactions to select for and differentiate between various microorganisms.

10.3.1.2   Elimination of product interference

Optical systems rely on a clear color change due to the microbial metabolism. Many product particles and colors, as well as the turbidity from the growth of microorganisms, can interfere with the optical system readings of color change. As a result, optical methods require a means of separation between the area containing the microorganisms and product and the reading zone to eliminate the masking effect of the optical signal. Various filtration techniques can be incorporated into sample preparation, such as membrane filtration for liquid samples or use of filtered stomacher bags for semisolid and solid samples. However, filtration techniques tend to be messy and time consuming and can lead to reduced sensitivity and accuracy. A better approach is to create a physical barrier to prevent the product and the microorganisms from penetrating the reading zone. The commercial systems in use today rely on this separation approach.

10.3.2   Commercial products

The Soleris System (Neogen Corp., Lansing, MI) has two commercial models: Soleris 128 and Soleris 32. The Soleris 128 model has four independent incubators. Each incubator can be set to a different temperature, which is important when simultaneously running tests with different temperature requirements (e.g., coliform at 35°C and mold at 25°C). The system has a single light-emitting diode (LED) and a photodetector placed on the opposite side of the LED to read the signal changes. The system has no moving parts and delivers accuracy and reproducibility similar to, or better than, results produced by standard methods. Each of the samples placed into the system is independent of the other. The Soleris 32 has a single temperature incubator, providing the same performance as the Soleris 128.

Test Vials: The Soleris system utilizes two types of disposable vials: a direct vial and a vial-in-a-vial (ViV). Both vials are ready to use.

Direct Vials: An agar plug is used to separate the liquid medium containing the product and microorganisms and the reading zone, thereby eliminating the masking of the optical pathway by the product or microbial turbidity. Changes in color, expressed as optical units, are sensed by the optical sensor and recorded in the computer. The sample is entered directly into the vial. Most of the direct vials monitor pH changes due to microbial metabolism.

Vial-in-a-Vial (ViV): The ViV technology is similar to the impedance indirect method. It consists of two vials: an external vial (similar to the vial used in the direct method) and a smaller internal vial that fits inside the external vial. A pH reagent is pipetted to the bottom of the external vial, and the sample is added to the inner vial containing the growth medium. The microorganisms grow in the liquid medium in the internal vial and generate CO_2 gas from their metabolic activity. The internal vial is placed unsealed in the external vial, allowing the generated gas released from the internal container to fill the head space of the external vial. The generated CO_2 interacts with the KOH compound to lower its pH, causing the dye indicator in the optical window to change from purple to colorless. The total volume of the internal vial is

3.5 mL, allowing for a sample size of up to 1.0 mL. Users of the instrument may note that the direct vials are easier to use and allow a larger sample size. However, the direct vials are unable to detect nonfermenting organisms in most cases.

The BioLumix instrument (BioLumix, Ann Arbor, MI) has a capacity of 32 simultaneous test samples with a single incubating temperature. One PC can control up to 32 instruments, resulting in the ability to monitor 1024 samples simultaneously and allowing the system to grow with the user's needs. The interlocking, front-loading design allows for safe stacking of up to three instruments, saving precious laboratory counter space. The multiplicity of temperatures is important whenever different tests requiring different temperatures are run simultaneously (e.g., total counts at 35°C and yeast and mold at 28°C). The optical system contains a yellow LED, a blue LED, and a fluorescent tube. This optical system allows detection of most color changes and fluorescent changes, resulting in a versatile system. It also allows detecting two reactions in a single vial (e.g., coliform and *E. coli*).

Test Vials: BioLumix has two types of ready-to-use disposable vials:

> *Membrane Vials:* The membrane vial has an embedded membrane filter that separates the area where the sample and microorganism are present (incubation zone) from the reading zone. The user introduces the sample by simply opening the screw cap and dropping the sample into the incubation zone. The reading zone, however, is not influenced by the sample since only small molecules and ions can diffuse through the filter. Therefore, the detection zone remains clear throughout the test. Both color and fluorescence signals can be monitored simultaneously, allowing two analytes to be monitored simultaneously (e.g., coliform and *E. coli*). The production of the yellow *ortho*-nitrophenyl-β-galactoside (ONPG) by coliform is monitored with a blue LED, while MUG utilization is detected by a fluorescence sensor.

> CO_2 *Vials:* A CO_2 sensor located at the bottom of the vial detects released CO_2. The transparent solid sensor changes its color whenever CO_2 diffuses into the sensor. Only gases can penetrate the sensor, blocking liquids, microorganisms, and particulate matter. The user introduces the sample by opening the screw cap and dropping the sample into the incubation zone. In the liquid, sample particles mixed with media yield a nonclear solution. The detection zone, however, is not influenced by the sample since only gases can diffuse through the sensor. The color of the sensor is dark in sterile vials. As microorganisms grow, the sensor turns yellow, indicating CO_2 production and metabolic

growth. The medium at the incubation zone does not contain any dye indicator.

10.3.3 Advantages and disadvantages of optical systems

The optical systems have an advantage over the impedance systems since they do not solely rely on selective media. Therefore, it tends to be easier to design media for optical systems. In addition, the color or fluorescent signals tend to be less susceptible to product interference due to the separation mechanisms described. Also, the disposables tend to be less complex as they do not require any metal electrodes.

10.4 Bioluminescence-based methods

10.4.1 Background and principles

The escalating economic costs of managing a food safety crisis resulting from microbiological contamination have driven the need for rapid response and preventive testing programs. Bioluminescence-based methods address speed and prevention and have broad applications in microbiological testing programs across a wide array of industry sectors. Primarily known as an accepted standard to monitor efficacy of cleaning and sanitizing procedures in the food (Leon and Albrecht 2007) and beverage industries, bioluminescence methods have expanded their use to assess surface hygiene in farm (Kelly et al. 2001), pharmaceutical, personal care, microelectronics, health care (Wren et al. 2008; Aycicek et al. 2006), and other industrial settings. In addition, bioluminescence methods have been adapted to rapidly determine microbiological keeping quality of dairy and dairy substitute end products (Bell et al. 1994) and have been employed as reactants in pathogen and pesticide (Saul et al. 1996) detection.

Hygienic monitoring of food- and beverage-processing environments is a significant challenge due to the high standards that need to be achieved, the potential for product residue to foster conditions that allow microorganisms to survive and contaminate products, and the "cleanability" of complex and intricate processing equipment. Bacteria can survive under extreme stress conditions even if cells are injured or are viable but nonculturable. Certain bacteria and molds can produce spores as a means to enhanced survivability in harsh conditions. Bioluminescence technology rapidly measures residual ATP (Kyriakides et al. 1991) from viable and nonviable cells. ATP is present in the cells of all organisms, which makes it an ideal biomarker of microbial and organic soil (Whitehead et al. 2008) contaminants. Although methods are available that are specific to microbial ATP (Siragusa et al. 1996) or somatic ATP (Frundzhyn et al. 2008), these methods have not yet been commercially viable. However, bioluminescence-based methods offer a broad, sensitive solution to measure residual levels of organic soil and microbial contamination. This nonspecificity is an advantage as organic product residues may provide a nutritious medium for microbial growth and act as barriers to the direct action of both sanitizers and disinfectants, which are both undesirable outcomes that would be important to detect in a sanitation program. Reduction or removal of either form of ATP as measured by bioluminescence provides a high level of confidence that the hygienic standards for safe, quality food processing have been met.

The general ATP procedure requires a single service swab device with premeasured reagents to support the luminescence reaction and a premoistened swab tip. The swab itself should be premoistened with a buffered solution to aid in dislodging cells if encrusted in biofilms or to disperse cell clumps. If biofilms are present on a surface, it can be difficult to detect ATP as extracellular material produced in biofilms literally "cements" the cells to the surface. Typically, a surface area of 10 cm by 10 cm is swabbed, reengaged in the swab device, and then inserted into a handheld luminometer. The light is generated when ATP is hydrolyzed in the presence of luciferin substrate and luciferase enzyme (Deluca 1976) (Figure 10.3).

Within seconds after the reaction, a relative light unit (RLU) value is displayed on the device. Most systems detect ATP in the 3- to 50-femtomole range. Photons of light are directly proportional to the amount of ATP present in extracellular and cellular bacterial cells. The greater the amount of light produced, the greater the bioburden will be.

Detection systems in luminometers utilize photodiodes, or more sensitive photomultiplier tubes. Sensitivity is determined by comparing RLU readings of a blank sample (containing no ATP, e.g., a single-use swab that has not touched a surface), and calibration is accomplished with standardized samples that contain predetermined ATP levels. Background RLU output from electrical noise, phosphorescence, or reagent chemistry is corrected during this calibration procedure.

Since ATP is deposited on surface samples by both nonmicrobial residual organic matter and microbial ATP, RLU levels do not correlate linearly to conventional plate count methods. However, ATP bioluminescence provides an instant assessment of the hygienic status of the surface. Ideally, ATP systems should be flexible to allow swabs to be sampled and collected for later readings. Methodology has recently been developed to detect lower levels of ATP (0.01 femtomoles or less) rapidly using enhanced luminescence chemistry and detection systems that amplify the light signal. This has provided a new strategy to improve allergen control (Jackson et al. 2008) through higher levels of cleaning. It also establishes new benchmarks for the level of cleaning required for aseptic manufacturing or clean room surfaces.

Prediction intervals for the linear regression analysis between total plate count (TPC) and ATP bioluminescence tests assayed on 25-cm^2

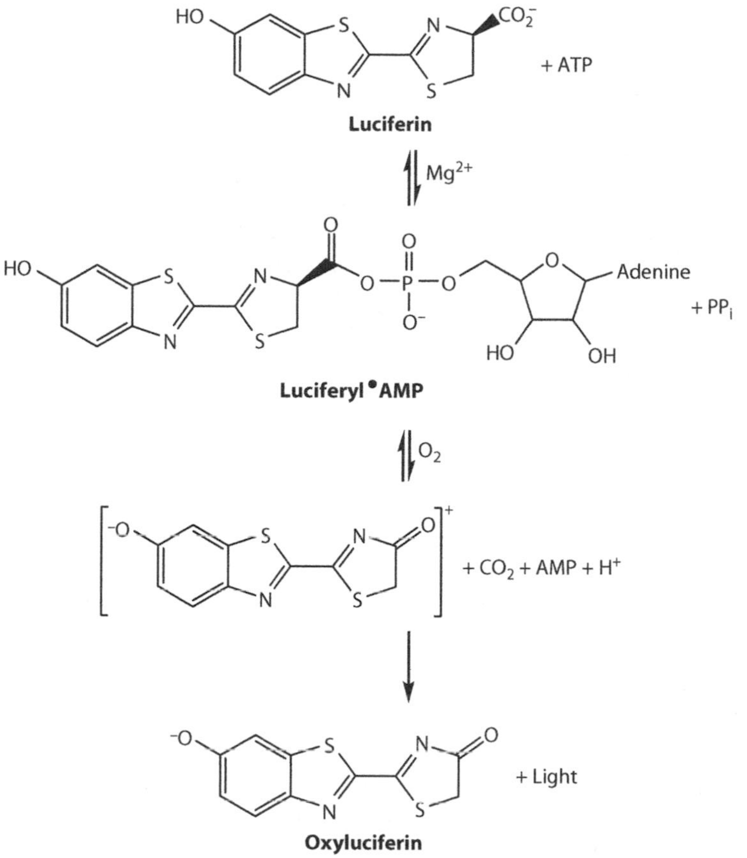

Figure 10.3 Generation of light when ATP is hydrolyzed in the presence of luciferin substrate and luciferase enzyme. (From Deluca, M. 1976. Firefly luciferase. In
Advances in enzymology and related areas of molecular biology, ed. A. Meister. With
permission from John Wiley & Sons, Ltd.)

plastic surfaces is shown in Figure 10.4. In this chart, $\log_{10}$ ATP is the
dependent variable, and $\log_{10}$ TPC is the response variable. The data are
for sanitation of plastic cutting boards (Leon and Albrecht 2007).

A reduction (McPherson and Savarino 2007) of RLUs as a percentage is
a more efficient method for determining the best cleaning and sanitizing
procedure. A simple formula for percentage RLU calculation is as follows:

$$\frac{\text{Initial RLUs} - \text{Post RLUs}}{\text{Initial RLUs}} = \% \text{ RLU reduction}$$

Initial – post = RLUs eliminated during the cleaning/sanitizing process

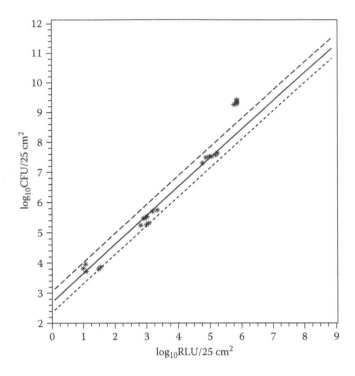

Figure 10.4 Prediction intervals for the linear regression analysis between total plate count (TPC) and ATP bioluminescence (ATP) tests assayed on 25-cm² plastic surfaces, where $\log_{10}$ ATP is the dependent variable and $\log_{10}$ TPC is the response variable. (Taken from Leon, M. B., and J. A. Albrecht. 2007. Comparison of adenosine triphosphate (ATP) bioluminescence and aerobic plate counts (APC) on plastic cutting boards. *Journal of Foodservice* 18 (4):145–152. With permission from John Wiley & Sons Ltd.)

The highest acceptable standard is 0 post-RLU and 100% RLU reduction. This can be realistically achieved through careful consideration to equipment design and diligent efforts to optimize clean room SSOPs (sanitation standard operating procedures). When the reduction of RLUs becomes a challenge in remediation, this can be used to initiate a complete disassembly and removal of equipment from the process environment for more thorough cleaning. A unique ATP threshold should be established by environmental surface type due to differing abilities of surfaces to retain soil or differences in extractability of ATP from different surfaces. This requires research on ATP levels on surface types under varying conditions, such as clean and sanitized preoperative, not clean postoperative, and degrees of cleanliness in between. If the limit is exceeded on a swab sampling point, the area should be cleaned and retested until an acceptable ATP level is achieved. Once an SSOP is optimized, adherence to its written form will demonstrate knowledge of a commitment to good sanitation and maintaining a safe production process. The SSOP must also document the monitoring and verification procedures used, including the frequency and record-keeping processes associated with monitoring and corrective action procedures. RLU readings that fail the luminometer threshold can be recleaned and resanitized immediately if undesirable results are visible in real-time during inspection. The RLU data are stored with customized software programs, and when integrated with conventional microbiological results, a macropicture of the overall success and lapses of hygiene standards is recorded. Correlations between RLU and microbial counts are possible, but only for certain food-processing environments.

10.4.2   Product testing applications

Bioluminescence technology is well established for sterility and postprocess microbial contamination testing of ESL (extended shelf life), UHT, and aseptic products. Consumer demand for "fresh-tasting" beverage products and a longer refrigerated shelf life (30–90 days) places the burden on ESL processors to manufacture a safe brand with timely product release into the distribution chain. The principles are similar to the bioluminescence hygiene platform but differ in that the end product is first preincubated at 30–32°C for 24–48 h in the original package. A small portion of the sample is removed, typically 50 μL, and added to a microwell. Reagents are automatically dispensed into microwells that are designed to eliminate all nonmicrobial product-sourced ATP, including ATP released from nonmicrobial cells. Afterward, only ATP from spoilage organisms, which were stimulated to grow in the preliminary stress incubation, is rapidly identified. The 96-well microplate and rapid analysis turnaround (<30 min) allow for higher sample throughput and several days' faster release of product when compared to conventional agar plating methods.

End product testing via bioluminescence has been successfully applied to dairy products, soy milk, rice milk, soups, and nutraceutical beverages. The pharmaceutical and personal care products industries also realize the benefits of the faster methods by validating bioluminescence for products like liquid cough medicines, shampoos, and fabric softeners. Microflora in such products are different from those found in foods, beverages, and ingredients. The products are shelf stable and contain preservatives, some of which are not permitted for human ingestion through food. To neutralize the inhibitors in product formulations and enable resuscitation of potential spoilage microorganisms, product samples are aseptically transferred to an ATP-free neutralizing broth, which is then incubated. Such neutralization may also prove useful for certain food products.

Enhanced bioluminescence-based methods can be coupled with membrane filtration to identify microorganisms in liquids such as process water, rinse water, clean-in-place (CIP) (Reinemann et al. 1998) water, emulsions, and high-purity water samples at low levels (<10 CFU/mL). Air samplers have also been integrated with ATP-based environmental monitoring systems. The air is first filtered to remove particles greater than 0.5 µm. Filters themselves or filtered collection sites can be swabbed directly to measure total ATP. Alternatively, samples can be treated with pyrase, an ATP-eliminating enzyme reagent, followed by a detergent that frees ATP from viable microorganisms for detection. With either approach, the amount of ATP detected provides a rapid indirect indicator of the bioburden of the air sample. Enhanced bioluminescence-based methods can also amplify the light signal through adenylate kinase (AK) and adenosine diphosphate (ADP), which allows AK to generate more ATP than the microorganism originally contained. The accelerated time to generate results through the AK/ADP (Albright 2008) platform provides earlier food product release if used for end-product hold-and-test situations.

When selecting a bioluminescence-based method, ease of use, data logging, and management are important, but careful consideration should also be given to the evaluation of system and reagent sensitivity, repeatability, ruggedness, and stability as systems vary widely in their ability to detect contamination from both free and intracellular ATP (Carrick et al. 2001). Validation research should accompany the implementation of ATP in food-, beverage-, and ingredient-processing quality systems monitoring. Since bioluminescence-based hygiene methods are used in both cold and hot environments, the luminometer should have a temperature sensor to compensate during the reading time as the bioluminescence reaction and detection system are sensitive to temperature changes. To meet good manufacturing practices (GMP), hazard analysis and critical control point (HACCP), and International Organization for Standardization (ISO) requirements, the bioluminescence-based method should have ATP performance controls and luminometer calibrators to ensure the system is working properly. Validation in end-product bioluminescence testing requires even greater scrutiny as parallel testing with existing conventional methods must show equivalent or better detection for the organisms of concern and over a prolonged period of time. However, assuming such validations are achieved, ATP bioluminescence offers a considerable benefit in that results are available quickly, if not in real time. This facilitates immediate action and better on-the-job training opportunities.

ATP measurement systems, while not replacing traditional microbiological techniques, actually complement them by providing an early warning capability to measure and document environmental surface cleanliness on a continuous basis and to evaluate the effectiveness of remediation programs. The flexibility to adapt enhanced sensitivity ATP systems for surface, air, and liquid monitoring ensures greater accuracy in hygienic monitoring. For sanitation programs, ATP measurement provides a true measure of "hygiene" and "cleanliness" by detecting both microorganisms and organic product residues present on surfaces. For end-product testing, ATP is measured from aerobes and anaerobes and from all species of bacteria, yeast, or mold, which distinguishes it from conventional methods. The use of ATP assays demonstrates due diligence and adherence to quality standards in cleaning and sanitizing, thereby providing documented evidence of any corrective actions that might have been taken in response to potential contamination. Good sanitation should be an ongoing strategic objective as improved hygiene and cleaning standards directly result in fewer microorganisms, increased shelf life, and ultimately a safer environment in which to produce food.

References

Albright, J. 2008. An analysis of AK-enhanced ATP bioluminescence for sterility testing using the Celsis Advance System. Paper presented at the annual meeting of the Parenteral Drug Association, Chicago, IL, October 20–23.

Aycicek, H., U. Oguz, and K. Karci. 2006. Comparison of results of ATP bioluminescence and traditional hygiene swabbing methods for the determination of surface cleanliness at a hospital kitchen. *International Journal of Hygiene and Environmental Health* 209:203–206.

Bell, C., P. A. Stallard, S. E. Brown, and J. T. E. Standley. 1994. ATP-bioluminescence techniques for assessing the hygienic condition of milk transport tankers. *International Dairy Journal* 4 (7):629–640.

Bishop, J. R., C. H. White, and R. Firstenberg-Eden. 1984. A rapid impedimetric method for determining the potential shelf-life of pasteurized milk. *Journal of Food Protection* 47:471–475.

Cady, P. 1975. Rapid automated bacterial identification by impedance measurement. In *New approaches to the identification of micro-organisms*, ed. C. Goranheden and T. Illeini. New York: Wiley, pp. 75–99.

Carrick, K., M. Barney, A. Navarro, and D. Ryder. 2001. The comparison of four bioluminometers and their swab kits for instant hygiene monitoring and detection of microorganisms in the brewery. *Journal of the Institute of Brewing* 107 (1):31–37.

Deak, T., and L. R. Buchat. 1993. Comparison of conductimetric and traditional plating technique for detecting yeast in fruit juices. *Journal of Applied Bacteriology* 75:546–550.

Deluca, M. 1976. Firefly luciferase. In *Advances in enzymology and related areas of molecular biology*, ed. A. Meister. Hoboken, NJ: Wiley.

Firstenberg-Eden, R., and G. Eden. 1984. *Impedance microbiology.* Letchworth, UK: Research Studies Press.

Frundzhyn, V. G., I. M. Parkhomenko, L. Y. Brovko, and N. N. Ugarova. 2008. Improved bioluminescent assay of somatic cell counts in raw milk. *Journal of Dairy Research* 75 (3):279–283.

Gibson, D. M., P. Coombs, and D. W. Pimbley. 1992. Automated conductance method for the detection of *Salmonella* in foods—collaborative study. *Journal of AOAC International* 75:292–302.

Jackson, L. S., F. M. Al-Taher, M. Moorman, J. W. DeVries, R. Tippett, K. M. J. Swanson, T.-J. Fu, R. Salter, G. Dunaif, S. Estes, S. Albillos, and S. M. Gendel. 2008. Cleaning and other control and validation strategies to prevent allergen cross-contact in food-processing operations. *Journal of Food Protection* 71:445–458.

Jørgenson, B. R., D. M. Gibson, and H. H. Huss. 1988. Microbiological quality and shelf-life prediction of chilled fish. *International Journal of Food Microbiology* 6:295–307.

Kelly, J. A., S. F. Amass, D. Ragland, P. M. Spicer, and R. Alvarez. 2001. Analysis of Lightning and BioClean tests for assessment of sanitation levels in pork production facilities. *Journal of Swine Health and Production* 9 (5):207–213.

Kyriakides, A. L., C. M. Costello, G. Doyle, M. C. Easter, and I. Johnson. 1991. Rapid hygiene monitoring using ATP-bioluminescence. In *Bioluminescence and chemiluminescence: current status*, ed. P. E. Stanley and L. J. Kricka. Chichester, UK: Wiley, pp. 519–522.

Leon, M. B., and J. A. Albrecht. 2007. Comparison of adenosine triphosphate (ATP) bioluminescence and aerobic plate counts (APC) on plastic cutting boards. *Journal of Foodservice* 18 (4):145–152.

Manafi, M. 2000. New developments in chromogenic and fluorogenic culture media. *International Journal of Food Microbiology* 60:205–218.

Manafi, M., W. Kneifel, and S. Bascomb. 1991. Fluorogenic and chromogenic substrates used in bacterial diagnostics. *Microbiology and Molecular Biology Review* 55 (3):335–348.

McPherson, J., and J. Savarino. 2007. Using bioluminescence technology to monitor kennel sanitizing procedures. *TechTalk* 12(6).

Ogden, I. D. 1986. Use of conductance method to predict bacterial counts in fish. *Journal of Applied Bacteriology* 61:263–268.

Perry, J. D., M. Ford, J. Taylor, A. L. Jones, R. Freeman, and F. K. Gould. 1999. ABC medium, a new chromogenic agar for selective isolation of *Salmonella* spp. *Journal of Clinical Microbiology* 37:766–768.

Reinemann, D. J., A. Grasshoff, A. C. L. Wong, and A. Muljadi. 1998. Efficacy assessment of CIP processes in milking machines. Paper presented at International Meeting, Fouling and Cleaning in Food Processing, Jesus College, Cambridge, UK. April 6–8.

Restaino, L., E. W. Frampton, W. C. Lionberg, and R. J. Becker. 2006. A chromogenic plating medium for the isolation and identification of *Enterobacter sakazakii* from foods, food ingredients, and environmental sources. *Journal of Food Protection* 69:315–322.

Russell, S. M. 1997. A rapid method for determination of shelf-life of broiler chicken carcasses. *Journal of Food Protection* 60:148–152.

Saul, S. J., E. Zomer, D. Puopolo, and S. E. Charm. 1996. Use of a new rapid bioluminescence method for screening organophosphate and N-methylcarbamate insecticides in processed baby foods. *Journal of Food Protection* 59 (3):306–311.

Siragusa, G. R., W. J. Dorsa, C. N. Cutter, L. J. Perino, and M. Koohmaraie. 1996. Use of a newly developed rapid microbial ATP bioluminescence assay to detect microbial contamination on poultry carcasses. *Journal of Bioluminescence and Chemiluminescence* 11 (6):297–301.

Steward, G. N. 1899. The change produced by the growth of bacteria in the molecular concentration and electrical conductivity of culture media. *Journal of Experimental Medicine* 4:235–245.

Timms, S. K., O. Colquhoun, and C. R Fricker. 1966. Detection of *Escherichia coli* in potable water using indirect impedance technology. *Journal of Microbiological Methods* 26:125–132.

Ur, A., and D. Brown. 1974. Rapid detection of bacterial activity by impedance measurement. In *New approach to the identification of microorganisms*, ed. C. Heden and T. Illeni. New York: Wiley, pp. 61–71.

White, C. H., J. Wilson, and M. W. Schilling. 2006. An investigation of the use of the MicroFoss as an indicator of the shelf life of pasteurized fluid milk. *Journal of Dairy Science* 89:2459–2464.

Whitehead, K. A., L. A. Smith, and J. Verran. 2008. The detection of food soils and cells on stainless steel using industrial methods: UV illumination and ATP bioluminescence. *International Journal of Food Microbiology* 127 (1–2):121–128.

Wren, M., M. Rollins, A. Jeanes, T. Hall, P. Coën, and V. Gant. 2008. Removing bacteria from hospital surfaces: a laboratory comparison of ultramicrofibre and standard cloths. *Journal of Hospital Infection* 70 (3):265–271.

chapter eleven

Nucleic acid-based methods for detection of foodborne pathogens

Bledar Bisha and Lawrence Goodridge

Contents

11.1 Introduction

The "gold standard" methods for detection of foodborne bacterial pathogens are still considered to be culture-based methods, which require the performance of numerous procedural steps, between preenrichment, enrichment, selective plating, identification, and confirmation. These lengthy procedures can take several days to carry out and are costly, requiring extensive manual labor and large amounts of media and reagents.

These cultural methods, while proven to be reliable, may not necessarily be optimal for detection of foodborne pathogens due to the current fast-paced production and distribution practices of food products by the food-processing industry. The rapidity of the food production system has actually helped spur the development and application of alternative molecular and more rapid methods for detection of foodborne pathogens. Early accurate detection of pathogens along the food production chain (from raw ingredients to finished product) can aid in avoiding the transmission of foodborne disease to consumers, thereby further protecting public health and lowering the financial burden on the health care system, as well as reducing economic losses to the food industry, which stem from costly product recalls, loss of reputation, and litigation costs.

Nucleic acid-based detection methods present a valid alternative to conventional detection methods, allowing for increasingly rapid, specific, and sensitive detection of pathogens in the food and processing environment. Methods such as polymerase chain reaction (PCR) are already finding wide acceptance across the food industry pathogen-testing settings, crossing the barrier from the research laboratories. Although new PCR-based methods are constantly being researched and developed, there are numerous PCR detection methods already widely used for routine pathogen testing and spoilage microorganism monitoring.

There is a continued need in food microbiological analysis for methods that can provide comparable results to traditional methods while permitting the possibility of speedier analysis with fewer preparation steps. This has led to extensive research in development of rapid tests, with nucleic acid-based approaches considered very important (Feng, 2007). The desired properties of these alternative rapid methods have been discussed previously and serve as guidelines for further development of such methods (Swaminathan and Feng, 1994; Fung, 1995; Notermans et al., 1997; de Boer and Beumer, 1999; Brehm-Stecher and Johnson, 2007; Feng, 2007). Such methods should possess the following qualities:

- Rapidity
- High accuracy (repeatability)
- High degree of specificity
- High level of sensitivity
- Reproducibility
- Compatibility with a wide range of food and environmental sample matrices
- Ease of use and easily automated
- Inexpensive (equipment cost and ongoing cost per sample)

In addition, new methods must be validated against standard methods, preferably with naturally contaminated samples. It is clear that such an "ideal" detection method does not yet exist; however, many of the alternative nucleic acid-based detection methods covered here fulfill several of these desired characteristics.

Nucleic acid-based detection methods employ techniques to specifically recognize and amplify nucleic acid material (DNA, ribosomal RNA [rRNA], messenger RNA [mRNA]) in the target cells that is unique to those cells (Brehm-Stecher and Johnson, 2007). Several important nucleic acid-based detection methods, such as PCR and its variations, nucleic acid sequence-based amplification (NASBA), DNA hybridization, fluorescence *in situ* hybridization (FISH), and others, are covered in this chapter.

11.2   Nucleic acid amplification techniques

11.2.1   Polymerase chain reaction

The basic mechanism of conventional PCR involves the use of a DNA polymerase enzyme and two oligonucleotide primers to amplify a specific region of DNA to multiple copies. This is done through a series of repeated cycles (denaturation at 95°C, annealing at 55°C, and extension at 72°C) in the presence of cofactors and Mg^{2+}. For conventional PCR, the results of the amplification are visualized by staining using agarose gel electrophoresis and fluorescent staining with DNA intercalating dyes. The presence or absence of bands or fragments (amplicons) of a particular size determines whether a positive or a negative result is obtained (Sanderson and Nichols, 2003; Fairchild et al., 2006; Mothershed and Whitney, 2006). Alternatively, enzyme-linked immunosorbent assay (ELISA) endpoint detection of the amplicons can be performed, obviating the need for agarose gel systems. This is done by attaching a label to the primers (e.g., digoxigenin), which is integrated into the synthesized amplification products, followed by immobilization of the amplicon on a microtiter plate by a capture probe and final colorimetric detection by an enzyme-antibody conjugate. Enzymes can be alkaline phosphatase or horseradish peroxidase, and an antidigoxigenin antibody can be used (Fairchild et al., 2006; Mothershed and Whitney, 2006)

Primers can be designed to target rRNA encoding genes and genes that encode proteins, including virulence factors, allowing the amplification of target sequences by 10^6-fold in a matter of hours (Fratamico and Bayles, 2005; Feng, 2007). For example, one of the most commonly used diagnostic targets for PCR of *Listeria monocytogenes* has been the *hlyA* gene, which encodes the protein listeriolysin (LLO) (Ryser et al., 1996; Blais et al., 1997; Norton et al., 2001). Other targets have included the *iap* gene, which encodes a protein associated with invasion (Cocolin et al., 2002; Schmid et al., 2003); 16S rRNA genes (Wang et al., 1992); and *inlB*, which encodes internalin B (Ingianni et al., 2001; Pangallo et al., 2001). For example, Jung et al. (2003) were able to detect spiked *L. monocytogenes* in frankfurters with a PCR protocol targeting the *inl*AB gene, following an enrichment of 16 h and reaching a detection sensitivity of 10 CFU (colony-forming units) 25 g^{-1} (with a limit of detection of 10^5 CFU mL^{-1} in pure culture). For *Salmonella* spp., the most common diagnostic target employed has been the *invA* gene, which is highly conserved within the genus (Rahn et al., 1992). For specific serovars, diagnostic targets have been sequences encoding antibiotic resistance (Carlson et al., 1999) or phage types (Hermans et al., 2005), as well as genes or sequences encoding surface antigens (Luk et al., 1993; Herrera-León et al., 2004).

Variations to conventional PCR include real-time PCR, conventional or real-time reverse transcriptase PCR (RT-PCR), and conventional or real-time multiplex PCR. Some of these adaptations are discussed in the following material.

11.2.2   Nested PCR

Nested PCR is a variant of conventional PCR that uses two different sets of primers to carry out the process; the first set is a so-called outer pair, which is employed to amplify a region of interest, followed by another amplification by an "internal" primer pair. Nested PCR has been reported to have an increased detection sensitivity compared to conventionally performed PCR (Sanderson and Nichols, 2003; Mothershed and Whitney, 2006).

11.2.3   Reverse transcriptase PCR

RT-PCR uses RNA, not DNA, as a target for amplification in a reaction catalyzed by the enzyme reverse transcriptase, producing complementary DNA (cDNA), which can then be amplified by standard PCR (Mothershed and Whitney, 2006; Theron et al., 2010). Such an approach can be useful in detecting viable bacteria since a molecule with a very short half-life, such as mRNA, can be targeted (Mustapha and Li, 2006) or be applied to detect foodborne viral pathogens containing RNA genetic material (Sobsey, 1994; Schwab et al., 1996; Huang et al., 2000; D'Souza and Jaykus, 2006; Theron et al. 2010).

Klein and Juneja (1997) used RT-PCR targeting the *iap* gene transcript to detect as few as 3 cells of *L. monocytogenes* per gram ground beef following a total procedure of 54 h. Morin et al. (2004) employed multiplex RT-PCR to detect *S.* Typhi, along with *Escherichia coli* O157:H7, with a sensitivity of 30 cells. RT-PCR was also successfully employed to detect live *S. enterica* in jalapeño and serrano peppers as well as tomatoes and spinach by targeting mRNA related to *inv*A (González-Escalona et al., 2009). RT-PCR and nested RT-PCR have received attention as valuable tools for detection of noroviruses from a variety of foods as well as hepatitis A virus, and successful detection has been demonstrated in a number of studies (D'Souza and Jaykus, 2006).

11.2.4   Real-time PCR and quantitative real-time PCR

Real-time PCR assays combine the amplification and detection in one step, allowing for online monitoring and thus speedier and more sensitive detection that requires fewer postamplification procedural steps to accomplish detection. This PCR variation is made possible through the use of a fluorescent dye that incorporates into the amplicon as it is being amplified. The fluorescent signal that is incorporated can be due to intercalating dyes such as SYBR® Green, hybridization probes (Light-Cycler), molecular beacons, Scorpions™, or dual-labeled probes (TaqMan®) (Fairchild et al., 2006; Mothershed and Whitney, 2006). A quantitative PCR (qPCR) assay is performed by employing serial dilutions of known target copies following exponential increase in initial target nucleic acid to determine the unknown amount of nucleic acids in the sample analyzed.

Real-time multiplex RT-PCR was found to be 10 times more sensitive than conventional PCR when used for the multiplex detection of enteroviruses, astroviruses, and noroviruses (Beuret, 2004). A multiplex real-time qPCR employing multiple TaqMan probes used three different probes targeting the genes for β-glucuronidase (*uid*A) in *Escherichia coli*, thermonuclease gene (*nuc*) in *Staphylococcus aureus*, and the origin of the replication sequence (*oriC*) gene of *Salmonella enterica*, allowing for detection of approximately 10^3 CFU g^{-1} of minimally processed vegetables (Elizaquível and Aznar, 2008; Dwivedi and Jaykus, 2010). A SYBR Green qPCR using primers against the *inv*A gene for *Salmonella* spp. detected 7.5×10^2 CFU 100 mL^{-1} in chicken carcass rinses and irrigation water (Wolffs et al., 2006). Approximately 10 CFU of *Vibrio vulnificus* in oysters and seawater were detected and quantified using qPCR targeting the *toxR* virulence gene of the pathogen (Takahashi et al., 2005).

11.2.5   Conventional or real-time multiplex PCR

Multiplex PCR allows the amplification of two or more target sequences via use of two or more primer sets, which permits detection of several bacterial cells in one reaction and can eventually supply more information on the sample analyzed (Fratamico and Bayles, 2005). To allow for identification of two or more different target amplicons, two or more primer sets are used to amplify exclusive fragment sizes, while the conditions of the reaction such as primer concentration, Mg^{2+}, free dNTPs (deoxyribonucleotide triphosphates), and the enzyme DNA polymerase are adjusted to conditions that will favor all reactions. This permits the identification of two or more microorganisms in one reaction while discounting time and labor needed to perform detection (Harmon et al., 1997; Houf et al., 2000; Fairchild et al., 2006).

Simultaneous detection of *L. monocytogenes, S.* Typhimurium, and *E. coli* O157:H7 was achieved in fresh produce following 30-h enrichment with a detection sensitivity of 1 cell mL^{-1} for *Salmonella* and *E. coli* O157:H7 and 10^2–10^3 cells mL^{-1} for *L. monocytogenes* (Bhagwat, 2003). A multiplex real-time PCR TaqMan assay that targeted four different genes was able to detect total and pathogenic *V. parahaemolyticus*, including the pandemic O3:K6 serotype in oysters. The gene targets included the *tdh* and *trh* genes (detection of pathogenic *V. parahaemolyticus*), ORF8 (detection of pandemic *V. parahaemolyticus* O3:K6), and *tlh* (detection of total *V. parahaemolyticus*) (Ward and Bej, 2006).

11.2.6   Nucleic acid sequence-based amplification

Several isothermal nucleic acid method variations have been reported in the scientific literature, with some of them finding use in clinical applications, including transcription-mediated amplification, strand displacement amplification, rolling circle amplification, cycling probe technology, and NASBA (Mothershed and Whitney, 2006). Of these approaches, NASBA has been more extensively researched as a detection method for microbial pathogens in food and is covered in more detail here.

NASBA, like PCR, is also applied to extracted nucleic acid, but differently from PCR, NASBA is an isothermal reaction that does not require going through amplification cycles and can be performed in tubes not requiring equipment such as a PCR heat block since it is conducted isothermically at 41°C. This method amplifies RNA from RNA or DNA by using transcriptive amplification via use of the enzymes avian myeloblastosis virus-reverse transcriptase (AMV-RT), ribonuclease H, and T7 RNA polymerase. The target sequences can be amplified to more than 10^9 copies in a period of 90 min. In addition, it offers the advantage of being applicable to mRNA targets, allowing for a realistic assessment of viability (Sanderson and Nichols, 2003; Churchill et al., 2006; Mothershed and Whitney, 2006; Fratamico and Bayles, 2005; Feng, 2007).

NASBA has been employed to detect *L. monocytogenes* in meats and seafood using 16S rRNA sequences as well as *hlyA* mRNA, allowing for a sensitivity of 10 CFU 60 g^{-1} of meat (Uyttendaele et al., 1995; Blais et al., 1997). This method was also used with the NucliSens kit to detect *S. enterica* in meat, poultry, and other foods by targeting mRNA transcribed from the *dnaK* gene (Simpkins et al., 2000; D'Souza and Jaykus, 2003). Multiplexed detection using NASBA of hepatitis A virus and noroviruses from lettuce and deli turkey has been reported (Jean et al., 2004).

11.3   Hybridization techniques

11.3.1   DNA hybridization

DNA hybridization is referred to here as the method that is applied to bacterial colonies following nucleic acid extraction and separation via electrophoresis (whole-cell detection is covered in the section about FISH). DNA hybridization (Southern blotting) is performed by transferring colonies to a solid support, followed by cell lysis, denaturation of the DNA, and hybridization. Annealing of single-stranded DNA counterparts occurs based on sequence similarities between the two strands and allows for nucleic acid target detection and quantification. An alternative format uses hybridization in solution, and changes in fluorescence signal a positive reaction. If mRNA or rRNA and not DNA is used as a target, the variation of the hybridization procedure is named Northern blotting (Sanderson and Nichols, 2003; Fratamico and Bayles, 2005). Oligonucleotide (<20 nucleotides long) or polynucleotide (>50 nucleotides long) probes can be used, with oligonucleotides finding more widespread use because they allow for single-mismatch discrimination (Theron et al., 2010). More details on types of probes and probe labeling used in nucleic acid hybridization are given in the section on FISH. Typically, low sensitivities of 10^4–10^5 bacterial cells have been reported for detection via DNA hybridization methods, raising the need for enrichment prior to the performance of the assays and increasing the time-to-detection timeframe (Swaminathan and Feng, 1994). An example of the DNA hybridization technique is given by the GeneTrack commercial kit, which uses a dipstick with a capture probe specifically targeting rRNA encoding genes and is available in *Salmonella* and *Listeria* formats as well as for *Campylobacter* and *E. coli* O157:H7 (Fung, 2002).

DNA hybridization detection methods have received attention early. An example can be given by the assay developed by Datta and Wentz (1989), who developed hemolysin gene-specific oligonucleotides for *L. monocytogenes* in a dot blot assay to differentiate it from nonpathogenic listeriae within a 4-day timeframe. Peterkin et al. (1992) combined digoxigenin-labeled oligonucleotides for DNA hybridization with a hydrophobic grid-membrane filter capture step to accomplish successful detection of *Listeria monocytogenes* in spiked soft cheese, raw milk, and chicken meat.

11.3.2   Fluorescence in situ *hybridization*

DeLong et al. (1989) introduced the use of fluorescently labeled probes that replaced the radioactively labeled probes for *in situ* detection of bacteria, hence the name fluorescence *in situ* hybridization (FISH). Fluorescent probes supplied a number of advantages compared to radioactively labeled probes, among which enhanced safety, improved resolution, reduction in detection steps, and because of the flexibility of using dyes of different excitation and emission spectra, the possibility of detection of several target sequences in one sample.

Oligonucleotide probes are designed in silico by aligning them with known target sequences in databases by using the probe-matching tools in resources such as, for example, the ARB database (Ludwig et al., 2004) or the Ribosomal Database Project II (RDP-II) (Maidak et al., 2001). Commonly, oligonucleotide probes that are generated in an automatic synthesizer are between 15 and 30 base pairs (bp) long (Moter and Göbel, 2000). For direct fluorescent labeling, probes are commonly labeled with fluorochromes chemically at the 5′ end via an amino linker or less often labeled at the 3′ end enzymatically using a terminal transferase. Typical fluorochromes used for probe labeling are fluorescein derivatives (e.g., fluorescein isothiocyanate or FITC), rhodamine dyes (e.g., Texas Red) or cyanine dyes like Cy3, Cy5, and Cy7, which typically produce brighter staining (Zimmer and Wähnert, 1986; Moter and Göbel, 2000).

The fundamental principle in FISH is the specific hybridization of complementary nucleic acid sequences within whole permeabilized cells via fluorescently labeled probes for detection (Moter and Göbel, 2000). FISH assays are overwhelmingly performed using rRNA as a target. This diagnostic target is desired because it is genetically stable, contains both highly conserved and variable regions, and is present in a high number of copies in each target cell (Woese, 1987; Moter and Göbel, 2000). The fact that rRNA contains both variable and conserved regions allows for development of probes that can differentiate target cells at different taxonomic levels. Probes can discriminate between large taxonomic groups (Archea, Bacteria, and Eukarya) or down to the specific genus or species (Moter and Göbel, 2000; Amann and Fuchs, 2008).

While 5S, 16S, and 23S of bacteria can be used as a diagnostic target for oligonucleotide probes, 16S rRNA has more commonly been used for such purposes. The reason for such an occurrence is the availability of large databases for 16S rRNA of bacteria that have been deposited following PCR amplification and sequencing during the recent two decades, covering basically about 8000 species of Bacteria and Archea and amounting to about 500,000 small subunit (SSU) rRNA entries (Amann and Fuchs, 2008). The 5S rRNA contains only about 120 nucleotides and does not provide enough variability for discrimination via hybridization with oligonucleotide probes; however, 16S rRNA, which contains about 1600 nucleotides, serves as a more suitable diagnostic target. When 16S rRNA does not provide enough variability to accomplish interspecies and intraspecies differentiation, 23S rRNA, which contains about 3000 nucleotides, can provide enough variability to allow for those discriminations. While the lack of extensive deposited 23S rRNA sequences has accounted for only limited use of 23S rRNA as a diagnostic target in environmental studies, 23S RNA of main foodborne bacterial pathogens has been validly described; thus, it can be successfully used as a target for FISH.

Accessibility of target regions of the 16S rRNA and 23S rRNA of *Escherichia coli* and 18S rRNA of *Sacharomyces cerevisiae* to oligonucleotide probes has been studied in three large systematic studies (Fuchs et al., 1998, 2001; Behrens et al., 2003). It was determined that the tertiary structure of the ribosomes did not affect the hybridization efficiency, probably due to the denaturation and destabilization caused by fixation and treatments at high temperatures during hybridization, but instead secondary structures were in fact the higher-order structures that affected accessibility to the oligonucleotide probes. Regions that provided hybridization efficiencies ranging from low to high were mapped in detail and can be taken into account when probes are designed. However, Yilmaz et al. (2006) designed a thermodynamic study relating the hybridization efficiency of 16S rRNA of *E. coli* to overall Gibbs free-energy change ($\Delta G^{\circ}_{overall}$) and determined that there are no truly inaccessible regions of 16S rRNA, but these regions can be made accessible by designing probes with $\Delta G^{\circ}_{overall}$ of less than 1.3 kcal/mol.

FISH-based assays are performed by following these common steps:

1. Fixation
2. Preparation
3. Hybridization
4. Washing
5. Visualization and documentation

Fixation is performed to permeabilize the cells, allowing for the entry of the probes as well as preventing nucleic acids from degradation. This procedure is carried on using cross-linking fixatives like aldehydes (e.g., formaldehyde or paraformaldehyde) or precipitating fixatives such as methanol or ethanol. Precipitating agents are the fixative of choice when fixation of cells possessing a thick cell wall such as Gram-positive bacteria is desired, while aldehyde-based fixatives perform very well with Gram-negative microorganisms. Fixation causes loss of cell viability, and even though FISH of live cells has been shown by Silverman and Kool (2005), the possibility of uptake of oligonucleotide probes by live cells was contradicted by later work (Amann and Fuchs, 2008).

The second step, sample preparation, may involve treatment of Gram-positive bacteria with compounds that improve the permeability of cells to probes (e.g., lysozyme or lysostaphin) (Schönhuber et al., 1997; Wagner et al., 1998; Moter and Göbel, 2000) or, for example, simply coating slides with gelatin (Amann et al., 1990) or other coating agents, ethanol dehydration of samples air-dried onto slides, or dewaxing of the paraffin-embedded preparations (Boye et al., 1998). The preparation step might not be necessary if no special treatments are needed to conduct a complete FISH procedure.

The third step is hybridization, which basically involves annealing of the probe to its target sequence in the cell rRNA. Hybridizations are conducted under stringent (high degree of homology between probe and target sequence) conditions. The stringency of hybridization can be varied by varying the formamide concentration in the hybridization buffer, the temperature of hybridization, or the salt concentration. To ensure a successful hybridization, the temperature of the hybridization must be maintained below the melting temperature of the probe for annealing of the probes to target sequences to occur. An empirical formula can be used for calculating the T_m of an oligonucleotide in relation to its GC content: T_m (in °C) = $2(A + T) + 4(G + C)$ (Suggs et al., 1981). Formamide in the hybridization buffer weakens the hydrogen bonds of the target sequence, practically lowering the T_m, while decreasing its concentration will increase T_m. Washing is performed in order to mediate removal of the unbound probes. It is carried out using wash buffer containing the same elements which were described above for hybridization buffers while keeping stringency under control.

Visualization can be performed by fluorescence microscopy or following mounting of the samples in a mounting medium with or without added agents, which prevent fading of the fluorochromes under intense illumination, or by cytometry.

Several variations to FISH have been described that have been used to improve the sensitivity of this method or increase the hybridization efficiency; some of them are covered here. One approach uses hybridization with several monolabeled oligonucleotides targeted to different sequences (Glöckner et al., 1996) to increase the signal intensity. Using oligonucleotides labeled with horseradish peroxidase followed by catalyzed reported deposition of fluorescently labeled tyramide (CARD-FISH) has been shown to improve signal intensity (Schönhuber et al., 1997); however, special and harsh pretreatment of whole cells to improve uptake of the large enzyme complexes is needed. DNA oligonucleotides can be modified by designing two complementary sequences on both sides of the probe sequence to permit formation of a stem-loop in solution. These structures are called molecular beacons (MBs), and they are labeled with a fluorochrome in one end and a quencher at the other end; thus, they only fluoresce when they are annealed to the target sequence but not when they are unbound, increasing the signal-to-noise ratio (Tyagi and Kramer, 1996; Xi et al., 2003; Lenaerts et al., 2007).

Peptide nucleic acids (PNAs) bring on another exciting development in probe technology. PNAs are molecules that mimic DNA but instead are in possession of an uncharged, achiral backbone consisting of repetitive units of N-(2-aminoethyl) glycine. These unique characteristics imparted by PNA molecules allow them to penetrate thick cell walls of Gram-positive cells better; they also improve hybridization kinetics; make their employment independent of salt concentration; make them resistant to nucleases, which can be present in food matrices; as well as increase the accessibility of regions of the ribosome that are inaccessible to DNA probes (Egholm et al., 1993; Demidov et al., 1994; Stender et al., 2002; Brehm-Stecher et al., 2005).

Potential problems with FISH might arise from the fact that 16S rRNA can be too conserved, not allowing intraspecies or interspecies differentiation (Fox et al., 1992); however, as mentioned, 23S rRNA can serve in these cases as a more suitable diagnostic target that provides more region variability (Amann and Ludwig, 2000). Generally, the possibility of hybridization of the specific probes with rRNA of unknown microorganisms that contain similar sequences with the target microorganisms exists (Amann and Ludwig, 2000); however, this is not likely when FISH of well-characterized bacterial pathogens such as *Listeria monocytogenes* and *Salmonella* spp. in their environmental niche is attempted. On designing probes in silico, they should be carefully checked against closely related microorganisms and microorganisms found in the same environmental niche to confirm that the hybridizations are indeed specific at the set hybridization conditions. Problems that occur with low signal intensity, due to the low number of ribosomes and target rRNA subsequent stress and injury (Amann et al., 1995), might be overcome by including brief resuscitation and nonselective enrichment steps. Problems with target accessibility have been discussed and should be taken into consideration when designing oligonucleotides for probing of bacteria.

FISH allows for specific and sensitive labeling of target whole cells in a relatively short time, which makes its use desirable in rapid microbiology. This method can be combined with endpoint analysis methods such as epifluorescence microscopy or flow cytometry to process and document a large number of target cells rapidly. Fang et al. (2003) used DNA probes targeting 23S rRNA of *Salmonella* spp. to detect this pathogen in 18 different kinds of foods with improved outcome compared to the culture method. Another study (Bisha and Brehm-Stecher, 2009a) used a cocktail of two DNA probes, providing coverage of both *S. enterica* and *S. bongori* to specifically hybridize with *Salmonella* spp. on the surface of tomatoes following extraction via adhesive tape and either direct detection via epifluorescence microscopy of approximately 10^3 CFU cm² or miniculture nonselective enrichment for 5 h and flow cytometric detection. *Salmonella*-specific DNA-FISH was conducted in another study to simultaneously detect and characterize the interactions of this pathogen in the alfalfa sprouts matrix via imaging cytometry (Bisha and Brehm-Stecher, 2009b). PNA-FISH for specific *Salmonella* detection was recently applied to a number of sample matrices, including powdered infant formula, and following enrichment for 8 h the method was able to detect as few as 1 CFU 10 g⁻¹ of the pathogen (Almeida et al., 2010).

Probes and protocols for FISH of other foodborne pathogens have been described. For example, another study was able to combine specific detection and quantitative enumeration of *Listeria* spp. by conducting FISH with a 23S rRNA DNA probe on hydrophilic membrane filters to achieve a comparable result with plate counts in a shorter time (Fuchizawa et al., 2008).

11.4 Aptamers, microarrays, and nucleic acid-based biosensors: alternative and emerging nucleic acid-based approaches for detection of foodborne pathogens

11.4.1 Aptamers

Aptamers are single-stranded DNA or RNA ligands folded into stable secondary structures and randomly selected for specificity to targets against a library of sequences. The selection process is named systematic evolution of ligands by exponential enrichment (SELEX) and entails a number of iterative selection cycles against a library of different oligonucleotide sequences (Ellington and Szostak, 1990; Tuerk and Gold, 1990; Tombelli et al., 2007). Compared to antibodies, aptamers have advantages as ligands since they are stable in higher temperatures and can regain their initial structure if exposed to denaturation, offer less batch-to-batch variation, and can be optimized in their affinity to target molecules by introducing modifications to their structure during chemical synthesis (Tombelli et al., 2007). The inherent vulnerability of aptamers to nucleases has been counteracted by modification of the ribose ring at the 2′-location, contributing to enhanced stability of these molecules (Pieken et al., 1991). In food microbiology applications, aptamers have been used mainly as capture molecules, as illustrated by one study in which DNA aptamers with affinity for outer membrane proteins of *Salmonella enterica* serovar Typhimurium were used for concentration of this pathogen from chicken rinses (Joshi et al., 2009). In another study, *Campylobacter jejuni* was detected at levels of 10–250 CFU mL⁻¹ in a variety of spiked foods following capture of the pathogen with magnetic beads linked to aptamers specific for surface proteins of the organism and final detection with a fluorescent quantum dot assay (Bruno et al., 2009). The potential of aptamers in direct detection of pathogens has been demonstrated, however, in detection of *Francisella tularensis*; a dot blot assay was developed involving the employment of aptamers labeled with biotin, which were then reacted with alkaline phosphatase labeled with streptavidin, with a positive reaction signaled by a color change (Vivekananda and Kiel, 2006; Dwivedi and Jaykus, 2010).

11.4.2   Nucleic acid-based arrays

Microarrays are constructed by immobilizing probes (DNA, cDNA, or oligonucleotides) at an ordered two-dimensional configuration to a solid support, which can be glass or nylon membranes (Theron et al., 2010). Typically, the nucleic acid target is labeled with an enzyme or fluorochrome and, following hybridization to the capture probes on the microarray, is washed away, so the unbound labeled nucleic acids do not contribute to detection by direct fluorescence scanning or enzyme-conferred signal analysis (de Boer and Beumer, 1999). Microarrays impart several advantages, including small size, possibility for multiplexing, and capacity to analyze multiple samples; however, they are prone to interference from the sample matrix and nonspecific binding (de Boer and Beumer, 1999; van Hal et al., 2000; Lemarchand et al., 2004; Theron et al., 2010). Since the limit of detection for DNA hybridization using microarrays is 10^4–10^5 copies, this method is yet dependent on cultural enrichment to detect pathogens in food sensitively (Dwivedi and Jaykus, 2010).

Even though microarrays remain very much in the research stage, they have been applied for detection of foodborne pathogens, including *S. enterica, Bacillus* spp., and *Campylobacter* spp. from milk (Cremonesi et al., 2009) and *Yersinia enterocolitica* from mung bean, alfalfa sprouts, mamey sapote, and cilantro (Siddique et al., 2009). Detection of multiple foodborne pathogens in one assay has been reported, as, for example, when Hong et al. (2004) detected 14 foodborne bacterial pathogens in the same format with a sensitivity of approximately 100 CFU mL^{-1} for each species.

11.4.3   Nucleic acid-based biosensors

Biosensors are devices consisting of two elements, a bioreceptor used to recognize the target and a transducer that translates the signal into a measurable electrical signal that can than be used for analysis and detection (Velusamy et al., 2010). A variety of biorecognition elements can be used in biosensors, including antibodies, enzymes, biomimetic compounds, and more; however, we only briefly cover nucleic acid-based biosensors. Both DNA and PNA (see section on FISH) have been reported as bioreceptors for biosensor development. Unique target sequences are recognized by the synthesized bioreceptor, allowing for rapid and specific detection of the pathogen. Nucleic acid-based biosensors are still in the research stage; nevertheless, successful detection of foodborne pathogens has been reported. To name a few, nucleic acid-based biosensor detection of *Escherichia coli* O157:H7 (Chen et al., 2008) and *Salmonella* spp. (Lermo et al., 2007) has been reported. In one study, multiplex detection of several foodborne pathogens in one assay was reported, including *Listeria* spp., *Campylobacter* spp., *Staphylococcus aureus* enterotoxin genes, *Clostridium perfringens* enterotoxin genes, and associated virulence factors (Sergeev et al., 2004; Velusamy et al., 2010).

11.5   Nucleic acid-based subtyping of foodborne pathogens

Extensive coverage of nucleic acid-based typing methods of bacterial pathogens for epidemiological surveillance is beyond the scope of this chapter; however, it is important to mention briefly chief typing methods that use nucleic acid-based approaches. Pulsed field gel electrophoresis (PFGE), repetitive element PCR (rep-PCR), restriction fragment length polymorphism (RFLP), randomly amplified polymorphic DNA (RAPD), amplified fragment length polymorphism (AFLP), ribotyping, and multilocus variable-number tandem-repeat analysis (MLVA) are available tools that can be used in epidemiological surveillance of foodborne pathogens (Churchill et al., 2006; Boxrud, 2010). PFGE is considered the gold standard for subtyping of foodborne pathogens (Boxrud, 2010). MLST (multiplelocus sequence typing) was found to compare favorably to PFGE for subtyping of this pathogen (Bender et al., 2001; Barrett and Gerner-Smidt, 2007). PFGE has also been the common subtyping method for *L. monocytogenes*, while riboprinting, AFLP, and the sequencing-based genotyping method MLST (including a modification that allows simultaneous subtyping of three virulence and three virulence-associated gene foci, MVLST) have also been used (Zhang et al., 2004; Barrett and Gerner-Smidt, 2007; Boxrud, 2010). MLST subtyping has been reported for *Campylobacter jejuni*, while single locus sequencing has been used for subtyping of noroviruses and *Staphylococcus aureus* (Mattison et al., 2009; McTavish et al., 2009; Grundmann et al., 2010; Boxrud, 2010).

11.6   Preanalytical sample preparation for nucleic acid-based detection

Sample preparation for rapid molecular detection is an important step that can ensure successful detection if performed properly or complicate the performance of the test. An example of such need for preprocessing of sample would be the employment of PCR to detect pathogenic bacteria in food, which is limited by the small sample size, the presence of inhibitors, and the need for concentration of the target organism (Bej and Mahbubani, 1992). Concentration, separation, and purification of bacterial cells from the complex food matrix in which they are found can effectively aid final detection via rapid detection methods and technologies by removing assay inhibitors or concentrating the target cells to detectable numbers (Swaminathan and Feng, 1994; Wilson, 1997; de Boer and Beumer, 1999; Jaykus, 2001; Stevens and Jaykus, 2004). Sharpe (2003) listed four different approaches for detection of bacterial pathogens in foods:

1. Extraction of whole cells with subsequent concentration and identification
2. Detection using cell phenotypes such as serological and enzymological characteristics
3. Chemical extraction, as is the case for extracted nucleic acids
4. Direct detection of pathogens in food following transfer of a property or a label to the target cells, which will allow for them to be detected against the background noise supplied by the food sample.

These approaches have to be considered and validated for each particular detection method and food matrix, taking into account the method and sample variability to enhance detection capabilities and sensitivity of the detection methods. Stevens and Jaykus (2004) pointed out three important issues that should be considered to overcome problems associated with practical employment of rapid molecular methods for detection of bacteria in food:

1. Pathogen separation from the particulate matter in samples
2. Effective removal of assay inhibitors from the sample
3. Reduction in sample size without compromising cell viability

Several strategies have been developed to address these issues; these strategies can be grouped into physical, chemical, physicochemical, and biological sample preparation methods (Stevens and Jaykus, 2004). An example of a physical method is given by the study of Niederhauser et al. (1992). This group used differential centrifugation to enhance the PCR detection of *L. monocytogenes* in meat by 1000-fold; basically, they centrifuged food samples at 100g to remove the larger particulate matter, followed by centrifugation at 3000g to harvest cells. Other physical sample preparation methods that have been developed include more centrifugation methods (high speed and density gradient), filtration, ultrasound, spraying, gas bubbles, and more.

Chemical methods of separation and concentration have also been used and range from adsorption and desorption methods to dielectrophoresis and biphasic partitioning (Sharpe, 2003; Stevens and Jaykus, 2004). Pedersen et al. (1998) used a biphasic partitioning method using 5% dextran and polyethylene glycol to effectively detect *L. monocytogenes* and

Salmonella enterica serovar Berta from smoked sausage via PCR without interference from inhibitors and particulate matter.

Physicochemical methods are a combination of chemical and physical methods for cell extraction and concentration. An example is given by the employment of metal hydroxides (Stevens and Jaykus, 2004). *Listeria monocytogenes* and *S. enterica* serovar Enteritidis have been effectively concentrated in reconstituted nonfat dry milk, increasing the detection sensitivity in a study involving endpoint detection via RT-PCR (Lucore et al., 2000).

Finally, biological methods that are based on immunoaffinity also supply a valid alternative for sample preparation (e.g., immunomagnetic separation [IMS]). Due to its high specificity, IMS has been incorporated in the official method for detection and identification of *Salmonella* spp. and *E. coli* O157:H7 in foods (Association of Analytic Chemists [AOAC], 1995; Stevens and Jaykus, 2004). Circulating IMS has been a recent development to the more traditional IMS methods, allowing the analysis of large sample sizes.

11.7   Concluding remarks

There are several potential advantages and disadvantages associated with use of nucleic acid-based methods. Among advantages, speed of analysis can be listed as the major benefit, allowing for a significant decrease in sample analysis time. Theoretically, methods such as PCR are able to amplify even one target sequence logarithmically to a detectable number of copies. Also, nucleic acid-based methods generally show good correlation with culture-based methods (Sanderson and Nichols, 2003). Disadvantages include the need for enrichment, which cannot be completely overcome, and the interference to the assays, such as PCR from inhibitors present in food such as phenolics, polysaccharides, glycogen, calcium ions, fat, and others (Rossen et al., 1992; Powell et al., 1994; Bickley et al., 1996; Liu, 2008). Inability to distinguish between live and dead cells has been listed as a potential disadvantage of nucleic acid-based methods such as PCR and NASBA; target nucleic acids have been shown to persist for 30 h following cell death by heat inactivation (Birch et al., 2001; Sanderson and Nichols, 2003). However, novel PCR approaches have circumvented the problem of viability by incorporating compounds such as ethidium bromide monoazide or propidium monoazide in the reaction. Such compounds only permeate membranes of dead cells; thus, additional valuable information on viability is obtained (Dwivedi and Jaykus, 2010). Other disadvantages include susceptibility to contamination characteristic of amplification-based techniques, degradation of RNA during extraction in RNA-based methods such as NASBA that lead to false-negative results, and the large number of manipulation steps needed to perform methods such as NASBA (Sanderson and Nichols, 2003).

Nucleic acid-based methods are continuously gaining importance as valid approaches for detection of foodborne pathogens beyond the microbial research lab. Such methods are becoming valuable tools to detect and track foodborne pathogens and specific spoilage microoorgansims in raw ingredients, food products, and associated environments. As researchers continue to develop and refine these methods, there is likely to continue to be a need for their deployment as part of food safety management systems.

References

Almeida, C., N.F. Azevedo, R.M. Fernandes, C.W. Keevil, and M.J. Viera. 2010. Fluorescence in situ hybridization method using a peptide nucleic acid probe for identification of *Salmonella* spp. in a broad spectrum of samples. *Applied and Environmental Microbiology* 76:4476–4485.

Amann, R., and B.M. Fuchs. 2008. Single-cell identification in microbial communities by improved fluorescence *in situ* hybridization techniques. *Nature Reviews Microbiology* 6:339–348.

Amann, R.I., L. Krumholz, and D.A. Stahl. 1990. Fluorescent-oligonucleotide probing of whole cells for determinative, phylogenic, and environmental studies of microbiology. *Journal of Bacteriology* 172:762–770.

Amann, R.I., and W. Ludwig. 2000. Ribosomal RNA-targeted nucleic acid probes for studies in microbial ecology. *FEMS Microbiology Reviews* 24:555–565.

Amann, R.I., W. Ludwig, and K.H. Schleifer. 1995. Phylogenetic identification and in situ detection of individual microbial cells without cultivation. *Microbiology Reviews* 59:143–169.

Association of Analytic Chemists (AOAC). 1995. *Official methods of analysis of AOAC International*. 16th ed. AOAC International, Arlington, VA.

Barrett, T.J., and P. Gerner-Smidt. 2007. Molecular sources tracking and molecular subtyping. In *Food microbiology fundamentals and frontiers*, 3rd ed., ed. M.P. Doyle and L.R. Beuchat. ASM Press, Washington, DC, pp. 987–1004.

Behrens, S., C. Rühland, J. Inácio, H. Huber, A. Fonseca, I. Spencer-Martins, B.M. Fuchs, and R. Amann. 2003. In situ accessibility of small-subunit rRNA of members of the domains Bacteria, Archaea, and Eucarya to Cy3-labeled oligonucleotide probes. *Applied and Environmental Microbiology* 69:1748–1758.

Bej, A.K., and M.H. Mahbubani. 1992. Applications of the polymerase chain reaction in environmental microbiology. *PCR Methods and Applications* 1:151–159.

Bender, J.B., C.W. Hedberg, D.J. Boxrud, J.M. Besser, J.H. Wicklund, K.E. Smith, and M.T. Ostreholm. 2001. Use of molecular subtyping in surveillance for *Salmonella enterica* serotype Typhimurium. *New England Journal of Medicine* 344:189–195.

Beuret, C. 2004. Simultaneous detection of enteric viruses by multiplex real-time RT-PCR. *Journal of Virological Methods* 115:1–8.

Bhagwat, A.A. 2003. Simultaneous detection of *Escherichia coli* O157:H7, *Listeria monocytogenes* and *Salmonella* strains by real-time PCR. *International Journal of Food Microbiology* 84:217–224.

Bickley, J., J.K. Short, D.G. McDowell, and H.C. Parkes. 1996. Polymerase chain reaction (PCR) detection *Listeria monocytogenes* in diluted milk and reversal of PCR inhibition caused by calcium ions. *Letters in Applied Microbiology* 22:153–158.

Birch, L., C.E. Dawson, J.H. Cornett, and J.T. Kerr. 2001. A comparison of nucleic acid amplification techniques for the assessment of bacterial viability. *Letters in Applied Microbiology* 33:296–301.

Bisha, B., and B.F. Brehm-Stecher. 2009a. Simple adhesive tape-based-sampling of tomato surfaces combined with rapid fluorescence in situ hybridization for *Salmonella* detection. *Applied and Environmental Microbiology* 75:1450–1455.

Bisha, B., and B.F. Brehm-Stecher. 2009b. Flow-through imaging cytometry for characterization of *Salmonella* subpopulations in alfalfa sprouts, a complex food system. *Biotechnology Journal* 4:880–887.

Blais, B.W., G. Turner, R. Sooknanan, and L.T. Malek. 1997. A nucleic acid sequence-based amplification system for detection of *Listeria monocytogenes hlyA* sequences. *Applied and Environmental Microbiology* 63:310–313.

Boxrud, D. 2010. Advances in subtyping methods of foodborne disease pathogens. *Current Opinion in Biotechnology* 21:137–141.

Boye, M., T.K. Jensen, K. Møller, T.D. Leser, and S.E. Jorsal. 1998. Specific detection of the genus *Serpulina*, *S. hyodysenteriae* and *S. pilosicoli* in porcine intestines by fluorescent rRNA in situ hybridization. *Molecular and Cellular Probes* 12:323–330.

Brehm-Stecher, B.F., J.J. Hyldig-Nielsen, and E.A. Johnson. 2005. Design and evaluation of 16S rRNA-targeted peptide nucleic acid probes for whole-cell detection of members of the genus *Listeria*. *Applied and Environmental Microbiology* 71:5451–5457.

Brehm-Stecher, B.F., and E.A. Johnson. 2007. Rapid detection of *Listeria*. In *Listeria, Listeriosis and Food Safety*, 3rd ed., ed. E. Marth and E. Ryser. Dekker, New York, pp. 257–282.

Bruno, J.G., T. Phillips, M.P. Carrillo, and R. Crowell. 2009. Plastic-adherent DNA aptamer-magnetic bead and quantum dot sandwich assay for *Campylobacter* detection. *Journal of Fluorescence* 19:427–435.

Carlson, S.A., L.F. Bolton, C.E. Briggs, H.S. Hurd, V.K. Sharma, P.J. Fedorka-Cray, and B.D. Jones. 1999. Detection of multiresistant *Salmonella typhimurium* DT104 using multiplex and fluorogenic PCR. *Molecular and Cellular Probes* 13:213–222.

Chen, S.H., V.C. Wu, Y.C. Chuang, and C.S. Lin. 2008. Using oligonucleotide-functionalized Au nanoparticles to rapidly detect foodborne pathogens on a piezoelectric biosensor. *Journal of Microbiological Methods* 73:7–17.

Churchill, R.L., H. Lee, and J.C. Hall. 2006. Detection of *Listeria monocytogenes* and the toxin listeriolysin O in food. *Journal of Microbiological Methods* 64:141–170.

Cocolin, L., K. Rantsiou, L. Iacumin, C. Cantoni, and G. Comi. 2002. Direct identification in food samples of *Listeria* spp. and *Listeria monocytogenes* by molecular methods. *Applied and Environmental Microbiology* 68:6273–6282.

Cremonesi, P., G. Pisoni, M. Severgnini, C. Consolandi, P. Moroni, M. Raschetti, and B. Castiglioni. 2009. Pathogen detection in milk samples by ligation detection reaction-mediated universal array method. *Journal of Dairy Science* 92:3027–3039.

Datta, A.R., and B.A. Wentz. 1989. Identification and enumeration of virulent *Listeria* strains. *International Journal of Food Microbiology* 8:259–264.

De Boer, E., and R.R. Beumer. 1999. Methodology for detection and typing of foodborne microorganisms. *International Journal of Food Microbiology* 50:119–130.

DeLong, E.F., G.S. Wickham, and N.R. Pace. 1989. Phylogenetic stains: ribosomal RNA-based probes for the identification of single cells. *Science* 243:1360–1363.

Demidov, V.V., V.N. Potaman, M.D. Frank-Kamenetskii, M. Egholm, O. Buchard, S.H. Sönnichsen, and P.E. Nielsen.1994. Suitability of peptide nucleic acids in human serum and cellular extracts. *Biochemical Pharmacology* 48:1310–1313.

D'Souza, D.H., and L.A. Jaykus. 2003. Nucleic acid sequence based amplification for the rapid and sensitive detection of *Salmonella enterica* from foods. *Journal of Applied Microbiology* 95:1343–1350.

D'Souza, D.H., and L.A. Jaykus. 2006. Molecular approaches for the detection of foodborne viral pathogens. In *PCR methods in food*, ed. J.J. Maurer. Springer Science, New York, pp. 91–117.

Dwivedi, H.P., and L.A. Jaykus. 2010. Detection of pathogens in foods: the current state-of-the art and future directions. *Critical Reviews in Microbiology* 37:40–63.

Egholm, M., O. Buchardt, L. Christensen, C. Behrens, S.M. Frier, D.A. Driver, R.H. Berg, S.K. Kim, B. Norden, and P.E. Nielsen. 1993. PNA hybridizes to complementary oligonucleotides obeying the Watson-Crick hydrogen bonding rules. *Nature* 365:566–568.

Elizaquível, P., and R. Aznar. 2008. A multiplex RTi-PCR reaction for simultaneous detection of *Escherichia coli* O157:H7, *Salmonella* spp. and *Staphylococcus aureus* on fresh, minimally processed vegetables. *Food Microbiology* 25:705–713.

Ellington, A.D., and J.W. Szostak. 1990. In vitro selection of RNA molecules that bind specific ligands. *Nature* 346:818–822.

Fairchild, A., M.D. Lee, and J.J. Maurer. 2006. PCR basics. In *PCR methods in food*, ed. J.J. Maurer. Springer Science, New York, pp. 1–25.

Fang, Q., S. Brockmann, K. Botzenhart, and A. Wiedenmann. 2003. Improved detection of *Salmonella* spp. in foods by fluorescent in situ hybridization with 23S rRNA probes: a comparison with conventional culture methods. *Journal of Food Protection* 66:723–731.

Feng, P. 2007. Rapid methods for the detection of foodborne pathogens. In *Food microbiology fundamentals and frontiers*, 3rd ed., ed. M.P. Doyle and L.R. Beuchat. ASM Press, Washington, DC, pp. 911–934.

Fox, G.E., J.D. Wisotzkey, and P. Jurtshuk, Jr. 1992. How close is close: 16S rRNA sequence identity may not be sufficient to guarantee species identity. *International Journal of Systematic Bacteriology* 42:166–170.

Fratamico, P.M., and D.O. Bayles. 2005. Molecular approaches for detection, identification, and analysis of foodborne pathogens. In *Foodborne pathogens: microbiology and molecular biology*, ed. P.M. Fratamico, A.K. Bhunia, and J.L. Smith. Caister Academic Press, Norfolk, UK, pp. 1–13.

Fuchizawa, I., S. Shimizu, Y. Kawai, and K. Yamazaki. 2008. Specific detection and quantitative enumeration of *Listeria* spp. using fluorescent in situ hybridization in combination with filter cultivation (FISHFC). *Journal of Applied Microbiology* 105:502–509.

Fuchs, B.M., K. Syutsubo, W. Ludwig, and R. Amann. 2001. In situ accessibility of *Escherichia coli* 23S rRNA to fluorescently labeled oligonucleotide probes. *Applied and Environmental Microbiology* 67:961–968.

Fuchs, B.M., G. Wallner, W. Beisker, I. Schwippl, W. Ludwig, and R. Amann. 1998. Flow cytometric analysis of the *in situ* accessibility of *Escherichia coli* 16S rRNA for fluorescently labeled oligonucleotide probes. *Applied and Environmental Microbiology* 64:4973–4982.

Fung, D.Y.C. 1995. What's needed in rapid detection of foodborne pathogens. *Food Technology* 49:64–67.

Fung, D.Y.C. 2002. Rapid methods and automation in microbiology. *Comprehensive Reviews in Food Science and Food Safety* 1:3–22.

Glöckner, F.O., R. Amann, A. Alfreider, J. Pernthaler, R. Psenner, K. Trebesius, and K.-H. Schleifer. 1996. An in situ hybridization protocol for detection of and identification of planktonic bacteria. *Systematic and Applied Microbiology* 19:403–406.

González-Escalona, N., T.S. Hammack, M. Russell, A.P. Jacobson, A.J. De Jesús, E.W. Brown, and K.A. Lampel. 2009. Detection of live *Salmonella* spp. cells in produce by a TaqMan-based quantitative reverse transcriptase real-time PCR targeting *invA* mRNA. *Applied and Environmental Microbiology* 75:3714–3720.

Grundmann, H., D.M. Aanensen, C.C. van den Wijngaard, B.G., Spratt, D. Harmsen, and A.W. Friedrich. 2010. Geographic distribution of *Staphylococcus aureus* causing invasive infections in Europe: a molecular-epidemiological analysis. *PLoS Medicine* 7:e1000215.

Harmon, K.M., G.M. Ransom, and I.V. Wesley. 1997. Differentiation of *Campylobacter jejuni* and *Campylobacter coli* by polymerase chain reaction. *Molecular and Cellular Probes* 11:195–200.

Hermans, A.P., T. Abee, M.H. Zwietering, and H.J. Aarts. 2005. Identification of novel *Salmonella enterica* serovar Typhimurium DT104-specific prophage and nonprophage chromosomal sequences among serovar Typhimurium isolates by genomic subtractive hybridization. *Applied and Environmental Microbiology* 71:4979–4985.

Herrera-León, S., J.R. McQuiston, M.A. Usera, P.I. Fields, J. Garaizar, and M.A. Echeita. 2004. Multiplex PCR for distinguishing the most common phase-1 flagellar antigens of *Salmonella* spp. *Journal of Clinical Microbiology* 42:2581–2586.

Hong, B.X., L.F. Jiang, Y.S. Hu, D.Y. Fang, and H.Y. Guo. 2004. Application of oligonucleotide array technology for the rapid detection of pathogenic bacteria of foodborne infections. *Journal of Microbiological Methods* 58:403–411.

Houf, K., A. Tutenel, L. De Zutter, J. Van Hoof, and P. Vandamme. 2000. Development of a multiplex PCR assay for the simultaneous detection and identification of *Arcobacter butzleri*, *Arcobacter cryaerophilus*, and *Arcobacter skirrowii*. *FEMS Microbiology Letters* 193:89–94.

Huang, P.W., D. Laborde, V.R. Land, D.O. Matson, A.W. Smith, and X. Jiang. 2000. Concentration and detection of caliciviruses in water samples by reverse transcription-PCR. *Applied and Environmental Microbiology* 66:4383–4388.

Ingianni, A., M. Floris, P. Palomba, M.A. Madeddu, M. Quartuccio, and R. Pompei. 2001. Rapid detection of *Listeria monocytogenes* in foods by a combination of PCR and DNA probe. *Molecular and Cellular Probes* 15:275–280.

Jaykus, L.A. 2001. Detection of human enteric viruses in foods. In *Foodborne diseases handbook, volume 2: viruses, parasites, pathogens and HACCP*, ed. Y.H. Hui, S. Sattar, K.D. Murrel, W.-K. Nip, and P.S. Stanfield. Marcel Dekker, New York, pp. 137–164.

Jean, J., D.H. D'Souza, and L.A. Jaykus. 2004. Multiplex nucleic acid sequence-based amplification (NASBA) for the simultaneous detection of enteric viruses in ready-to-eat food. *Applied and Environmental Microbiology* 70:6603–6610.

Joshi, R., H. Janagama, H.P. Dwivedi, T.M. Senthil Kumar, L.A. Jaykus, J. Schefers, and S. Sreevatsan. 2009. Selection, characterization, application of DNA aptamers for the capture and detection of *Salmonella enterica* serovars. *Molecular and Cellular Probes* 23:20–28.

Jung, Y.S., J.F. Frank, R.E. Brackett, and J. Chen. 2003. Polymerase chain reaction detection of *Listeria monocytogenes* in frankfurters using oligonucleotide primers targeting the genes encoding internalin AB. *Journal of Food Protection* 66:237–241.

Klein, P.G., and V.K. Juneja. 1997. Sensitive detection of viable of *Listeria monocytogenes* by reverse transcription-PCR. *Applied and Environmental Microbiology* 63:4441–4448.

Lemarchand, K., L. Masson, and R. Brousseau. 2004. Molecular biology and DNA microarray technology for microbial quality monitoring of water. *Critical Reviews in Microbiology* 30:145–172.

Lenaerts, J., H.M. Lappin-Scott, and J. Porter. 2007. Improved fluorescent in situ hybridization method for detection of bacteria from activated sludge and river water by using DNA molecular beacons and flow cytometry. *Applied and Environmental Microbiology* 73:2020–2023.

Lermo, A., S. Campoy, J. Barbé, S. Hernández, S. Alegret, and M.I. Pividori. 2007. *In situ* DNA amplification with magnetic primers for the electrochemical detection of food pathogens. *Biosensors and Bioelectronics* 22:2010–2017.

Liu, D. 2008. Preparation of *Listeria monocytogenes* specimens for molecular detection and identification. *International Journal of Food Microbiology* 122:229–242.

Lucore, L.A., M.A. Cullison, and L.A. Jaykus. 2000. Immobilization with metal hydroxides as a means to concentrate food-borne bacteria for detection by cultural and molecular methods. *Applied and Environmental Microbiology* 66:1769–1776.

Ludwig, W, O. Strunk, R. Westram, L. Richter, H. Meier, et al. 2004. ARB: a software environment for sequence data. *Nucleic Acids Research* 32:1363–1371.

Luk, J.M., U. Kongmuang, P.R. Reeves, and A.A. Lindberg. 1993. Selective amplification of abequose and paratose synthase genes (*rfb*) by polymerase chain reaction for identification of *Salmonella* major serogroups (A, B, C2, and D). *Journal of Clinical Microbiology* 31:2118–2123.

Maidak, B.L., J.R. Cole, T.G. Lilburn, C.T. Parker, Jr., P.R. Saxman, R.J. Farris, et al. 2001. The RDP-II (Ribosomal Database Project). *Nucleic Acids Research* 29:173–174.

Mattison, K., E. Grudeski, B. Auk, H. Charest, S.J. Drews, A. Fritzinger, et al. 2009. Multicenter comparison of two norovirus ORF2-based genotyping protocols. *Journal of Clinical Microbiology* 47:3927–3932.

McTavish, S.M., C.E. Pope, C. Nicol, D. Campbell, N. French, and P.E. Carter. 2009. Multilocus sequence typing of *Campylobacter jejuni*, and the correlation between clonal complex and pulsed-field gel electrophoresis macrorestriction profile. *FEMS Microbiology Letters* 298:149–156.

Morin, N.J., Z. Gong, and X.F. Li. 2004. Reverse transcription-multiplex PCR assay for simultaneous detection of *Escherichia coli* O157:H7, *Vibrio cholerae* O1, *Salmonella* Typhi. *Clinical Chemistry* 50:2037–2044.

Moter, A., and U.B. Göbel. 2000. Fluorescence in situ hybridization (FISH) for direct visualization of microorganisms. *Journal of Microbiological Methods* 41:85–112.

Mothershed, E.A., and A.M. Whitney. 2006. Nucleic acid-based methods for the detection of bacterial pathogens: present and future considerations for the clinical laboratory. *Clinica Chimica Acta* 363:206–220.

Mustapha, A., and Y. Li. 2006. Molecular detection of foodborne bacterial pathogens. In *PCR methods in food*, ed. J.J. Maurer. Springer Science, New York, pp. 69–90.

Niederhauser, C., U. Candrian, C. Höfelein, M. Jermini, H.P. Bühler, and J. Lüthy. 1992. Use of polymerase chain reaction for detection of *Listeria monocytogenes* in food. *Applied and Environmental Microbiology* 58:1564–1568.

Norton, D.M., M.A. McCamey, K.L. Gall, J.M. Scarlett, K.J. Boor, and M. Wiedmann. 2001. Molecular studies on the ecology of *Listeria monocytogenes* in the smoked fish processing industry. *Applied and Environmental Microbiology* 67:198–205.

Notermans, S., R.R. Beumer, and F.M. Rombouts. 1997. Detecting foodborne pathogens and their toxins, conventional versus rapid and automated methods. In *Food microbiology, fundamentals and frontiers*, ed. M.P. Doyle, L.R. Beuchat, and T.J. Montville. ASM Press, Washington, DC, pp. 697–709.

Pangallo, D., E. Kaclíková, T. Kuchta, and H. Drahovská. 2001. Detection of *Listeria monocytogenes* by polymerase chain reaction oriented to the *inlB* gene. *Microbiologica* 24:333–339.

Pedersen, L.H., P. Skouboe, L. Rossen, and O.F. Rasmussen. 1998. Separation of *Listeria monocytogenes* and *Salmonella berta* from a complex food matrix by aqueous polymer two-phase partitioning. *Letters in Applied Microbiology* 26:47–50.

Peterkin, P.I., E.S. Idziak, and A.N. Sharpe. 1992. Use of a hydrophobic grid-membrane filter DNA probe method to detect *Listeria monocytogenes* in artificially-contaminated foods. *Food Microbiology* 9:155–160.

Pieken, W., D.B. Olsen, F. Benseler, H. Aurup, and F. Eckstein. 1991. Kinetic characterization of ribonuclease-resistant 2'-modified hammerhead ribozymes. *Science* 253:314–317.

Powell, H.A., C.M. Gooding, S.D. Garrett, B.M. Lund, and R.A. Mckee. 1994. Proteinase inhibition of the detection of *Listeria monocytogenes* in milk using the polymerase chain reaction. *Letters in Applied Microbiology* 18:59–61.

Rahn, K., S.A. Grandis, R.C. Clarke, S.A. McEwen, J.E. Galán, C. Ginocchio, R. Curtiss III, and C.L. Gyles. 1992. Amplification of an *invA* gene sequence of *Salmonella typhimurium* by polymerase chain reaction as a specific method of detection of *Salmonella*. *Molecular and Cellular Probes* 6:271–279.

Rossen, L., P. Nørskov, K. Holmstrøm, and O.F. Rasmussen. 1992. Inhibition of PCR by components of food samples, microbial diagnostic assays and DNA-extraction solutions. *International Journal of Food Microbiology* 17:37–45.

Ryser, E.T., S.M. Arimi, M.M. Bunduki, and C.W. Donnelly. 1996. Recovery of different *Listeria* ribotypes from naturally contaminated, raw refrigerated meat and poultry products with two primary enrichment media. *Applied and Environmental Microbiology* 62:1781–1787.

Sanderson, K., and D. Nichols. 2003. Genetic techniques: PCR, NASBA, hybridisation and microarrays. In *Detecting pathogens in food*, ed. T.A. McMeekin. Woodhead and CRC Press, Boca Raton, FL, pp. 259–270.

Schmid, M., M. Walcher, A. Bubert, M. Wagner, M. Wagner, and K.H. Schleifer. 2003. Nucleic acid-based, cultivation-independent detection of *Listeria* spp. and genotypes of *Listeria monocytogenes*. *FEMS Immunology and Medical Microbiology* 35:215–225.

Schönhuber, W., B. Fuchs, S. Juretschko, and R. Amann. 1997. Improved sensitivity of whole-cell hybridization by combination of horseradish peroxidase-labeled oligonucleotides and tyramide signal amplification. *Applied and Environmental Microbiology* 63:3268–3273.

Schwab, K.J., R. DeLeon, and M.D. Sobsey. 1996. Immunoaffinity concentration and purification of waterborne enteric viruses for detection by reverse transcriptase PCR. *Applied and Environmental Microbiology* 62:2086–2094.

Sergeev, N., M. Distler, S. Courtney, S.F. Al-Khaldi, D. Volokhov, V. Chizhikov, et al. 2004. Multipathogen oligonucleotide microarray for environment and biodefense applications. *Biosensors and Bioelectronics* 20:684–698.

Sharpe, A.N. 2003. Separation and concentration of samples. In *Detecting pathogens in food*, ed. T.A. McMeekin. CRC Press, Boca Raton, FL.

Siddique, N., D. Sharma, and S.F. Al-Khaldi. 2009. Detection of *Yersinia enterocolitica* in alfalfa, mung bean, cilantro, mamey sapote (*Pouteria sapota*) food matrices using DNA microarray chip hybridization. *Current Microbiology* 59:233–239.

Silverman, A.P., and E.T. Kool. 2005. Quenched autoligation probes allow discrimination of live bacterial species by single nucleotide differences in rRNA. *Nucleic Acids Research* 33:4978–4986.

Simpkins, S.A., A.B. Chan, J. Hays, B. Pöpping, and N. Cook. 2000. An RNA transcription-based amplification technique (NASBA) for the detection of viable *Salmonella enterica*. *Letters in Applied Microbiology* 30:75–79.

Sobsey, M.D. 1994. Molecular methods to detect viruses in environmental samples. In *Rapid methods and automation in microbiology and immunology*, ed. R.C. Spencer, E.P. Wright, and S.W.B. Newsom. Intercept, Andover, Hampshire, UK, pp. 387–400.

Stender, H., M. Fiandaca, J.J. Hyldig-Nielsen, and J. Coull. 2002. PNA for rapid microbiology. *Journal of Microbiological Methods* 48:1–17.

Stevens, K.A., and L.A. Jaykus. 2004. Bacterial separation and concentration form complex sample matrices: a review. *Critical Reviews in Microbiology* 30:7–24.

Suggs, S.V., T. Hirose, T. Miyake, E.H. Kawashima, M.J. Johnson, K. Itakura, and R.B. Wallace. 1981. Use of synthetic oligodeoxynucleotides for the isolation of specific cloned DNA sequences. In *Developmental biology using purified genes*, ed. D.B. Brown and C.F. Fox. Academic Press, New York, pp. 683–693.

Swaminathan, B., and P. Feng. 1994. Rapid detection of food-borne pathogenic bacteria. *Annual Review of Microbiology* 48:401–426.

Takahashi, H., Y. Hara-Kudo, J. Miyasaka, S. Kumagai, and H. Konuma. 2005. Development of a quantitative real-time polymerase chain reaction targeted to the *toxR* for detection of *Vibrio vulnificus*. *Journal of Microbiological Methods* 61:77–85.

Theron, J., T. Eugene Cloete, and M. deKwaadsteniet. 2010. Current molecular and emerging nanobiotechnology approaches for the detection of microbial pathogens. *Critical Reviews in Microbiology* 36:318–339.

Tombelli, S., M. Minunni, and M. Mascini. 2007. Aptamers-based assays for diagnostics, environmental and food analysis. *Biomolecular Engineering* 24:191–200.

Tuerk, C., and L. Gold. 1990. Systematic evolution of ligands by exponential enrichment: RNA ligands to bacteriophage T4 DNA polymerase. *Science* 249:505–510.

Tyagi, S., and F.R. Kramer. 1996. Molecular beacons: probes that fluoresce upon hybridization. *Nature Biotechnology* 14:303–308.

Uyttendaele, M., R. Schukkink, B. van Gemen, and J. Debevere. 1995. Development of NASBA, a nucleic acid amplification system, for identification of *Listeria monocytogenes* and comparison to ELISA and modified FDA method. *International Journal of Food Microbiology* 27:77–89.

Van Hal, N.L., O. Vorst, A.M. Van Houwelingen, E.J. Kok, A. Peijnenburg, A. Aharoni, A.J. van Tunen, and J. Keijer. 2000. The application of DNA microarrays in gene expression analysis. *Journal of Biotechnology* 78:271–280.

Velusamy, V., K. Arshak, O. Korostynska, K. Oliwa, and C. Adley. 2010. An overview of foodborne pathogen detection: in the perspective of biosensors. *Biotechnology Advances* 28:232–254.

Vivekananda, J., and J.L. Kiel. 2006. Anti-*Francisella tularensis* DNA aptamers detect tularemia antigen from different subspecies by aptamer-linked immobilized sorbent assay. *Laboratory Investigation* 86:610–618.

Wagner, M., M. Schmid, S. Juretschko, K.H. Trebesius, A. Bubert, W. Goebel, and K.H. Schleifer. 1998. In situ detection of a virulence factor mRNA and 16S rRNA in *Listeria monocytogenes*. *FEMS Microbiology Letters* 160:159–168.

Wang, R.F., W.W. Cao, and M.G. Johnson. 1992. 16S rRNA-based probes and polymerase chain reaction method to detect *Listeria monocytogenes* cells added to foods. *Applied and Environmental Microbiology* 58:2827–2831.

Ward, L.N., and A.K. Bej. 2006. Detection of *Vibrio parahaemolyticus* in shellfish by use of multiplexed real-time PCR with TaqMan fluorescent probes. *Applied and Environmental Microbiology* 72:2031–2042.

Wilson, I.G. 1997. Inhibition and facilitation of nucleic acid amplification. *Applied and Environmental Microbiology* 63:3741–3751.

Woese, C.R. 1987. Bacterial evolution. *Microbiological Reviews* 51:221–271.

Wolffs, P.F., K. Glencross, R. Thibaudeau, and M.W. Griffiths. 2006. Direct quantitation and detection of salmonellae in biological samples without enrichment, using two-step filtration and real-time PCR. *Applied and Environmental Microbiology* 72:3896–3900.

Xi, C., M. Balberg, S.A. Boppart, and L. Raskin. 2003. Use of DNA and peptide nucleic acid molecular beacons for detection and quantification of rRNA in solution and in whole cells. *Applied and Environmental Microbiology* 69:5673–5678.

Yilmaz, L.S., H.E. Okten, and D.R. Noguera. 2006. Making all parts of the 16S rRNA of *Escherichia coli* accessible in situ to single DNA oligonucleotides. *Applied and Environmental Microbiology* 77:733–744.

Zhang, W., B.M. Jayarao, and S.J. Knabel. 2004. Multi-virulence-locus sequence typing of *Listeria monocytogenes*. *Applied and Environmental Microbiology* 70:913–920.

Zimmer, C., and U. Wähnert. 1986. Nonintercalating DNA-binding ligands: specificity of the interaction and their use as tools in biophysical, biochemical and biological investigation of the genetic material. *Progress in Biophysics and Molecular Biology* 47:31–112.

chapter twelve

Reporting food microbiology research outcomes

Mark Carter

Contents

12.1 Introduction

Communication is the activity of conveying *meaningful information*. Communication requires a sender, a message, and an intended recipient, although the receiver need not be present or aware of the sender's intent to communicate at the time of communication; thus, communication can occur across vast distances in time and space. Communication requires that the communicating parties share an area of communicative commonality. The communication process is complete once the receiver has understood the sender.[*]

Without much notice, society has become accustomed to the daily research reports that show up through all types of media and distribution channels. Information is more prevalent and accessible than ever; therefore, the new challenge is not finding information but sorting through it and extracting what is usable. Research produced by academic, governmental, and industrial institutions is used for a plethora of purposes. Report after report is paraded out and communicated to a wide audience with varying degrees of effectiveness. Usually, the reports have

---

[*] The Wikipedia entry for "Communication" was used as a reference for this paragraph (http://en.wikipedia.org/wiki/Communication).

some significant point or finding that has wide-ranging implications for society or can be an incremental addition or correction to existing scientific knowledge. No matter the depth or complexity of the research, its outcomes have impact only if they are communicated in clear, concise messages. Clarity and conciseness should be the researchers' goal when communicating study outcomes to maximize impact. The delivery of a clear, concise message that accurately describes significant points to the appropriate audience is a goal of research reporting.

Food microbiology research can encompass a wide variety of techniques and applications. Research can be used for the purpose of gaining basic knowledge of food systems and their intrinsic characteristics, or it can be applied to develop very specific knowledge of a unique item or process. Food-related technologies such as antimicrobial ingredients, sanitizers, and microbial diagnostics can be researched and developed for many purposes. Every day in labs throughout the world, millions of analyses are performed to generate data that will in some way, shape, or form become a research report. The previous chapters of this book have covered specific areas of microbiological research and development. This chapter gives an overview of the reporting process and how to communicate research and development results effectively.

12.2   The basics

The typical research report can take on many forms. We normally think of a research report as being divided into clear, concise sections that describe the research process and its outcomes. Many times, the report includes sections such as background, objective(s), materials and methods, results, and finally a discussion, followed by references and acknowledgment of funding sources or nonauthor contributors to the work. This format is very similar to a journal article's format and is commonly used, understood, and accepted in scientific circles. Depending on the length and complexity of the report, other sections can be added, such as an executive summary, which is typically more suitable for reporting beyond the scientific community. Summary sections augment the traditional format and serve to condense and summarize the key items that can be buried within a research report. In many instances, an abstract can serve the same function as the executive summary. In either case, the persons writing these summaries must overcome any bias to make sure that what is included and what is omitted does not falsely slant the perception of the bulk of the report. Executives reading these summaries will likely make decisions based on them and not the body of the report. Similarly, researchers will decide whether to retrieve the entire article based largely on the abstract.

The research report is the cornerstone of numerous decisions related to food quality and safety. The challenging process of reporting food microbiology research and development outcomes can be seen as a process that begins with the project design and ends with the communication of the outcomes to various target audiences. Developing the correct message for each audience is ultimately important to the scientist, researcher, and any research stakeholders. The term *stakeholder* is used because it normally encompasses individuals with a diverse group of backgrounds and functions who all have a vested interest in the research. It can include a very small group of individuals inside a department or a large group that includes large business units or an even larger group of consumers. For example, a government agency may fund research on a particular pathogen in foods, with the funding awarded to an academic institution and its outcome having an impact on private industry and consumers. In such an example, the stakeholders have vested interests in the findings for differing reasons: The government agency wants to make sure that their funding actually resulted in tangible new knowledge from which to base policy decisions; the academic institution wants to promote how their research program has an impact in society; the industry is interested because the findings will help it assess and manage risk and gauge what will be needed for compliance; and consumers want to see that they are informed about risk and are properly protected.

Typically, the communication process follows several routes and includes multiple forms of media. Many of us have been trained to communicate as scientists through traditional means, channels, and directions. For the purpose of this chapter, let us list the potential types of communicative interactions:

1. Peer to peer (individual or group)
2. Researcher to subordinate (individual or group)
3. Researcher to superior (individual or group)
4. Researcher to cross functional (subordinate, peer, superior)

There are also four basic types of communication:

1. Nonverbal
2. Visual
3. Oral
4. Written

As we move forward in the chapter, keep in mind how communication needs change based on the target audience. While peer-to-peer communication can be the most natural and easy of any to the scientist, group cross-functional communication can be the most challenging of all inter-actions. We illustrate why this is so further in the chapter.

12.3   Driven by purpose

The reporting style of an outcome is driven by the purpose of the research. Here is an example of a purpose statement (i.e., objective) that could potentially be in a report or some form of communication after the research is completed:

> "The purpose of this research project is to determine the effect of com-pounds A, B, and C on the shelf life of sliced luncheon meat."

This is a simple statement that describes the project and sets the premise for the final report. The report content will be driven by study design and execution. As scientists, you know you are testing a hypothesis for which the outcome of the research can be something that is highly anticipated or something that contradicts conventional thought and wisdom. On the surface, this purpose statement seems clear enough, which should make communicating the outcome simple. Can the state-ment be improved? If so, what do we want to add? At what point is the statement confusing?

Here is another version of the purpose statement for the same study:

> "The purpose of this research project is to determine the effect of 1, 5, and 10% solutions of compounds A, B, and C on the shelf life of sliced luncheon meat held at 5, 15, and 35°C."

A little more information goes a long way. Suddenly, we have a much better idea of the study design and purpose and just enough more infor-mation to decide whether to continue reading. As a scientist, you can begin to understand with this added bit of detail just what kinds of data will be generated. You might even begin to formulate ideas about poten-tial results. It is hoped the study has been designed in such a way that produces statistically valid results. The previous statement can be a part of a larger introduction section. Background information could act as a primer for the purpose statement. The communication that flows from this point should focus on explaining the study outcomes as they relate to the purpose statement.

As stated, developing clear, concise communication that is targeted to the appropriate audience is the goal of a reporting situation. Writing the final report *after* completing research is something scientists and researchers have been trained to do. If it can be accomplished, writing the introduction and materials and methods *before* completing the research should be considered a best practice. At times, it is difficult to translate the report so that it is understood by the appropriate target audience. In the next section are factors to consider when developing communications tools beyond the final research report.

12.4   Are we speaking the same language?

What does it mean to speak the same language? Normally, this discussion leads to conversations about the number of different languages or dialects of English that are spoken within a given group.* We normally assume that all of us use a common language for oral and written communications. In many instances, this is very true even in global organizations spanning regions with different languages. One of the most overlooked instances of language differences occurs between and among functional groups. Microbiology has a language all its own, but many dialects are spoken within the science. Because of our many peer-to-peer conversations, communicating research results becomes infiltrated by myriad acronyms, initials, and slang that make it easy for scientific peers to communicate. While these habits are effective for peer-to-peer information exchange, they may impede group cross-functional communication.

Research must often be communicated to subordinates or superiors within the same technical function. Most likely, the message is slightly modified (e.g., with more or less detail) for the recipient, but it is rare to have to significantly alter the message because it is being delivered in a common (technical) language. The effectiveness of the communication is dependent on the content and the quality of the research. Your peer group's common interests and language make this process easier to manage. The challenges begin as we move beyond our "tribe" of peers, subordinates, and superiors to communicate research results with other groups. The sense of speaking the same language begins to change. Imagine a research setting where two microbiologists from different microbiological disciplines with different training and backgrounds are present for the same research review. A food microbiologist with a background in aseptic thermal processing has a slightly different "dialect" than a food microbiologist with a background in pathogens and meat microbiology. However, it is not out of the realm of possibility that the two could easily be on the same project team with both expected to understand a problem in the same way. After all, to the business they are both technical—they

---

* Editor's note: For those whose first language is English, it is fortuitous and convenient that English is the most published and the predominant common language in science and business communications. However, Mandarin is the most spoken language on the planet, followed by English, then Spanish and Hindi, which are also spoken by hundreds of millions. Multilingual writing adds value and scope of impact to research reports.

are both microbiologists—but this does not mean that they will both understand the nuances of each other's subdiscipline.

This issue of language commonality only increases as we begin to include more people from diverse scientific and nonscientific backgrounds. Let us continue with the research purpose statement we used previously. A project team has been assembled, and it is time to present findings from the luncheon meat study. For the sake of discussion, we are going to imagine a project review where the following people are present:

1. Sensory scientist
2. Process engineer
3. Food microbiologist
4. Product manager
5. Quality director

Each participant will have a different overview of the project and different responsibilities and vested interests. Each participant has been trained in a different language than the food microbiologist who is responsible for performing the research and communicating the results. Often, the researcher will come to the review with the materials he or she has prepared and be ready to give the results of the project from his or her point of view and written in his or her language. The issue is the group's composition. The research report and presentations have been written in "food microbiologese." The group interprets the report in the language of their respective schooling and professional experience. The process engineer could be thinking of the ways to apply the specific compound to the product. The sensory scientist is likely wondering which of the compounds will impart an acceptable flavor profile. The product manager is considering product placement within the market, label cleanliness, and material cost impact on profitability. The quality manager is thinking about measurement processes to verify proper production and safety. Each stakeholder comes to the review with a different motivation and perspective. Each will hear you (the food microbiologist) in a different language. A well-done research report will address all the technical outcomes of the experiments and will attempt to cater to all the main stakeholders. The research report represents the first of many forms of communicating the results. Many of us, through time and experience, have learned to communicate in and across multiple languages. Being able to report research results fluently in other languages is a key factor when it comes to communicating technical results.

Figure 12.1 represents a simple way to think of the communication matrix by audience. Use it as a simple reminder of what languages you may

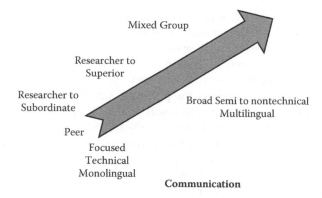

Figure 12.1 Communication matrix for matching report detail, jargon, and language to target audience.

encounter and what cultural differences are present. Scientific communication is challenging. It is imperative to develop a clear audience-based message of scientific results. From my own experience, this becomes even more difficult as you truly begin to cross cultural and language boundaries. Always remember that what you thought you said and what your audience heard can be miles (or kilometers) apart.

12.5 Intuitive leaps (How did we get here?)

All research reports have data that have to be compiled and communicated to a group of stakeholders. In many cases, the initial communications are between peers or subordinates. Data and results are exchanged between people who have a common knowledge of the project. Usually, the participants have very similar backgrounds and experiences, which simplifies the exchanges. In my experience, this type of conversation is usually characteristic of 80% of the interactions experienced within any group.

A pitfall that commonly occurs when reporting research results is the intuitive leap. The goal of the research report is to guide a reader through the research process and deliver the reader to a conclusion. Often in peer-to-peer situations, intuitive leaps become part of the conversation and communication and are hardly noticed. Shared backgrounds and knowledge make this possible with minor consequence. Many people have seen a document on the Internet in which words are spelled without vowels and the reader has no problem reading and understanding the content. This works because our brains allow us to make leaps and see what the whole word should be. Our common knowledge and language allow us to be successful even though key information is absent.

In a written research report as well as any oral or visual presentations, this can cause major problems. Many times, we start an experiment with a design that we feel will answer the question at hand only to look at the results and find out we may have missed something. Even in this case, the outcome of the research can seem perfectly logical because we can make an intuitive leap over the potential small data gap. Notice I say gap; I have seen cavernous pits that screamed, "Stop and do this again!" It is possible for the researcher to move ahead with a report based on the data and can make a small leap to conclusions. Depending on the reader or audiences, that small gap may be the cavernous pit.

If we think of our research example and the process of reporting to a diverse group, it becomes evident that some of the stakeholders will need to be walked through the process, and others may be able to make the leaps with you. The bottom line is that at some time every researcher will be faced with reporting data that have a gap, and some leap has to be made. In a perfect world, there would never be a gap or cavern, and the path through research projects would be simple. Our challenge is to recognize that there will be gaps in research that will need to be reported and to be properly prepared to explain the gaps. Do not make people jump the cavern. You want the recipient of the message to understand how they (and you) got where you are.

12.6   A picture tells a thousand words

As simple as it sounds, one of the most overlooked and ignored tools to communicate research data is that a picture tells a thousand words. Visual communication is not normally a scientist's forte, but it is essential for non-peer-to-peer forms of communication. It is also often the best choice for presentations. Everyone has been to the presentation where you hear the dreaded phrase, "I put this slide up, and I'm sorry you can't read it," or "This slide is a little crowded." The idea of a clear, concise message is immediately lost. A simple chart can illustrate results quickly and easily, and the details should be explained verbally.

Again, returning to our example used throughout the chapter, our research is complete, and the final report is ready. The data from a trial of our 1% solution could look like Figure 12.2. This simple chart is an effective way to present the results of our research. The results from each compound at each temperature are clearly displayed. We can scan each line for the data and relate them back to the corresponding compound.

The same data can be represented graphically. For most, this format will convey the message in an even simpler way (Figure 12.3). This is a simplified example of how a picture can represent data at a glance. As data become more complex, it becomes even more important to be able to use graphs to represent and communicate results. As simple and obvious

	Compounds		
Temperatures	A	B	C
5	60	60	60
15	30	35	40
25	15	25	30

Figure 12.2 Example of simple chart depicting results of the theoretical experiment of effects of compounds A, B, and C on the shelf life days for sliced luncheon meat at 5, 15, and 25°C.

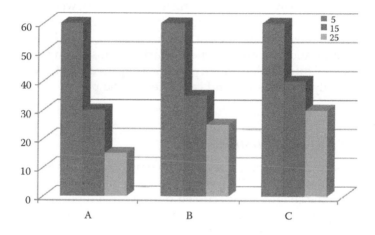

Figure 12.3 Example of simple graph depicting results of the theoretical experiment of effects of compounds A, B, and C on the shelf life days for sliced luncheon meat at 5, 15, and 25°C.

as this concept seems, its power is often overlooked. Visual communication style is usually consistent within a group. It is interesting to observe students present research at technical conferences because it is reflective of a style preference within their department or developed by their advisor. Many companies develop a style of visual communication that helps the companies' brand identity.

Figures 12.2 and 12.3 are examples from our theoretical research project. They both communicate the same information. Remember that a fairly diverse group of individuals is seeing the data presentation, and these individuals all speak different languages. Some have a technical knowledge base sufficient to understand details, while others will not. Often, details must be omitted to cater to those without such knowledge. The technical people can be counted on to make intuitive leaps to fill in the details or to ask you follow-up questions about details. Keeping it simple is better than asking nontechnical audience members to comprehend substantial detail. Our goal is to convey the information as simply and succinctly as possible. We have three choices to communicate the data:

1. Text: bullet points explaining results or a written report
2. Table: formatted text with categorized and sorted data
3. Chart: pictorial or graphic representation of data

All three forms are perfectly fine to use. Tailor the message and type of visual communication to the audience and situation. Our stakeholders' backgrounds will play a role in interpretation, but a picture can cross boundaries more easily than words. Charts always work well if they accurately tell the story. Just because something is in a chart does not necessarily mean it is simpler to interpret, but if done correctly, it can be a powerful tool. Communications with muddled charts or misleading scales and poorly arranged information have a powerful negative effect because they become visually unappealing and can threaten the perceived credibility of the work.

12.7   Getting the business

Any discussion of reporting research outcomes would have a huge cavernous gap if we do not discuss reporting to the business functions. Previously in the chapter we discussed our research report meeting, and it included a product (marketing) manager. For many who are responsible for research within a company or for a company client, interaction with the "business" is inevitable, if not essential. We have discussed differences in language and culture and how they affect the research reporting process. Researchers are often very well prepared to communicate research outcomes within scientific communities but not necessarily to business professionals. Those who can bridge these communication gaps between science and business stand to benefit.

Research performed by or for corporations will at some point be reviewed or used by managers in business functions to make decisions that could generate significant costs or significant value and profit. Business functions normally use multiple financial tools to assess the following two parameters:

1. Cost of research
2. Value of research

The cost of the research is exactly that: "cost." Researchers should be able to account for and communicate research costs. Simple accounting measures can be used to monitor the costs associated with a project. Learning the language of the business group is a difficult but necessary step for many researchers. Costs will be analyzed to determine the overall return on investment (ROI). Many companies will compute this in different ways, but it serves the researcher to have a grasp of the process and terms. This is part of becoming fluent in another "language." It is also one of the most difficult languages for researchers to master unless they had some business administration training at some point along their career path. The cost of research leads us directly to the "value," and the two should not be confused.

Value is driven by many things, but every company wants to understand its ROI. As scientists and researchers, our perception of the value of research is probably measured in scientific terms. Corporations look at many things to determine where to invest research money, and their philosophies are as diverse as the products they produce. Companies invest precious dollars in basic and applied research. Patents and trade secrets are interesting research outcomes, even assets, that can add significant value to a company by giving a competitive advantage or protection. Measuring and calculating value happens in many ways and is perceived differently depending on a company's culture.

Each not-for-profit funding body, like trade associations, special-interest groups, or government agencies, perceives value differently from corporations and differently from each other depending on mission. Many will include project-specific value metrics. To even be considered for project funding, researchers will already have had to explain how the approach would meet these value criteria. Since this work is done a priori, communicating that added value back to the funding body should simply be a matter of filling in the details, assuming the study went as planned.

If we use our own research example, we can look at the business parameters that can be discussed from a research standpoint. Our project reviewed three compounds at three different concentrations on sliced luncheon meat. The process of developing a research report is focused on the technical aspects and benefits of each compound. At some point, the business functions will consider whether to proceed with a compound based on the research results and subsequent communications. If the research finding is of perceived or measured value, the business functions will now take those data and begin the process to implement the changes in their systems. The research outcomes can have a tremendous value to the organization. The process of extending the shelf life of a product can create a great deal of value. Many times, the business analysis will touch many areas: marketing, sales, procurement, operations, distribution, quality, and many more. Being fluent in the languages and measures of the business can help broaden the appeal and comprehension of a research report, leading to greater impact and in turn greater potential for future research funding.

12.8   Summary

Reporting research outcomes is a process that requires thought and preparation. The ability to communicate accurately and precisely in-depth technical data to diverse groups is becoming a key factor for success and promotion of ideas. Academia, government, and private industry generate data, journal articles, and reports at a stunning pace. The quality and quantity of reports vary greatly, but reporting outcomes will continue to be a part of every scientist's job. It is necessary to keep in mind the type of communication that will be most effective in a given situation. In many situations, the old adage of keeping it simple still applies to communicating research outcomes no matter how detailed the study. Otherwise, great research outcomes can be lost in a muddled message.

Changes in media and distribution channels may well change the way research reporting occurs. Publishing and peer review have become easier due to the Web. Many journals require no page charges, many have open access, and many are archived online going back 20 to 50 years. We can only speculate what the use of YouTube, Skype, or other social or scientific media outlets will have on the future delivery of research reports and the subsequent outcomes. As long as we have research occurring, scientists will look for ways to distribute and communicate what they have learned. The bottom line is that the same fundamentals for effective reporting will not change regardless of the distribution mechanism.

Index

Page references in **bold** refer to tables.

Printed and bound by CPI Group (UK) Ltd, Croydon, CR0 4YY

21/10/2024

01777105-0010